BUSINESS/SCIENCE/TECHNOLOGY DIVISION
CHICAGO PUBLIC LIBRARY
400 SOUTH STATE STREET
CHICAGO, IL 60605

QH
545
.T4
F56
2001

HWLCSC

"Fingerprints" of climate change : adapt

R0400371439

Form 178 rev. 11-00

"Fingerprints" of Climate Change

Adapted Behaviour and Shifting Species Ranges

"Fingerprints" of Climate Change

Adapted Behaviour and Shifting Species Ranges

Edited by

G.-R. Walther
University of Hannover
Hannover, Germany

C. A. Burga
University of Zurich-Irchel
Zurich, Switzerland

and

P. J. Edwards
Swiss Federal Institute of Technology
Zurich, Switzerland

Kluwer Academic / Plenum Publishers
New York, Boston, Dordrecht, London, Moscow

Proceedings of the International Conference "Fingerprints" for Climate Change: Adapted Behaviour and Shifting Species Ranges, held February 23–25, 2001, at Ascona, Switzerland

ISBN 0-306-46716-X

©2001 Kluwer Academic/Plenum Publishers, New York
233 Spring Street, New York, New York 10013

http://www.wkap.nl

10 9 8 7 6 5 4 3 2 1

A C.I.P. record for this book is available from the Library of Congress

All rights reserved

No part of this book may be reproduced, stored in a retrieval system, or transmitted in any form or by any means, electronic, mechanical, photocopying, microfilming, recording, or otherwise, without written permission from the Publisher

Printed in the United States of America

Preface

This volume includes contributions to the international conference *"FINGERPRINTS" of CLIMATE CHANGE*, held from February $23^{rd} - 25^{th}$ 2001 at the International Conference Centre *"Centro Stefano Franscini"* at Ascona – Monte Verità (Switzerland). This conference was also the fifth meeting of the working group *"Biomonitoring/Global Change"* of the *Reinhold-Tüxen-Gesellschaft für Vegetationskunde*, Hannover. This international working group was founded by Conradin A. Burga in 1996 and comprises mainly botanists and ecologists, but also scientists from other fields. The program of this conference covered a wider range of topics than usual, and included several speakers from overseas.

We would like to thank to all the participants and invited speakers for coming to Monte Verità. The invited speakers were: Professor Ove Hoegh-Guldberg (University of Queensland, Australia), Professor Christian Körner (University of Basel, Switzerland), Dr. Camille Parmesan (University of Texas, Austin, USA), Dr. Peter Convey (British Antarctic Survey, Cambridge, GB) and Dr. Brendan Kelly (University of Alaska, Juneau, USA).

This conference was the first international meeting dedicated to the biological signals of recent climate warming. By this that means that the presentations were not based only on computer models or on predictions for the future, but reported changes which are actually occurring in the biosphere. These include shifts in the range of species and changes in the behaviour of organisms. During this conference 21 contributions reported research findings from a wide range of habitats, including coral reefs in the tropics, Mediterranean regions, coastal areas, temperate grass- and woodlands and alpine and arctic areas on five continents.

This conference volume is divided in two parts. The first part starts with an introductory overview on studies concerned with the impacts of recent climate warming on the phenology and physiology of organisms as well as the range and distribution of species (Walther). In the following, examples of adapted behaviour and shifting ranges are presented for Antarctic species (Convey), Arctic mammals (Kelly), butterflies (Parmesan and Hill *et al.*) and dragonflies (Ott). Findings of plant phenological changes are presented (Defila and Clot and Menzel and Estrella), as well as reports on shifts in the ranges of plant species from various habitats (Pauli *et al.*, Walther *et al.*, Sobrino-Vesperinas *et al.* and Metzing and Gerlach). Furthermore, the impact of climate warming on tropical marine ecosystems, especially coral bleaching, is discussed (Hoegh-Guldberg and Tsimilli-Michael and Strasser).

In the second part of the volume different methodological aspects are discussed, such as monitoring permanent plots (Stampfli), the experimental approach (Erschbamer) and the effectiveness of indicator values (Pignatti *et al.*). The final and synoptic chapter summarises the findings of the conference and draws some general conclusions (Körner and Walther).

We thank the following persons for their critical comments and suggestions to the submitted manuscripts: A. Barker, D. Bergstrom, G. Carraro, T. Denk, H. Dierschke, P. Dustan, A. Erhardt, W.K. Fitt,. H. Freund, F. Gugerli, S. Harrison, M.P. Lesser, C. Nellemann, J. Plötz, C. Schneider and N.R. Webb.

We are very grateful to the ETH Zurich and to the Centro Stefano Franscini Ascona – Monte Verità for the financial and organisational support of this conference and to Kluwer Academic / Plenum Publishers London for publication of these proceedings.

Conradin A. Burga

Contents

Adapted behaviour and shifting ranges of species – a result of recent climate warming?
   G.-R. Walther   1

Terrestrial ecosystem responses to climate changes in the Antarctic
   P. Convey   17

Climate change and ice breeding pinnipeds
   B. P. Kelly   43

Detection of range shifts: General methodological issues and case studies using butterflies
   C. Parmesan   57

Climate and recent range changes in butterflies
   J. K. Hill, C. D. Thomas and B. Huntley   77

Expansion of Mediterranean Odonata in Germany and Europe – consequences of climatic changes
   J. Ott   89

Phytophenological trends in different seasons, regions and altitudes in Switzerland
   C. Defila and B. Clot   113

Plant phenological changes
A. Menzel and N. Estrella — 123

High summits of the Alps in a changing climate
H. Pauli, M. Gottfried and G. Grabherr — 139

Evergreen broad-leaved species as indicators for climate change
G.-R. Walther, G. Carraro and F. Klötzli — 151

The expansion of thermophilic plants in the Iberian Peninsula as a sign of climatic change
E. Sobrino Vesperinas, A. González Moreno, M. Sanz Elorza, E. Dana Sánchez, D. Sánchez Mata and R. Gavilán — 163

Climate change and coastal flora
D. Metzing and A. Gerlach — 185

Sizing the impact: Coral reef ecosystems as early casualties of climate change
O. Hoegh-Guldberg — 203

Fingerprints of climate changes on the photosynthetic apparatus' behaviour, monitored by the JIP-test.
M. Tsimilli-Michael and R. Strasser — 229

Did recent climatic shifts affect productivity of grass-dominated vegetation in southern Switzerland?
A. Stampfli — 249

Responses of some Austrian glacier foreland plants to experimentally changed microclimatic conditions
B. Erschbamer — 263

Reliability and effectiveness of Ellenberg's indices in checking flora and vegetation changes induced by climatic variations
S. Pignatti, P. Bianco, G. Fanelli, R. Guarino, J. Petersen and P. Tescarollo — 281

Fingerprints of climate change – concluding remarks
Ch. Körner and G.-R. Walther — 305

Index — 317

"Fingerprints" of Climate Change

Adapted Behaviour and Shifting Species Ranges

Adapted behaviour and shifting ranges of species – a result of recent climate warming?
An introduction to "FINGERPRINTS" of CLIMATE CHANGE

GIAN-RETO WALTHER
Institute of Geobotany, University of Hannover, Nienburger Str. 17, 30167 Hannover, Germany

Abstract: The Earth's climate has warmed by a mean of 0.6 °C over the last 100 years. The observed change in environmental conditions has promoted the re-evaluation of long-term data sets. The studies demonstrate that there has been systematic change in the abundance and distribution of a broad range of species, and provide convincing evidence that recent climate warming has affected biological systems. In this introduction, numerous examples of adapted behaviour and shifting ranges of species related to recent climate change are provided. These observations are neither derived from computer models nor considered as predictions for potential future impacts, but document changes which are actually occurring in the biosphere. They provide ecological evidence that organisms are responding to the recent warming trend of the past three decades and thus represent the biological "fingerprints" of climatic change.

Since 1856, the global mean temperature has warmed by a mean of 0.6 ± 0.2 °C; regional temperature changes have varied, ranging from increases of greater than 0.6 °C to cooling in some regions (IPCC 2001a). The recent upward trend in temperatures began in the 1970s, with increasingly pronounced warming towards the end of the 20th century (Figure 1). The ten warmest years have all occurred since 1983, with eight of these occurring since 1990 (IPCC 2001a); this period includes a string of 16 consecutive months (from May 1997 to August 1998) where each monthly mean temperature broke all of the previous records (Karl et al. 2000). The year 2000 has continued the run of warm years in spite of the persistent cooling influence of the tropical Pacific La Niña (WMO 2001).

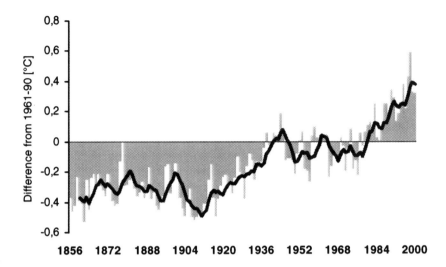

Figure 1. Annual land air and sea surface temperature anomalies. Grey: Temperature anomalies, Black: Smoothed curve (Data from Jones et al. 2000, WMO 2001).

Knowing that the mean global temperature has increased tells us little about how temperatures will change in a particular location, or how the ecosystem in that location will respond. From an ecological point of view, particular variables such as the annual minimum temperature (Alward et al. 1999, Godefroid & Tanghe 2000, Walther 2000) or temperature in specific months (Beebee 1995, Sparks & Yates 1997, Brown et al. 1999) may be more important than the annual mean. Many biological processes undergo sudden shifts at particular threshold values (Allen & Breshears 1998, Hoegh-Guldberg 2001; cf. also Easterling et al. 2000a,b, Parmesan et al. 2000).

Various approaches have been followed in investigating the potential impacts of climate change upon biological systems. Up to the mid 1990s, many publications concerned with the ecological consequences of global climate change focused upon the response of organisms to past climatic changes, or were based on models to predict future impacts (e.g. Peters & Lovejoy 1992, Woodward 1992, Gates 1993, Graves & Reavey 1996, IPCC 1996). However, models are usually quite limited in their spatial resolution and often do not include all the relevant feedback effects and other processes such as migration, competition, or human disturbance of the landscape (Ennis & Marcus 1996, cf. also Davis et al. 1998). Another approach to investigate climate change impacts was to manipulate temperatures using a variety of methods to heat the experimental ecosystems. These differing approaches have revealed some consistent patterns of response among the ecosystems studied, as well as puzzling inconsistencies (see e.g. Walker et al. 1999, Shaver et al. 2000 and cited literature). However, it has been argued that the ecological effects of global change are "too complex to be comprehensively addressed by experimental approaches, ..." (Ringold & Groffman 1997).

In recent years, evidence has increased that the warmer period since the 1970s has influenced the behaviour and ranges of many species. An increasing number of studies have documented observed impacts of regional climatic changes – primarily rising temperature – for a broad range of taxonomic groups with diverse geographical distributions (Table 1). Because these studies are based on observed changes rather than models, they can be considered as "fingerprints" of climate change and the result of a widespread trend toward warmer global temperatures. A selection of studies is provided in Figure 2, whereas Figure 3 gives the geographic references to the localities of the studies listed in Table 1.

Figure 2. Headlines of some recently published papers (citations: Grabherr et al. 1994, Holden 1994, Parmesan 1996, Brown et al. 1999, Menzel & Fabian 1999, Parmesan et al. 1999, Pockley 1999).

Evidence from both plants and animals indicate that the period of thirty years of warmer temperatures at the end of the 20th century has affected the phenology and physiology of organisms as well as the range and distribution of species. It has also led to changes in the structure and dynamics of ecosystems (cf. also IPCC 2001b). Table 1 gives an overview of studies reporting biological responses to climate change.

Table 1. Examples for adapted behaviour and shifting ranges of species related to recent climate change. The numbers given to the references refer to the geographic locations in Figure 3 (see also Kapelle et al. 1999, Hughes 2000, Wuethrich 2000, McCarty 2001, Lozan et al. 2001, Walther et al. 2001).

PLANTS
Phenology [18]Oglesby & Smith 1995, [108]Omasa et al. 1996, [96]Walkovsky 1998, [100]Ahas 1999, [15]Bradley et al. 1999, [73]Menzel & Fabian 1999, [57]Post & Stenseth 1999, [101]Ahas et al. 2000, [6]Beaubien & Freeland 2000, [74]Menzel 2000, [11]Schwartz & Reiter 2000, [19]Abu-Asab et al. 2001, [10]Cayan et al. 2001, [21]Fitzjarrald et al. 2001, [103]Gurung unpubl. see also: Menzel & Estrella 2001, Puhe & Ulrich 2001 **Growth rates** [17]Hamburg & Cogbill 1988, [13]Alward et al. 1999, [95]Hasenauer et al. 1999, [77]Pretzsch 1999, [1]Barber et al. 2000, [87]Paulsen et al. 2000. see also: Phillips 1996, Fung 1997, Myeni et al. 1997, Puhe & Ulrich 2001 **Phytoplankton/Algae** [97]Chisholm et al. 1995, [69]Kesel & Gödeke 1996, [52]Nehring 1998, [51]Reid et al. 1998, [25]Sagarin et al. 1999 **Vegetation shifts** [111]Wardle & Coleman 1992, [81]Landolt 1993, [116]Fowbert & Smith 1994, [89]Grabherr et al. 1994, [23]Peterson 1994, [117]Smith 1994, [114]Kennedy 1995, [2]Stuart Chapin III et al. 1995, [91]Klötzli et al. 1996, [90]Pauli et al. 1996, [22]Brown et al. 1997, [115]Grobe et al. 1997, [89]Walther 1997, [13]Alward et al. 1999, [92/93]Carraro et al. 1999a, b, [84]Klötzli & Walther 1999, [82]Meduna et al. 1999, [70]Frahm & Klaus 2000, [88]Keller et al. 2000, [68]Kesel 2000, [99]Meshinev et al. 2000, [104]Vlasenko 2000, [94]Walther 2000, [59]Kullmann 2001, [83]Landolt 2001, [86]Walther & Grundmann 2001 see also: Guisan et al. 1995, Theurillat et al. 1998, Serreze et al. 2000 Burga & Kratochwil 2001
ANIMALS
Corals/Zooplankton/marine Invertebrates [26]Barry et al. 1995, [31]Roemmich & McGowan 1995, [110]Hoegh-Guldberg 1999, [25]Sagarin et al. 1999, [106]Reaser et al. 2000, [107]Edwards et al. 2000, [109]Loya et al. 2001. Coral bleaching: see also Lough 2000, Hoegh-Guldberg 2001 **Insects** [39]Dennis & Shreeve 1991, [75]Ott 1996, [27]Parmesan 1996, [64]Ellis 1997, [65]Ellis et al. 1997, [22]Brown et al. 1997, [7]Stewart et al. 1997, [61]De Jong & Brakefield 1998, [105]Epstein et al. 1998, [43]Hill et al. 1999, [47]McEwen 1999, [28]Parmesan et al. 1999, [67]Handke 2000, [60]Lindgren et al. 2000, [76]Ott 2000, [44]Roy & Sparks 2000, [63]Visser & Holleman 2001, [103]Gurung unpubl. see also: Dennis 1993, Harrington & Stork 1995

Introduction to "FINGERPRINTS" of CLIMATE CHANGE

Amphibians & Reptiles
[49]Beebee 1995, [45]Forchhammer et al. 1998, [42]Reading 1998, [33]Pounds et al. 1999, [20]Corser 2001, [9]Kiesecker et al. 2001, [34]Pounds 2001

Fishes
[16]Drinkwater 1997, [29]Holbrook et al. 1997, [50]von Westernhagen 1998, [100]Ahas 1999, [98]Nieder et al. 2000, [53]O'Brian et al. 2000, [8]Dempson et al. 2001

Birds
[112]Cunningham & Moors 1994, [79]Bezzel & Jetz 1995, [4]LaRoe & Rusch 1995, [48]Mason 1995, [18]Oglesby & Smith 1995, [40]Crick et al. 1997, [66]Ludwichowski 1997, [30]Veit et al. 1997, [71]Bairlein & Winkel 1998, [78]Berthold et al. 1998, [45]Forchhammer et al. 1998, [46]McCleery & Perrins 1998, [54]Prop et al. 1998, [100]Ahas 1999, [80]Bergmann 1999, [15]Bradley et al. 1999, [32]Brown et al. 1999, [37]Crick & Sparks 1999, [5]Dunn & Winkler 1999, [33]Pounds et al. 1999, [118]Smith et al. 1999, [41]Sparks 1999, [38]Thomas & Lennon 1999, [12]Inouye et al. 2000, [58]Saether et al. 2000, [36]Stevenson & Bryant 2000, [72]Bairlein & Winkel 2001, [62]Both & Visser 2001, [113]Barbraud & Weimerskirch 2001, [35]Moss et al. 2001, [24]Oedekoven et al. 2001, [102]Yom-Tov 2001
see also: Burton 1995

Mammals
[22]Brown et al. 1997, [55]Post et al. 1997, [56/14]Post et al. 1999a, b, [57]Post & Stenseth 1999, [12]Inouye et al. 2000, [3]Kelly 2001

ECOLOGICAL CONSEQUENCES

Aebischer et al. 1990, Brown et al. 1997, Loeb et al. 1997, Veit et al. 1997, McGowan et al. 1998, Visser et al. 1998, Harrington et al. 1999, Harvell et al. 1999, Post et al. 1999b, Both & Visser 2001, Thomas et al. 2001

In general, biological responses to climate change are complex and trends may be masked by the effects of multiple causal factors. This makes the attribution of causality difficult or impossible within the confines of an individual study. Although the causal link to climate cannot be rigorously demonstrated, the existence of consistent patterns strongly suggest that biological systems are responding now to climate change, and that this is not simply a possibility for the future (cf. Bradley 2001, McCarty 2001). Most of the observed changes, such as poleward and elevational shifts, earlier development in the spring, or growing season extending into the autumn, are consistent with a response to warmer temperatures. The probability that all these observed impacts could occur by chance alone is considered negligible by the IPCC (2001b).

Ecological changes in response to climate change are expected to occur everywhere. The lack of observed impacts in a particular region, may be simply because climate change has not yet reached critical thresholds for such effects. However, "fingerprints" of climate change in terms of adapted behaviour or shifting species ranges are expected to be detectable first in

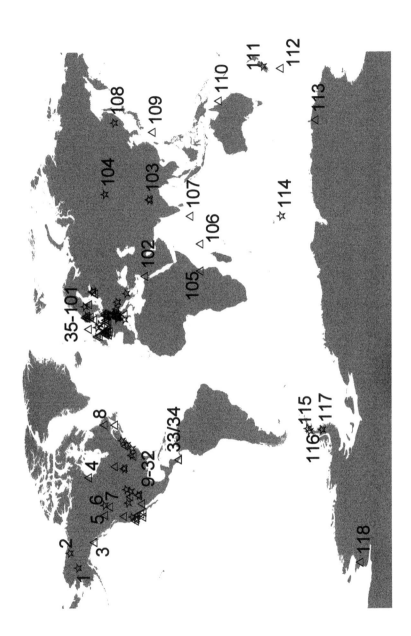

Figure 3. Global distribution of reported "fingerprints" of climate change (☆ = plants, Δ = animals; the numbers coincide with the references given in Table 1)

particularly sensitive areas, such as ecotones or at the distribution boundaries of species. The examples reported here tend to be in regions where recent regional warming has crossed a critical threshold (for example, the absence of winter frost) relevant to particular organisms.

The predicted further increases in mean temperature will increase the likelihood of crossing further thresholds, and we can expect more "fingerprints" of climate change in other areas and for other species in the relatively near future. Thus, the observed changes reported in this book may foreshadow the types of impacts likely to become more frequent and widespread with continued warming.

NOTE

Additional references of studies with reports on observed changes in the behaviour or shifts in the ranges of species related to recent climate warming are welcome.

ACKNOWLEDGMENTS

Many thanks to Peter J. Edwards for proof-reading and Regula Langenauer and Dave Dokken for providing relevant literature.

REFERENCES

Abu-Asab M.S., Peterson P.M., Shetler S.G. & Orli S.S., 2001, Earlier plant flowering in spring as a response to global warming in the Washington, DC, area. *Biodivers. Conserv.* 10(4): 597-612.

Aebischer N.J., Coulson J.C. & Colebrook J.M., 1990, Parallel long-term trends across four marine trophic levels and weather. *Nature* 347: 753-755.

Ahas R., 1999, Long-term phyto-, ornitho- and ichthyophenological time-series analyses in Estonia. *Int. J. Biometeorol.* 42: 119-123.

Ahas R., Jaagus J. & Aasa A., 2000, The phenological calendar of Estonia and its correlation with mean air temperature. *Int. J. Biometeorol.* 44: 159-166.

Allen C.D. & Breshears D.D., 1998, Drought-induced shift of a forest-woodland ecotone: Rapid landscape response to climate variation. *Proc. Nat. Acad. Sci. USA* 95: 14839-14842.

Alward R.D., Ketling J.K. & Milchunas D.G., 1999, Grassland vegetation changes and nocturnal global warming. *Science* 283: 229-231.

Bairlein F. & Winkel W., 1998, Vögel und Klimaänderungen. In: J.L. Lozan, H. Grassl & P. Hupfer (eds.) *Warnsignal Klima*. Wissenschaftliche Auswertungen, Hamburg, pp. 281-285.

Bairlein F. & Winkel W., 2001, Birds and climate change. In: J.L. Lozan, H. Grassl & P. Hupfer (eds.) *Climate of the 21st Century: Changes and Risks*. Wissenschaftliche Auswertungen, Hamburg, pp. 278-282.

Barber V.A., Juday G.P. & Finney B.P., 2000, Reduced growth of Alaskan white spruce in the twentieth century from temperature-induced drought stress. *Nature* 405: 668-673.

Barbraud C. & Weimerskirch H., 2001, Emperor penguins and climate change. *Nature* 411: 183-186.

Barry J.P., Baxter C.H., Sagarin R.D. & Gilman S.E., 1995, Climate-related long-term faunal changes in a California rocky intertidal community. *Science* 267: 672-675.

Beaubien E.G. & Freeland H.J., 2000, Spring phenology trends in Alberta, Canada: links to ocean temperature. *Int. J. Biometeor.* 44(2): 53-59.

Beebee T.J.C., 1995, Amphibian breeding and climate. *Nature* 374: 219-220.

Bergmann F., 1999, Long-term increase in numbers of early-fledged Reed Warblers (*Acrocephalus scirpaceus*) at Lake Constance (Southern Germany). *J. Ornithol.* 140(1): 81-86.

Berthold P., Fiedler W., Schlenker R. & Querner U., 1998, 25-year study of the population development of central European songbirds: A general decline, most evident in long-distance migrants. *Naturwissenschaften* 85: 350-353.

Bezzel E. & Jetz W., 1995, Delay of the autumn migratory period in the blackcap (*Sylvia atricappila*) 1966-1993: a reaction to global warming? *J. Ornithol.* 136: 83-87.

Both C. & Visser M.E., 2001, Adjustment to climate change is constrained by arrival date in a long-distance migrant bird. *Nature* 411: 296-298.

Bradley N.L., Leopold A.C., Ross J. & Huffaker W., 1999, Phenological changes reflect climate change in Wisconsin. *Proc. Nat. Acad. Sci USA* 96(17): 9701-9704.

Bradley R.S., 2001, Many citations support global warming trend. *Science* 292: 2011.

Brown J.H., Valone T.J. & Curtin C.G., 1997, Reorganization of an arid ecosystem in response to recent climate change. *Proc. Natl. Acad. Sci. USA* 94: 9729-9733.

Brown J.L., Li S.H. & Bhagabati N., 1999, Long-term trend toward earlier breeding in an American bird: A response to global warming? *Proc. Nat. Acad. Sci. USA* 96: 5565-5569.

Burga C.A. & Kratochwil A., 2001, *Biomonitoring: General and applied aspects on regional and global scales*. Tasks for vegetation science 35, Kluwer Academic Publ., Dordrecht.

Burton J.F., 1995, *Birds and climate change*. Christopher Helm, London.

Carraro G., Klötzli F., Walther G.-R., Gianoni P. & Mossi R., 1999, *Observed changes in vegetation in relation to climate warming*. Final report 31, vdf, Zürich, 87p. + Annex.

Carraro G., Gianoni P. & Mossi R., 1999b, Climatic influence on vegetation changes: a verification on regional scale of the laurophyllisation. In: F. Klötzli & G.-R. Walther (eds.) *Conference on recent shifts in vegetation boundaries of deciduous forests, especially due to general global warming*. Monte Verità; Proceedings of the Centro Stefano Franscini, Ascona. Birkhäuser, Basel, pp. 31-51.

Cayan D.R., Kammerdiener S.A., Dettinger M.D., Caprio J.M. & Peterson D.H., 2001, Changes in the onset of spring in the western United States. *Bull. Am. Meteor. Soc.* 82(3): 399-415.

Chisholm J.R.M., Joubert J.M. & Giaccone G., 1995, Caulerpa taxifolia in the northwest Mediterranean: Introduced species or migrant from the Red Sea? *C. R. Acad. Sci., Ser. III Sci Vie/Life Sci.* 318(12): 1219-1226.

Corser J.D., 2001, Decline of disjunct green salamander (*Aneides aeneus*) populations in the southern Appalachians. *Biol. Conserv.* 97: 119-126.

Crick H.Q.P., Dudley C., Glue D.E. & Thomson D.L., 1997, UK birds are laying eggs earlier. *Nature* 388: 526.

Crick H.Q.P. & Sparks T.H., 1999, Climate change related to egg-laying trends. *Nature* 399: 423-424.
Cunningham D.M. & Moors P.J., 1994, The decline of rockhopper penguins (*Eudyptes chrysocome*) at Campbell Island, Southern Ocean and the influence of rising sea temperatures. *Emu* 94: 27-36.
Davis A.J., Jenkinson L.S., Lawton J.H., Shorrocks B. & Wood S., 1998, Making mistakes when predicting shifts in species range in response to global warming. *Nature* 391: 783-786.
De Jong P.W. & Brakefield P.M., 1998, Climate and change inclines for melanism in the two-spot ladybird, *Adalia bipunctata* (Coleoptera: Coccinellidae). *Proc. Royal Soc. London B* 265: 39-43.
Dennis R.L.H., 1993, *Butterflies and climate change*. Manchester University Press, New York.
Dennis R.L.H. & Shreeve T.G., 1991, Climatic change and the British butterfly fauna: opportunities and constraints. *Biol. Conserv.* 55: 1-16.
Dempson J.B., O'Connell M.F. & Cochrane N.M., 2001, Potential impact of climate warming on recreational fishing opportunities for Atlantic salmon, *Salmo salar* L., in Newfoundland, Canada. *Fisheries Management and Ecology* 8: 69-82.
Drinkwater K.F., 1997, Impacts of climate variability on Atlantic Canadian fish and shellfish stocks. In: R.W. Shaw (ed.) *Climate variability and climate change in Atlantic Canada.* Volume VI of the Canada Country Study: climate impacts and adaptation. Atmospheric Science Division, Environment Canada, Atlantic Region, pp. 38-50.
Dunn P.O. & Winkler D.W., 1999, Climate change has affected the breeding date of tree swallows throughout North America. *Proc. R. Soc. Lond., Ser. B.: Biol. Sci.* 266: 2487-2490.
Edwards A.J., Clark S., Zahir H., Rajasuriya A., Naseer A. & Rubens J., 2001, Coral bleaching and mortality on artificial and natural reefs in Maldives in 1998, sea surface temperature anomalies and initial recovery. *Mar. Pollut. Bull.* 42: 7-15.
Ellis W.N., 1997, Recent shifts in phenology of Microlepidoptera, related to climatic change (Lepidoptera). *Entomol. Ber. (Amst.)* 57: 66-72.
Ellis W.N., Donner J.H. & Kuchlein J.H., 1997, Recent shifts in distribution of Microlepidoptera in The Netherlands. *Entomol. Ber. (Amst.)* 57: 119-125.
Ennis C.A. & Marcus N.H., 1996, *Biological consequences of global climate change.* University Science Books, Sausalito, California.
Easterling D.R., Evans J.L., Groisman P.Ya., Karl T.R., Kunkel K.E. & Ambenje P., 2000a, Observed variability and trends in extreme climate events: a brief review. *Bull. Am. Meteor. Soc.* 81(3): 417-425.
Easterling D.R., Meehl G.A., Parmesan C., Changnon S.A., Karl T.R. & Mearns L.O., 2000b, Climate extremes: Observations, modeling, and impacts. *Science* 289: 2068-2074.
Epstein P.R., Diaz H.F., Elias S., Grabherr G., Graham N.E., Martens W.J.M., Mosley-Thompson E. & Susskind J., 1998, Biological and physical signs of climate change: Focus on Mosquito-borne diseases. *Bull. Am. Meteor. Soc.* 79(3): 409-417.
Fitzjarrald D.R., Acevedo O.C. & Moore K.E., 2001, Climatic consequences of leaf presence in the eastern United States. *J. Clim.* 14(4): 598-614.
Forchhammer M.C., Post E. & Stenseth N.C., 1998, Breeding phenology and climate... *Nature* 391: 29-30.
Fowbert J.A. & Smith R.I.L., 1994, Rapid population increases in native vascular plants in the Argentine Islands, Antarctic peninsula. *Arct. Alp. Res.* 26(3): 290-296.
Frahm J.-P. & Klaus D., 2000, Moose als Indikatoren von rezenten und früheren Klimafluktuationen in Mitteleuropa. *NNA-Ber.* 2/2000: 69-75.

Fraser W.R., Trivelpiece W.Z., Ainley D.G. & Trivelpiece S.G., 1992, Increase in Antarctic penguin populations: reduced competition with whales or a loss of sea ice due to environmental warming? *Polar Biol.* 11: 525-531.

Fung I., 1997, A greener north. *Nature* 386: 659-660.

Gates D.M., 1993, *Climate change and its biological consequences.* Sinauer, Sunderland, Massachusetts.

Gatter W., 1992, Timing and patterns of visible autumn migration: can effects of global warming be detected? *J. Ornithol.* 133: 427-436.

Godefroid S. & Tanghe M., 2000, Influence of small climatic variations on the species composition of roadside grasslands. *Phytocoenologia* 30(3-4): 655-664.

Gonzalez P., 2001, Desertification and a shift of forest species in the West African Sahel. *Clim. Res.* (in press).

Grabherr G., Gottfried M. & Pauli H., 1994, Climate effects on mountain plants. *Nature* 369: 448.

Graves J. & Reavey D., 1996, *Global environmental change – plants, animals and communities.* Longman, Essex.

Grobe C.W., Ruhland C.T. & Day T.A., 1997, A new population of *Colobanthus quitensis* near Arthur Harbor, Antarctica: Correlating recruitment with warmer summer temperatures. *Arct. Alp. Res.* 29(2): 217-221.

Guisan A., Holten J.I., Spichiger R. & Tessier L. (eds.) 1995, *Potential ecological impacts of climate change in the Alps and Fennoscandian Mountains.* Publ. hors-série n° 8 des Conservatoire et Jardin botaniques de la ville de Genève, 194pp.

Gurung J.B., Global warming signs found in the Himalayan region, unpubl. data.

Hamburg S.P. & Cogbill C.V. 1988, Historical decline of red spruce populations and climatic warming. *Nature* 331: 428-431,

Handke K., 2000, Veränderungen in der Insektenfauna der Bremer Flussmarschen 1982-1999 – Zeichen eines Klimawandels? *NNA-Ber.* 2/2000: 37-54.

Harrington R. & Stork N.E. (eds.) 1995, *Insects in a changing environment.* Academic Press, San Diego, California, USA.

Harrington R., Woiwod I. & Sparks T., 1999, Climate change and trophic interactions. *Trends Ecol. Evol.* 14(4): 146-150.

Harvell C.D., Kim K., Burkholder J.M., Colwell R.R., Epstein P.R., Grimes D.J., Hofmann E.E., Lipp E.K., Osterhaus A.D.M.E., Overstreet R.M., Porter J.W., Smith G.W. & Vasta G.R., 1999, Emerging marine diseases – climate links and antropogenic factors. *Science* 285: 1505-1510.

Hasenauer H., Nemani R.R., Schadauer K. & Running S.W., 1999, Forest growth response to changing climate between 1961 and 1990 in Austria. *For. Ecol. Manag.* 122: 209-219.

Hill J.K., Thomas C.D. & Huntley B., 1999, Climate and habitat availability determine 20th century changes in a butterfly's range margins. *Proc. R. Soc. Lond., Ser. B.: Biol. Sci.* 266: 1197-1206.

Hoegh-Guldberg O., 1999, Climate change, coral bleaching and the future of the world's coral reefs. *Mar. Freshwater Res.* 50: 839-866.

Hoegh-Guldberg O., 2001, Sizing the impact: Coral reef ecosystems as early casualties of climate change. In: G.-R. Walther, C.A. Burga & P.J. Edwards (eds.) *"Fingerprints" of Climate Change – Adapted behaviour and shifting species ranges.* Kluwer Academic Publ., New York and London, 205-230.

Holbrook S.J., Schmitt R.J. & Stephens J.S., 1997, Changes in an assemblage of temperate reef fishes associated with a climate shift. *Ecol. Appl.* 7(4): 1299-1310.

Holden C. (ed.) 1994, Greening of the Antarctic Peninsula. *Science* 266: 35.

Hughes L., 2000, Biological consequences of global warming: is the signal already apparent? *Trends Ecol. Evol.* 15(2): 56-61.

Inouye D.W., Barr B., Armitage K.B. & Inouye B.D., 2000, Climate change is affecting altitudinal migrants and hibernating species. *Proc. Natl. Acad. Sci. USA* 97(4): 1630-1633.

IPCC 1996, *Climate change 1995: Impacts, adaptations and mitigation of climate change: Scientific-technical analyses.* Contribution of working group II to the second assessment report of the intergovernmental panel on climate change. Cambridge Univ. Press, Cambridge.

IPCC 2001a, *Climate change 2001: The scientific basis.* A report of working group I of the Intergovernmental Panel on Climate Change. Cambridge University Press, Cambridge.

IPCC 2001b, *Climate change 2001: Impacts, Adaptation, and Vulnerability.* A report of working group II of the Intergovernmental Panel on Climate Change. Cambridge University Press, Cambridge.

Jones P.D., Parker D.E., Osborn T.J. & Briffa K.R., 2000, Global and hemispheric temperature anomalies--land and marine instrumental records. In: *Trends: A Compendium of Data on Global Change.* Carbon Dioxide Information Analysis Center, Oak Ridge National Laboratory, U.S. Department of Energy, Oak Ridge, Tennessee, U.S.A.

Kapelle M., van Vuuren M.M.I. & Baas P., 1999, Priorities in research focusing at the effects of global climate change on biodiversity. *Biodivers. Conserv.* 8: 1383-1397.

Karl T.R., Knight R.W. & Baker B., 2000, The record breaking global temperatures of 1997 and 1998: Evidence for an increase in the rate of global warming? *Geophys. Res. Lett.* 27(5): 719-722.

Keller F., Kienast F. & Beniston M., 2000, Evidence of response of vegetation to environmental change on high-elevation sites in the Swiss Alps. *Reg. Environ. Change* 1(2): 70-77.

Kelly B.P., 2001, Climate change and ice breeding pinnipeds. In: G.-R. Walther, C.A. Burga & P.J. Edwards (eds.) *"Fingerprints" of Climate Change – Adapted behaviour and shifting species ranges.* Kluwer Academic Publ., New York and London, 45-57.

Kennedy A.D., 1995, Antarctic terrestrial ecosystem response to global environmental change. *Annu. Rev. Ecol. Syst.* 26: 683-704.

Kesel R., 2000, Auswirkungen der Klimaänderung auf Flora und Vegetation in Nordwestdeutschland. *NNA-Ber.* 2/2000: 2-12.

Kesel R. & Gödeke T., 1996, *Wolffia arrhiza, Azolla filiculoides, Lemna turionifera* und andere wärmeliebende Pflanzen in Bremen – Boten eines Klimawandels? *Abh. Naturwiss. Verein Bremen* 43(2): 339-362.

Kiesecker J.M., Blaustein A.R. & Belden L.K., 2001, Complex causes of amphibian population declines. *Nature* 410: 681-684.

Klötzli F., Walther G.-R., Carraro G. & Grundmann A., 1996, Anlaufender Biomwandel in Insubrien. *Verh. Ges. Ökol.* 26: 537-550

Klötzli F. & Walther G.-R., 1999, Recent vegetation shifts in Switzerland. In: F. Klötzli & G.-R. Walther (eds.) *Conference on recent shifts in vegetation boundaries of deciduous forests, especially due to general global warming.* Monte Verità; Proceedings of the Centro Stefano Franscini, Ascona. Birkhäuser, Basel, pp. 15-29.

Kullmann L., 2001, 20th century climate warming and tree-limit rise in the southern Scandes of Sweden. *Ambio* 30(2): 72-80.

Landolt E., 1993, Über Pflanzenarten, die sich in den letzten 150 Jahren in der Stadt Zürich stark ausgebreitet haben. *Phytocoenologia* 23: 651-663.

Landolt E., 2001, *Flora der Stadt Zürich.* Birkhäuser, Basel.

LaRoe E.T. & Rusch D.H., 1995, Changes in nesting behaviour of arctic geese. In: National Biological Service, U.S. Dept. of the Interior, *Our living resources.* A report to the Nation of the distribution, abundance and health of U.S. plants, animals and ecosystems. Washington, DC, pp. 388.

Lindgren E., Tälleklint L. & Polfeldt T., 2000, Impact of climatic change on the northern latitude limit and population density of the disease-transmitting European tick *Ixodes ricinus*. *Environ. Health Perspect.* 108(2): 119-123.

Loeb V., Siegel V., Holm-Hansen O., Hewitt R., Fraser W., Trivelpiece W. & Trivelpiece S., 1997, Effects of sea-ice extent and krill or salp dominance on the Antarctic food web. *Nature* 387: 897-900.

Lough J.M., 2000, 1997-98: Unprecedented thermal stress to coral reefs? *Geophys. Res. Lett.* 27(23):3901-3904.

Loya Y., Sakai K., Yamazato K., Nakano Y., Sambali H. & van Woesik R., 2001, Coral bleaching: the winners and the losers. *Ecol. Lett.* 4(2): 122-131.

Lozan J.L., Grassl H. & Hupfer P., 2001, *Climate of the 21st century: Changes and risks.* Wissenschaftliche Auswertungen, Hamburg.

Ludwichowski I., 1997, Long-term changes of wing-length, body mass and breeding parameters in first time breeding females of goldeneyes (*Bucephala changula changula*) in northern Germany. *Vogelwarte* 39: 103-116.

Mason C.F., 1995, Long-term trends in the arrival dates of spring migrants. *Bird Study* 42: 182-189.

McCarty J.P., 2001, Ecological consequences of recent climate change. *Conserv. Biol.* 15(2): 320-331.

McCleery R.H. & Perrins C.M., 1998, ...temperature and egg-laying trends. *Nature* 391: 30-31.

McEwen P., 1999, *Global warning: summer superbugs on the increase.* Insect Investigations Ltd., Cardiff University, unpubl. report.

McGowan J.A., Cayan D.R. & Dorman L.M., 1998, Climate-ocean variability and ecosystem response in the northeastern Pacific. *Science* 281: 210-217.

Meduna E., Schneller J.J. & Holderegger R., 1999, Prunus laurocerasus L., eine sich ausbreitende nichteinheimische Gehölzart: Untersuchungen zu Ausbreitung und Vorkommen in der Nordschweiz. *Z. Ökologie u. Naturschutz* 8: 147-155.

Menzel A., 2000, Trends in phenological phases in Europe between 1951 and 1996. *Int. J. Biometeor.* 44(2): 76-81.

Menzel A. & Fabian P., 1999, Growing season extended in Europe. *Nature* 397: 659.

Menzel A. & Estrella N., 2001, Plant phenological changes. In: G.-R. Walther, C.A. Burga & P.J. Edwards (eds.) *"Fingerprints" of Climate Change – Adapted behaviour and shifting species ranges.* Kluwer Academic Publ., New York and London, 125-139.

Meshinev T., Apostolova I. & Koleva E., 2000, Influence of warming on timberline rising: a case study on *Pinus peuce* Griseb. in Bulgaria. *Phytocoenologia* 30(3-4): 431-438.

Moss R., Oswald J. & Baines D., 2001, Climate change and breeding success: decline of the capercaillie in Scotland. *J. Anim. Ecol.* 70(1): 47-61.

Myeni R.B., Keeling C.D., Tucker C.J., Asrar G. & Nemani R.R., 1997, Increased plant growth in the northern high latitudes from 1981 to 1991. *Nature* 386: 698-702.

Nehring S., 1998, Establishment of thermophilic phytoplankton species on the North Sea: biological indicators of climatic changes? *ICES J. Mar. Sci.* 55: 818-823.

Nieder J., La Mesa G. & Vacchi M., 2000, Blenniidae along the Italian coasts of the Ligurian and the Tyrrhenian Sea: Community structure and new records of *Scartella cristata* for northern Italy. *Cybium* 24(4): 359-369.

O'Brian C.M., Fox C.J., Planque B. & Casey J., 2000, Climate variability and North Sea cod. *Nature* 404: 142.

Oedekoven C.S., Ainley D.G. & Spear L.B., 2001, Variable responses of seabirds to change in marine climate: California Current, 1985-1994. *Mar. Ecol. Prog. Ser.* 212: 265-281.

Oglesby R.T. & Smith C.R., 1995, Climate change in the Northeast. In: National Biological Service, U.S. Dept. of the Interior, *Our living resources*. A report to the Nation of the distribution, abundance and health of U.S. plants, animals and ecosystems. Washington, DC, pp. 390-391.

Omasa K., Kai K., Taoda H., Uchijima Z. & Yoshino M. (eds.) 1996, *Climate change and plants in East Asia*. Springer, Heidelberg.

Ott J., 1996, Zeigt die Ausbreitung der Feuerlibelle in Deutschland eine Klimaänderung an? *Nat.schutz Landsch.plan.* 28(2): 53-61.

Ott J., 2000, Die Ausbreitung mediterraner Libellenarten in Deutschland und Europa – die Folge einer Klimaänderung? *NNA-Ber.* 2/2000: 13-36.

Parmesan C., 1996, Climate and species' range. *Nature* 382: 765-766.

Parmesan C., Ryrholm N., Stefanescu C., Hill J.K., Thomas C.D., Descimon H., Huntley B., Kaila L., Kullberg J., Tammaru T., Tennent W.J., Thomas J.A. & Warren M., 1999, Poleward shifts in geographical ranges of butterfly species associated with regional warming. *Nature* 399: 579-583.

Parmesan C., Root T.L. & Willig M.R., 2000, Impacts of extreme weather and climate on terrestrial biota. *Bull. Am. Meteor. Soc.* 81(3): 443-450.

Paulsen J., Weber U.M. & Körner C., 2000, Tree growth near treeline: Abrupt or gradual reduction with altitude? *Arct. Antarct. Alp. Res.* 32: 14-20.

Pauli H., Gottfried M. & Grabherr G., 1996, Effects of climate change on mountain ecosystems – upward shifting of alpine plants. *World Res. Rev.* 8(3): 382-390.

Peters R.L. & Lovejoy, T.E., 1992, *Global warming and biodiversity*. Yale University Press, New Haven.

Peterson D.L., 1994, Recent changes in the growth and establishment of subalpine conifers in western North America. In: M. Beniston (ed.) *Mountain environments in changing climates*. Routledge, London, pp. 234-243.

Phillips O.L., 1996, Long-term environmental change in tropical forests: increasing tree turnover. *Environ. Conserv.* 23(3): 235-248.

Pockley P., 1999, Global warming 'could kill most coral reefs by 2100'. *Nature* 400: 98.

Post E., Stenseth N.C., Langvatn R. & Fromentin J.M., 1997, Global climate change and phenotypic variation among red deer cohorts. *Proc. R. Soc. Lond., Ser. B.: Biol. Sci.* 264: 1317-1324.

Post E., Forchhammer M.C., Stenseth N.C. & Langvatn R., 1999a, Extrinsic modification of vertebrate sex ratios by climatic variation. *Am. Nat.* 154(2): 194-204.

Post E., Peterson R.O., Stenseth N.C. & McLaren B.E., 1999b, Ecosystem consequences of wolf behavioural response to climate. *Nature* 401: 905-907.

Post E. & Stenseth N.C., 1999, Climatic variability, plant phenology, and northern ungulates. *Ecology* 80(4): 1322-1339.

Pounds J.A., 2001, Climate and amphibian declines. *Nature* 410: 639-640.

Pounds J.A., Fogden M.P.L. & Campbell J.H., 1999, Biological response to climate change on a tropical mountain. *Nature* 398: 611-615.

Pretzsch H., 1999, Changes in forest growth. *Forstwiss. Centralbl.* 118(4): 228-250.

Prop J., Black J.M., Shimmings P. & Owen M., 1998, The spring range of barnacle geese *Branta leucopsis* in relation to changes in land management and climate. *Biol. Conserv.* 86: 339-346.

Puhe J. & Ulrich B., 2001, *Global climate change and human impacts on forest ecosystems*. Ecological Studies 143, Springer, Berlin Heidelberg.

Reading C.J., 1998, The effect of winter temperatures on the timing of breeding activity in the common toad *Bufo bufo*. *Oecologia* 117: 469-475.

Reaser J.K., Pomerance R. & Thomas P.O., 2000, Coral bleaching and global climate change: Scientific findings and policy recommendations. *Conserv. Biol.* 14(5): 1500-1511.

Reid P.C., Edwards M., Hunt H.G. & Warner A.J., 1998, Phytoplankton change in the North Atlantic. *Nature* 391: 546.

Ringold P.L. & Groffman P.M., 1997, Inferential studies on climate change. *Ecol. Appl.* 7(3): 751-752.

Roemmich D. & McGowan J., 1995, Climatic warming and the decline of zooplankton in the California current. *Science* 267: 1324-1326.

Roy D.B. & Sparks T.H., 2000, Phenology of British butterflies and climate change. *Global Change Biol.* 6(4): 407-416.

Saether B.-E., Tufto J., Engen S., Jerstad K., Rostad O.W. & Skatan J.E., 2000, Population dynamical consequences of climate change for a small temperate songbird. *Science* 287: 854-856.

Sagarin R.D., Barry J.P., Gilman S.E. & Baxter C.H., 1999, Climate-related change in an intertidal community over short and long time scales. *Ecol. Monogr.* 69(4): 465-490.

Schwartz M.D. & Reiter B.E., 2000, Changes in North American spring. *Int. J. Climatol.* 20(8): 929-932.

Serreze M.C., Walsh J.E., Chapin F.S.III, Osterkamp T., Dyurgerov M., Romanovsky V., Oechel W.C., Morison J., Zhang T. & Barry R.G., 2000, Observational evidence of recent change in the northern high-latitude environment. *Clim. Change* 46: 159-207.

Shaver G.R., Canadell J., Chapin F.S.III, Gurevitch J., Harte J., Henry G., Ineson P., Jonasson S., Melillo J., Pitelka L. & Rustad L., 2000, Global warming and terrestrial ecosystems: a conceptual framework for analysis. *BioScience* 50(10): 871-882.

Smith R.C., Ainley D., Baker K., Domack E., Emslie S, Fraser B., Kennett J., Leventer A., Mosley-Thompson E., Stammerjohn S. & Vernet M., 1999, Marine ecosystem sensitivity to climate change. *BioScience* 49(5): 393-404.

Smith R.I.L., 1994, Vascular plants as bioindicators of regional warming in Antarctica. *Oecologia* 99(3-4): 322-328.

Sparks T.H., 1999, Phenology and the changing pattern of bird migration in Britain. *Int. J. Biometeorol.* 42: 134-138.

Sparks T.H. & Yates T.J., 1997, The effect of spring temperature on the appearance dates of British butterflies 1883-1993. *Ecography* 20: 368-374.

Stevenson I.R. & Bryant D.M., 2000, Climate change and constraints on breeding. *Nature* 406: 366-367.

Stewart R., Wheaton E. & Spittlehouse D., 1997, Climate change: Implications for the boreal forest. Implications of climate change: What do we know? Cited in: Brydges T., Ecological change and the challenges for monitoring. *Environ. Monit. Assess.* 67: 89-95.

Stuart Chapin III, F., Shaver G.R., Giblin A.E., Nadelhoffer K.J. & Laundre J.A., 1995, Responses of arctic tundra to experimental and observed changes in climate. *Ecology* 76(3): 694-711.

Theurillat J.-P., Felber F., Geissler P., Gobat J.-M., Fierz M., Fischlin A., Küpfer P., Schlüssel A., Velluti C., Zhao G.F. & Williams J., 1998, Sensitivity of plant and soil ecosystems of the Alps to climate change. In: P. Cebon, U. Dahinden, H. Davies, D.M. Imboden & C.C. Jäger (eds.) *View from the Alps – Regional perspectives on climate change*. MIT Press, Massachusetts, pp. 225-308.

Thomas C.D. & Lennon J.J., 1999, Birds extend their ranges northwards. *Nature* 399: 213.

Thomas D.W., Blondel J., Perret P., Lambrechts M.M. & Speakman J.R., 2001, Energetic and fitness costs of mismatching resource supply and demand in seasonally breeding birds. *Science* 291: 2598-2600.

Veit R.R., McGowan J.A., Ainley D.G., Wahls T.R. & Pyle P., 1997, Apex marine predator declines ninety percent in association with changing oceanic climate. *Global Change Biol.* 3: 23-28.

Visser M.E., van Noordwijk A.J., Tinbergen J.M. & Lessells C.M., 1998, Warmer springs lead to mistimed reproduction in great tits (*Parus major*). *Proc. R. Soc. Lond., Ser. B.: Biol. Sci.* 265: 1867-1870.

Visser M.E. & Holleman L.J.M., 2001, Warmer springs disrupt the synchrony of oak and winter moth phenology. *Proc. R. Soc. Lond., Ser. B.: Biol. Sci.* 268: 289-294.

Vlasenko V.I., 2000, The mapping of vegetation cover dynamics in the Sayan-Shushensky reserve. *Geobot. Mapping* 1998-2000. Komarov Bot. Inst., Acad. Sci. Russia.

von Westernhagen H. 1998, Klima und Fischerei. In: J.L. Lozan, H. Grassl & P. Hupfer (eds.) *Warnsignal Klima*. Wissenschaftliche Auswertungen, Hamburg, pp. 286-291.

Walker B., Steffen W., Canadell J. & Ingram J. (eds.), 1999, *The terrestrial biosphere and global change*. IGBP book series 4, Cambridge University Press, Cambridge.

Walkovsky A., 1998, Changes in phenology of the locust tree (*Robinia pseudoacacia* L.) in Hungary. *Int. J. Biometeorol.* 41: 155-160.

Walther G.-R., 1997, Longterm changes in species composition of Swiss beech forests. *Ann. Bot. (Roma)* 55: 77-84.

Walther G.-R., 2000, Climatic forcing on the dispersal of exotic species. *Phytocoenologia* 30(3-4): 409-430.

Walther G.-R. & Grundmann A., 2001, Trends of vegetation change in colline and submontane climax forests in Switzerland. *Bull. Geobot. Inst. ETH* 67, in press.

Walther G.-R., Burga C.A. & Edwards P.J. (eds.) 2001, *"Fingerprints" of Climate Change – Adapted behaviour and shifting species ranges*. Kluwer Academic Publ., Dordrecht.

Wardle P. & Coleman M.C., 1992, Evidence for rising upper limits of four native New Zealand forest trees. *N. Z. J. Bot.* 30(3): 303-314.

WMO 2001, *WMO statement on the status of the global climate in 2000*. WMO-No. 920, World Meteorological Organization, Geneva.

Woodward F.I. (ed.), 1992, *The ecological consequences of global climate change*. Advances in ecological research 22. Academic Press, London, 337pp.

Wuethrich B., 2000, How climate change alters rhythms of the wild. *Science* 287: 793/795.

Yom-Tov, Y., 2001, Global warming and body mass decline in Israeli passerine birds. *Proc. R. Soc. Lond., Ser. B.: Biol. Sci.* 268: 947-952.

Terrestrial ecosystem responses to climate changes in the Antarctic

PETER CONVEY
British Antarctic Survey, Natural Environment Research Council, High Cross, Madingley Road, Cambridge CB3 0ET, United Kingdom

Abstract: Parts of Antarctica, particularly the Antarctic Peninsula region and sub-Antarctic Islands, are experiencing rapid changes in climate, particularly temperature, precipitation/hydration and irradiation, although it is becoming clear that many of these are driven by regional rather than global processes. Terrestrial ecosystems of this remote region provide a "natural experiment" in which to identify biological responses (at scales between cell biochemistry and whole ecosystem) to changing climate variables, both in isolation and combination. The conclusions drawn may be applied to more complex lower latitude ecosystems, where change is perceived to have more direct relevance to Mankind. This paper gives an overview of recent and continuing studies of Antarctic terrestrial biology, assessing these in the context of existing predictive literature. The importance of flexibility (physiological and ecological) and resilience of existing taxa in the face of change are highlighted. In the long-term, large-scale changes in ecosystem structure, complexity and diversity are likely as a consequence of long-distance colonisation by exotic species. However, in the shorter term, geographical isolation will limit responses to those of existing terrestrial biota. In contrast with some earlier predictions of wide-ranging deleterious effects, these now appear likely to be subtle and multifactorial in origin.

INTRODUCTION

Terrestrial ecosystems of Antarctica are remote from those inhabited and exploited by Man. Therefore they, their climatic characteristics, and the processes of change that they may be undergoing, are unfamiliar to and indeed generally ignored by most biologists, politicians and the general public. One of the aims of this paper is, therefore, to highlight the direct relevance of studies of these high latitude southern ecosystems to the wider climate change debate. To enable this, it is necessary first to provide a background overview of the terrestrial biota of Antarctica, highlighting some relevant biological and ecological characteristics. By doing so, the main climatic influences on this biota are identified. With a firm biological background, the processes of change in the physical environment of Antarctica are then summarised, thereby providing the means to identify areas of biology which should be targeted in any search for "fingerprints of change". Studies of biological change in Antarctica are still at an early stage, therefore this paper will draw on existing predictive literature, as well as observational and experimental studies, to outline the current state of knowledge of climate change and its biological consequences in Antarctica.

ANTARCTIC TERRESTRIAL ECOSYSTEMS

Antarctica is the one continent on Earth which has no history of permanent human occupation and environmental manipulation. Indeed, the history of its discovery and exploration occupies little more than the last century, while the current transient human population only numbers in the thousands, mostly related to the presence of scientific bases operated by the 30+ signatories of the Antarctic Treaty. It is a continent of extremes. With a larger surface area than either Australia or Western Europe, it is the world's coldest, windiest, highest and most isolated continent. Less than 1 % of its area is clear, even seasonally, of permanent snow or ice, while much of the continent is technically a frigid desert with negligible precipitation. It is not surprising that its terrestrial biology is often, at best, ignored!

Covering such a large area, it is simplistic to consider the biota of Antarctica as a single entity. Conventionally, three biogeographical zones are recognised in the Antarctic (see Smith 1984, Longton 1988), separated on the basis of clear biological and climatological differences. Although terminology varies in the literature, these will be referred to as the sub-, maritime and continental Antarctic zones here, after Smith (1984).

The sub-Antarctic consists of a ring of isolated islands and archipelagos at high latitudes (50-55° S) in the Southern Ocean, including South Georgia, Îles

Kerguelen and Crozet, Prince Edward Is., Heard I. and Macquarie I. All are remote from continental landmasses and most, with the exceptions of South Georgia and Heard Is., lie north of the oceanic Southern Polar Front. They have a cold maritime climate, with seasonal variation damped by the strong oceanic influence (Convey 1996a, Danks 1999) and little or no impact of winter pack ice. Although hosting large breeding populations of marine vertebrates (birds and seals) which provide considerable input of nutrients to the terrestrial environment, the only true non-marine vertebrates present in the sub-Antarctic are a single insectivorous passerine and two species of duck, all with very limited distributions. Otherwise, the terrestrial fauna is dominated by invertebrate taxa, the most visible representatives of which are Diptera, Coleoptera, Acari and Collembola. With a single exception, woody plants are absent, and the flora consists of lichens, bryophytes, a limited range of ferns and flowering plants. In the absence of grazing vertebrates, the scale of development of a "megaherb" flora is unique (Block 1984, Smith 1984, Convey 2001, provide summaries of Antarctic biodiversity).

The maritime Antarctic includes the western side of the Antarctic Peninsula, and the offshore island groups of the South Shetland, South Orkney and South Sandwich Is. and Bouvetøya. The climate is more strongly seasonal than that of the sub-Antarctic, although again damped by the maritime influence, especially in the summer months. In particular, mean air temperatures are negative for most of the year, only reaching 0-2 °C for 1-4 months in summer. The terrestrial biota is much more restricted and is dominated by soil arthropods (Acari and Collembola) and invertebrates (Nematoda, Tardigrada, Rotifera), with only two higher insects (both Diptera) present. Likewise, the importance of higher plants in the flora is much diminished, with only two species present, and a fellfield vegetation of bryophytes and lichens is particularly well developed. Again, marine vertebrates may provide considerable nutrient input and habitat disturbance locally.

The continental Antarctic is by far the largest zone in terms of area, including the entire East Antarctic landmass and the east coast of the Antarctic Peninsula (the west Antarctic land mass is therefore divided between the continental and maritime zones). With a considerably more extreme climate than the maritime zone, mean monthly temperatures are rarely positive for a single month even at coastal locations, while inland they may rise only to -10 to -30 °C in summer. Flora and fauna are correspondingly more depauperate, represented at higher taxonomic levels by bryophytes, lichens and soil arthropods and invertebrates. At more extreme locations, visible vegetation disappears, leaving a microbial flora including algae, cyanobacteria, bacteria and fungi, while arthropod groups are similarly lost from the fauna (Friedmann 1993, Freckman & Virginia 1997, Wynn-Williams 1996a). At the extreme, detectable life retreats to the endolithic niche, with microbial communities existing in the pores between (usually sandstone) rock crystals (Friedmann 1982).

Perhaps surprisingly, these three zones do not have close analogues in northern polar latitudes. The Arctic consists of continental landmasses surrounding a relatively small, shallow ocean. The climatic consequences of this are considerable, with the most extreme terrestrial habitats of the High Arctic experiencing average temperatures well above zero (5-10 °C) for significant periods in summer, while even the sub-Arctic suffers a continental climate during winter, with temperatures well below zero. There is no northern region with conditions comparable to the continental Antarctic, whilst those of the maritime Antarctic may only be shared with the Greenland icecap, whose terrestrial biology has received little study.

As well as having reduced representation at higher taxonomic levels, Antarctic terrestrial communities (both plant and animal) typically have low species diversity even in those groups which are represented with, additionally, fewer trophic links and interactions than those of Arctic, temperate and tropical latitudes. Low diversity does not equate automatically, however, to low abundance or productivity - population densities, biomass and/or production of several Antarctic taxa (e.g. species of mite, springtail, bryophyte and grass) may be comparable or even greater than lower latitude representatives of the same groups (Convey 1996b). The very simplicity of these communities, in terms of species richness and trophic complexity, is often advanced as an advantage in terms of potential to understand ecosystem structure and functioning. This argument can now be extended to the identification of critical responses of both species and communities to aspects of environmental change. Additionally, communities of some of the most extreme Antarctic environments represent the first stages of natural colonisation and/or succession, or "life at the limits" - the current edge of life on earth, limited purely by physical/abiotic conditions. These are likely to be particularly sensitive to changes in climatic variables, while also providing the opportunity to separate the direct effects of abiotic change from those that act via other biotic processes.

Life history and ecophysiological studies of Antarctic terrestrial organisms have largely failed to identify convincing evidence of features truly indicative of novel adaptation to extreme environmental conditions. Rather, existing physiological capacities ("pre-adaptations"), particularly relating to cold and desiccation tolerance, have been refined and are well represented (see reviews by Longton 1988, Cannon & Block 1988, Block 1990, Sømme 1995, Convey 1996b, 1997a, 2000). Life history strategies (after Southwood 1977, 1988, Greenslade 1983, Grime 1988) are generally "adversity" selected, with high investment in features relating to stress tolerance, low investment in somatic growth, reproduction or dispersal abilities, and poor competitive abilities. One feature of many groups that is of particular relevance to their response to environmental change is an inherent flexibility in many life history and physiological parameters. This is clearly advantageous in the rapidly and unpredictably changing Antarctic terrestrial environment (in direct contrast to the extreme stability displayed in Antarctic marine habitats).

Finally, the extreme isolation of terrestrial locations of the Antarctic, combined with the relative uniformity of communities, at least within each of the three biogeographical zones stands as an advantage in studies of climate change. Although isolation is one cause of their low diversity relative to the Arctic, and increased colonisation is itself a predicted outcome of climate amelioration (see below), this very isolation may permit separation of the direct consequences of change within existing communities from those relating to increased diversity through colonisation. Further, the presence of specific species and communities across wide latitudinal and environmental gradients, particularly in the maritime and sub-Antarctic zones where the gradient extends up to 20 degrees of latitude, provides a "proxy" or natural experiment unique worldwide, which can be used to examine their biology across a wide gradient of natural conditions (cf. Addo-Bediako et al. 2000, Chown & Clarke 2000).

ASPECTS OF CLIMATE IMPORTANT TO ANTARCTIC TERRESTRIAL ORGANISMS

Four aspects of climate are expected to be of direct importance in the biology of Antarctic terrestrial organisms, temperature, water relations, radiation climate, and atmospheric CO_2 concentrations. Only the first three will be considered here. In common with low latitudes, increasing levels of atmospheric CO_2 exist in the Antarctic, however no attempts have been made to identify direct impacts of this on Antarctic terrestrial biota (see Oechel et al. 1997, for an introduction to the effects of changing CO_2 on Arctic ecosystems and Norby et al. 2001 for a recent wider commentary).

As already implied above, low thermal energy input is a feature of Antarctic terrestrial habitats. In terms of potential impacts on biology, temperature variation may act at several scales. While upper and lower extremes may clearly be limiting, other aspects such as diurnal and annual ranges, means and rates of change will also be influential. In the context of climate warming, the very low summer temperatures experienced in all Antarctic zones, even relative to the Arctic, are of great importance, as they are likely to be near minimum threshold temperatures for many physiological processes. For an organism with a specified threshold temperature, but occurring in a range of environments, a small temperature increment will therefore have a relatively greater biological impact in the more extreme environment.

The importance of water relationships in polar biology has been re-emphasised by Kennedy (1993), Sømme (1995) and Block (1996). It should be realised that water availability in these environments is not only governed by precipitation patterns, but also by thawing. Hence the presence of free water is normally separated temporally from the precipitation event, and may also be separated spatially. In the case of the water source being glacial melt, both temporal and physical scales may be large.

Two aspects of the radiation climate may be of particular significance to Antarctic terrestrial biology in the context of climate change. Firstly, although the annual pattern of variation in day length with latitude is obviously constant, exposure to direct insolation potentially varies as a function of meteorological variables such as cloud and snow cover, with clear implications for levels of primary production. Any biological consequences of variation may also be magnified in Antarctica, as the main autotrophic elements of the flora (lichens, bryophytes, microbial groups) inherently show photosynthetic saturation at relatively low levels of PAR, and reduced production at higher levels (Longton 1988, Post et al. 1990). Secondly, changes in exposure to ultra-violet radiation mediated by the Antarctic spring ozone hole have the potential to influence biological systems.

Finally, it is often simplistic to consider the actions of these major variables in isolation. Their interactions are likely to be important on both macro and micro temporal and spatial scales, while the predictability of their patterns of variation is also likely to have biological significance (Convey 1996b). Understandably, macroclimatic data across an area as large as Antarctica is recorded at a very coarse resolution. While the lack of a close relationship between coarse macroclimatic and fine-scale microclimatic data is well appreciated (e.g. Walton 1984, Smith 1988a), and a number of short-term data sets exist within the Antarctic which highlight the large and rapid changes possible in microclimatic conditions (e.g. Davey et al. 1992, Kennedy 1994, Smith 1988a, Walton 1982), there remains a paucity of studies in which climatic variables are described at temporal and physical scales relevant to the organisms under study.

CLIMATE CHANGE IN THE ANTARCTIC

The widely-held perception that Antarctica has, until recent years, remained under a permanent and stable icecap is clearly incorrect. Even throughout the Pleistocene, wide variations in climate and the extent of ice cover have been identified, analogous to the Ice Age advances and retreats of the northern hemisphere. These have included periods even during the last 1,000 to 50,000 years when local ice cover and shelf thickness (or existence) has been considerably less than present now (Sugden & Clapperton 1977, Clapperton & Sugden 1982, 1988, Lorius et al. 1985, Smith 1990, Pudsey & Evans in press). Thus, even in the Antarctic, climate change processes are not a new phenomenon.

Contemporary rapid trends of temperature increase are now well-documented in the maritime Antarctic region (Smith 1990, 1994, Fowbert & Smith 1994, King 1994, King & Harangozo 1998, Skvarca et al. 1998) with several sites reporting increases in annual air temperatures of at least 1 °C over the last 30-50 years, among the most rapid rates observed worldwide. Similar reports are available from a range of sub- and continental Antarctic sites (e.g. Adamson et al. 1988, Smith & Steenkamp 1990, Gordon & Timmis 1992, Frenot et al. 1997, Bergstrom & Chown 1999). In detail, the maritime Antarctic pattern is driven by a strong warming trend in the winter months, which is much weaker during summer (King 1994, King & Harangozo 1998). These temperature trends have been linked to decreases in winter sea ice extent and are thought to be regional rather than global processes, with a possible relationship to El Niño Southern Oscillation (ENSO) events in the southern Pacific Ocean (Cullather et al. 1996). Indeed, although many global circulation models do predict relatively high rates of warming at higher latitudes, the data from this region are of much greater magnitude than expected in current models. Nevertheless, irrespective of cause, current rates of warming in the maritime and sub-Antarctic in particular provide an unrivalled model system in which to search for its biological consequences.

A second general prediction of climate change models is that worldwide precipitation patterns will change, although such predictions currently remain at a very coarse level. Changes in precipitation will almost certainly be linked to changes in patterns of insolation, cloud cover and wind speed, and hence to temperature, especially at the microclimatic scale. Increases in precipitation have been predicted in the Antarctic coastal zone (Budd & Simmonds 1991) and more recently documented in the maritime Antarctic (Turner et al. 1997). This has been tentatively linked to a change in atmospheric circulation patterns bringing an increase in depressions approaching the Antarctic Peninsula from more northerly latitudes, which again may be linked to the ENSO. Decreased precipitation may also occur, as has been reported at Marion I. and Îles Kerguelen since the early 1950's (Frenot et al. 1997, Bergstrom & Chown 1999).

Water input to Antarctic terrestrial habitats is governed by snow and glacial melt as well as direct precipitation, indeed this is the only source of liquid water in the continental Antarctic. There have been several high-profile reports of ice shelf collapse on both east and west coasts of the Antarctic Peninsula in recent years (Doake & Vaughan 1991, Vaughan & Doake 1996, Pudsey & Evans in press), which again appear to be related to regional rather than global processes. These have no direct impact on terrestrial biota except by bringing maritime climatic influences closer to coastal terrestrial sites. Of more relevance are the rapid rates of glacial retreat and loss of "permanent" snow cover observed at a range of maritime and sub-Antarctic sites (Smith 1990, Gordon & Timmis 1992, Fowbert & Smith 1994, Frenot et al. 1997, Pugh & Davenport 1997, Fox & Cooper 1998), amounting to as much as 40 % of cover on Signy Island (South Orkney Islands) in the last 50 years. A more subtle effect relates to the formation of a sub-snow ice layer on the ground surface during brief periods of winter thaw, a feature observed in both polar regions. Warming may increase the frequency of thaws and hence encourage formation of an ice layer, with apparently negative effects on some soil faunal communities (Coulson et al. 2000).

Finally, the potential consequences of the annual formation in the austral spring of the Antarctic ozone hole have received much attention. This feature, formed as a result of winter concentration of anthropogenic pollutants at high altitudes over southern polar latitudes, has existed only since the early 1980's (Farman et al. 1985) and, as yet, shows no sign of recovery despite anti-pollution measures beginning to be put in place by some of the world's governments. At its peak, approximately two-thirds of the protective ozone layer is lost over high southern latitudes, which has the direct consequence of allowing greater penetration of lower wavelength UV-B radiation to the Earth's surface while leaving UV-A and PAR levels unaffected (hence the ratio of incident PAR:UV-B is altered). Even when the ozone hole is open, maximum levels of UV-B received are little different from normal mid-summer maxima. Rather, the significance of these levels is twofold: they occur at a much earlier point in the season than is normal, and lower wavelengths (within the UV spectrum) penetrate to the Earth's surface than would be normal at this time of year. Although ozone depletion is now being reported over the Arctic (Müller et al. 1997), and also over industrialised areas at lower latitudes, the scale and predictability of the Antarctic ozone hole result in the areas exposed to its effects remaining the best sites worldwide in which to study the potential impacts of changes in UV-B exposure.

PREDICTIVE LITERATURE

The Antarctic terrestrial environment has lagged behind much of the rest of the world in terms of examination of the possible consequences of climate change, despite the magnitude of changes experienced (Roberts 1989, Smith

& Steenkamp 1990, Voytek 1990). However, over the last decade, a predictive literature has been developed encompassing both the justification for using Antarctic terrestrial biology as a focus of climate change studies, and a wide range of predictions of potential consequences of change (e.g. Wynn-Williams 1994, 1996b, Kennedy 1995a, Convey 1997b, Walton et al. 1997, Bergstrom & Chown 1999). Considering each of the three major climatic variables separately, these general predictions are summarised in Table 1. Together, these lead to a number of large scale biological predictions, including an increase in rate of colonisation by species new to the Antarctic, thereby increasing diversity, biomass, trophic complexity and habitat structure, with possible loss of existing Antarctic species and communities largely through increased competition.

Table 1. Summary of general predictions of the consequences of changes in major climatic variables on Antarctic terrestrial biota.

Variable	Temperature	Water	UV-B radiation
Predicted consequences	*increase:* speed up development and life cycle; alter feeding preferences; population growth; extend active season; expand range limits (upper thermal limits may restrict distributions); exotic colonisation;	*increase:* extend active season; expose new ground for colonisation; *decrease:* reduce active season; local extinction;	target autotrophs and saprotrophs; change resource allocation strategies (screening, quenching, avoidance, repair); range limitation; local extinction; impact on food chain;

TERRESTRIAL BIOLOGICAL CHANGE IN ANTARCTICA

Responses within natural populations

Much of the evidence currently available for the effects of climate amelioration in the Antarctic biogeographical zones is observational or circumstantial. Among the most visually striking biological observations are those reported by Fowbert & Smith (1994), Smith (1994) and Grobe et al. (1997), relating to colonisation and rapid population expansion of the two flowering plants, *Deschampsia antarctica* and *Colobanthus quitensis* at sites on the Antarctic Peninsula (c. 65° S). These species are widely, but sparsely, distributed throughout the maritime Antarctic but, at a local scale, have shown rapid population expansion both in terms of number of plants and the area covered, consistent with warming and snow melt trends at this site.

These species are clearly already able to survive the environmental stresses imposed within the maritime Antarctic, and thus local population expansion is not surprising under climate amelioration. Although both are capable of vegetative population expansion, the most important source of new plants is likely to be seed production. Studies in the early 1970's indicated that production of mature seeds was unusual in most summers for both species (Edwards 1974). This situation appears to have reversed by the early-mid 1990's, certainly for *Deschampsia antarctica*, with seed produced in the majority of seasons (Convey 1996c). As well as giving immediate potential for population expansion, seeds may also remain dormant in soil propagule banks (McGraw & Day 1997), providing a pre-existing source of colonising propagules in the event of local climate amelioration, thereby accelerating the process.

Local and long distance colonisation

Similar rapid expansion has been observed in bryophyte ground coverage at sites in the maritime Antarctic. Signy Island in the South Orkney Islands is particularly well-studied in this context (Smith 1990), providing a model for other sites in the maritime Antarctic, which share very similar floras. Bryophytes are well-suited to take advantage of improved environmental conditions, possessing a range of both sexually and asexually-produced propagules which may be transported over all scales of distance, from millimeters to bipolar scales (Longton 1988). Sexually-produced spores, as well as being transported long distances at high altitude in the air column, may also remain dormant in the soil propagule bank for many years (Smith & Coupar 1986, Smith 1987, 1993, During 1997, 2001). Recent bryophyte population changes as described on Signy Island have, as is intuitively reasonable, involved mainly locally-occurring species – with their close physical proximity to sources of propagules they predominate in cultures derived from propagule banks. However, examples of species new to the island are also present (Convey & Smith 1993), indicative of the processes of long-distance colonisation and local climate amelioration acting together to allow successful establishment. Similarly, aerobiological studies have recently given clear demonstration of the transfer of biological material into the Antarctic from southern South America, including an estimate of the frequency of meteorological patterns leading to such events (less than once per year, Marshall 1996), but these studies have not yet identified any bryophyte (or lichen) propagules of species not already existing locally (Marshall &

Convey 1997). That such events do occur is shown clearly by the presence of lower latitude species associated with geothermal activity at several Antarctic locations (Smith 1991, Kennedy 1996, Convey et al. 2000a). However, even with climate amelioration, the frequency of such events will remain very low through the physical isolation of Antarctica.

Isolation similarly limits the potential for unassisted colonisation of Antarctic sites by invertebrate groups and, as yet, there are no examples of species becoming established in either maritime or continental Antarctic zones. However, known migratory species, particularly of Lepidoptera, are documented relatively frequently at several sub-Antarctic islands (Bonner & Honey 1987, Chown & Avenant 1992, Chown & Language 1994, Greenslade et al. 1999), with the moth *Plutella xylostella* recently becoming established on Marion I. Although known to be a strongly migratory species, even in this case the possibility of human assistance cannot be discounted. It is also not possible to demonstrate a causal link between climate amelioration and colonisation success in such isolated events, even though the observation is consistent with predictions (Chown & Language 1994).

It is clear both from early deliberate transplant experiments (Edwards & Greene 1973, Edwards 1980) and accidental or deliberate human-mediated introductions, that a wide range of floral and faunal taxa can establish and multiply, when the problems of transport to the Antarctic are overcome (Smith 1996, Bergstrom & Chown 1999). These taxa already possess appropriate ecophysiological features allowing them to survive the harsher environmental conditions at higher latitudes, which is an indication of the importance of geographical isolation in limiting current Antarctic biodiversity. Relaxation of environmental stress constraints (which help determine geographical distributions worldwide (Chown & Clarke 2000)) through climate amelioration will further assist colonists from lower latitudes to become established.

Direct human-mediated impacts

Alien colonisation via human influence is particularly noticeable in the sub-Antarctic. For instance, roughly half the vascular flora of South Georgia consists of grasses and weeds introduced from the Falkland Islands, South America and Europe during the operation of shore-based whaling stations in the first half of the Twentieth Century. Some of these, such as the grass *Poa annua* now cover large areas of ground, having displaced native taxa (Smith 1996). Similar consequences of range expansion by introduced *Agrostis stolonifera* on Marion Island have been described (Gremmen 1997, Gremmen & Smith 1999). Man has also introduced vertebrates either deliberately (reindeer, cattle, sheep, cats, rabbits) or accidentally (rats, mice) to most sub-

Antarctic islands which, in many cases, have naturalised successfully, leading both to major alterations of vegetation structure (Leader-Williams 1988, Chapuis et al. 1994, Smith 1996), and reductions or destruction of invertebrate (Vogel et al. 1984, Chown & Block 1997) and ground-nesting bird populations.

A wide range of invertebrate introductions to sub-Antarctic sites have been documented, including beetles (Ernsting 1993), flies (Chevrier et al. 1997), springtails (Deharveng 1981, Greenslade 1990, Convey et al. 1999) and mites (Pugh 1994) (see also Crafford et al. 1986). While the impact of these may not stand out visually, detailed studies have already demonstrated clear consequences on the trophic dynamics of existing communities (Smith & Steenkamp 1990, Chown & Smith 1993, Ernsting et al. 1995, 1999), or proposed displacement of native species (Convey et al. 1999). Examples of successful establishment in the more extreme maritime Antarctic are fewer, and have yet to be documented in the continental Antarctic. However, one species each of dipteran and enchytraeid worm were accidentally introduced to Signy Island during 1960's plant transplant experiments, and remain there today with small but gradually expanding ranges (Block et al. 1984, Convey 1992, Convey & Block 1996), while the grass *Poa pratensis* has become established at a site in the northern Antarctic Peninsula (Smith 1996). Alien invertebrates are frequently reported from various Antarctic stations, often associated with the import of cargo and foodstuffs. While a large majority of these are unable to survive away from active human habitation, a proportion clearly can. The accidental transfer of "resident" organisms between and even within Antarctic zones also remains a clear danger (e.g. Convey et al. 2000b).

While none of these examples can be traced directly to climate amelioration, their significance in this context lies in demonstrating (a) the potential of many existing "exotic" species to establish in the Antarctic given opportunity, which is likely to increase under scenarios of amelioration and (b) the great importance that human vectors are likely to assume in the colonisation process, especially given the current rate of increase in both scientific and recreational access to the Antarctic (Pugh 1994, Smith 1996). The proven impacts of some of these artificially-introduced alien taxa on the biology of naturally-occurring Antarctic species and communities are also likely to mirror closely the consequences of natural colonisation events, and hence are of relevance to this discussion of the potential consequences of climate change in the region.

Human exploitation of marine resources may also have secondary consequences leading to environmental change in terrestrial ecosystems. Over-exploitation of marine mammal populations (seals and whales) between the Eighteenth and Twentieth Centuries led to drastic reductions in numbers. In recent decades, the population of Antarctic Fur Seal (*Arctocephalus gazella*) has been able to recover rapidly, in the absence of feeding competition from whales, to levels probably greater than those of pre-exploitation. This has resulted in range expansion, with moulting animals now having a destructive impact on terrestrial ecosystems at a number of sites in the maritime Antarctic previously unoccupied by the species (Smith 1988b, Hodgson & Johnston 1997, Hodgson et al. 1998).

Natural models and biological correlates of climate change predictions

Biologists have yet to take advantage of the potential of the environmental gradient extending through the maritime Antarctic and into the sub-Antarctic to be used as a proxy natural experiment to test the predictions of climate change. A single study of reproductive investment in the oribatid mite, *Alaskozetes antarcticus* (Convey 1998) provides tantalising evidence of alterations in resource allocation strategy consistent with these predictions, with significantly greater investment in eggs in the milder sub-Antarctic compared with populations from the maritime zone. Such a pattern might be expected to apply more widely amongst Antarctic invertebrates, as their inherently very flexible physiological and life history responses to environmental stress should permit changes in resource allocation patterns. For instance, while high levels of investment on seasonal timescales in physiological strategies to counteract exposure to cold and desiccation is well-demonstrated in various groups (Block 1990, 1996, Ring & Danks 1994, Wharton 1995, Worland 1996, Worland et al. 1998, Bayley & Holmstrup 1999), recent research has also identified much more rapid responses, on timescales of hours, indicating that a great deal more flexibility in resource usage may exist than previously considered (Worland & Convey, in press).

Other studies of widely varying groups have demonstrated aspects of their biology with clear potential to respond to the opportunity presented by climate amelioration. The incidence of successful sexual reproduction in many species of moss is limited by temperature (Longton 1988), and thus any changes in the distribution of this form of reproduction may provide a sensitive indicator of local change, while also altering species colonisation potential (Frahm & Klaus 2001). However, many factors influence the success of this mode of reproduction in mosses and its associated investment patterns, with the limited data available in this field to date equivocal (Convey 1994a).

Life history flexibility, which is typical of many Antarctic terrestrial invertebrates (Convey 1994b, 1997a, 2000), provides the potential for changes in development rates and shortening of life cycle duration in response to climate amelioration. As yet, observational evidence of these events occurring in the field is lacking, while there are few experimental demonstrations. The sub-Antarctic diving beetle, *Langustis angusticollis*, provides a clear example of a species likely to show a rapid response (Arnold & Convey 1998). Unusually amongst Antarctic invertebrates, it possesses an environmentally-cued resting stage, with development arrested below a temperature of 7.3 °C. Thermal energy input to its lake environment currently limits this species to a two-year life cycle. However, an increase of only 1 °C would allow completion of an annual life cycle, leading to the potential of a very rapid increase in population and, as the beetle is the top predator in the lake ecosystem, a large impact on local trophic dynamics. An analogous experimental demonstration of very rapid change in life cycle structure and population increase in response to realistic temperature manipulation has been demonstrated in field studies of the Arctic aphid *Acyrthosiphon svalbardicum* (Strathdee et al. 1993, 1995).

Springtails are among the best-represented amongst arthropod groups of Antarctic terrestrial ecosystems. A feature of their physiology is that gas exchange occurs across the cuticle, rather than relying on a tracheal system, which therefore reduces the animal's ability to control its body water status via cuticular adaptations. Springtails are therefore likely to be vulnerable to changes in the water status of their microenvironment. In particular, through purely physical processes, they are vulnerable to drying stresses while trapped (inactive) between ice crystals in frozen soil (Worland 1996), and even when soil humidity drops below c. 98 % (Bayley & Holmstrup 1999). These properties have recently been used to propose that individual body water content can be used as a proxy measure of microhabitat water status. Using a detailed multivariate analysis, Block & Harrisson (1995) postulated that a significant upward trend in the body water content of *Cryptopygus antarcticus* at Signy Island over the period 1984-87 could be related to an interaction between increased insolation and wind speed over the same period, which resulted in greater rates of ice loss from local glaciers and snowfields, and hence increased availability of liquid water at microhabitat level. Recent analysis of a longer run of data from the same collection site (1984-95) has shown the earlier overall increase to be a short-lived feature, but has identified other long-term trends in body water content (Block & Convey, in press), including upward trends in water content during the late autumn and early spring, with some evidence of a downward trend during part of summer. These data are consistent with regional climate change observations, in that they indicate a lengthening of the summer season, with earlier water availability in spring and later availability in autumn; additionally, the decreasing trend in part of summer may indicate local exhaustion of water supply. However, intuitively attractive though this interpretation is, a causal link has yet to be proved with any specific environmental factor or combination.

The foregoing example highlights the danger of assuming that "climate amelioration" will automatically lead to reduced stress on organisms (see also Kennedy 1995a). Exhaustion of water supplies, effectively through increased temperatures, obviously results in desiccation stress and, at the extreme, may result in local range limitation. Indeed, summer die-back through desiccation has been reported in the two maritime Antarctic flowering plants (Komarkova et al. 1985). Coulson et al. (2000) have recently demonstrated that increased formation of an ice layer on the soil surface, as would happen with increased frequency of temporary thaws through winter warming, leads to large reductions in soil Collembola populations at an Arctic site. The impact of such effects are likely to be highly variable spatially through local topographical control of melt flow patterns. Finally, the possibility exists that microhabitat warming may exceed the upper thermal limits of some invertebrate species (van der Merwe et al. 1997), a feature which shows little geographical variation, at least in insects (Addo-Bediako et al. 2000).

In situ field manipulation experiments

The use of a range of field manipulation methodologies to mimic predicted changes in (some) environmental variables has become widespread in both Antarctic and Arctic studies (e.g. Wynn-Williams 1992, 1996b, Coulson et al. 1993, Strathdee et al. 1993, 1995, Henry 1997, Webb et al. 1998; see also Kennedy 1995b, Table 1). A range of "greenhouse" methodologies have been utilised (see Kennedy 1995c), all of which involve placing variations on the theme of chambers or screens over selected areas of habitat, with more labour-intensive methodologies including water or nutrient amendments. The attraction of such methodologies is clear, in that they are technologically simple and can be left in place for at least several months over the austral summer, or for longer periods. These are very real practical advantages at the harsh and generally inaccessible sites available for study. Greenhouse methodologies clearly impact temperature, but also affect humidity, precipitation, exposure to wind abrasion and substrate stabilisation. In studies relating to the impact of changed levels of UV-B radiation, the use of different types of screen material also allows selective absorption or transmission of specific wavelength ranges and may also be combined with UV supplementation using fluorescent lamps. Many experiments based on greenhouse methodologies are vulnerable to at least two serious criticisms, (i) such manipulations often affect several different environmental variables and their interactions, and it is not possible to separate these effects, and (ii) in detail, the environmental variation achieved may not match closely the predictions of the general models being tested. An elegant critique of these methods is provided by Kennedy (1995b,c), while more recent studies (e.g. Day et al. 1999) have developed methodologies to minimise such difficulties.

On maritime Antarctic Signy Island (c. 60° S), greenhouse methodologies have led to spectacular responses in studies of microbial and bryophyte community development, and growth of higher plants (Smith 1990, Wynn-Williams 1993, 1996b, Kennedy 1996). Existing species show greater ground coverage, lusher growth forms and greater reproductive output. Manipulation of visually bare ground has led, within a few years, to complete coverage by moss turf (Smith 1990), further emphasising the likely importance of local propagule banks in colonisation processes. Arthropod communities associated with the same manipulations have also shown rapid expansion (Kennedy 1994), in this case indicating an increase in reproductive output rather than de novo colonisation. More recent and ongoing studies at a much more extreme site on southern Alexander Island (c. 72° S) have demonstrated similar large increases in microbial populations (Wynn-Williams, unpublished data) over timescales of as little as 1-2 years. Nematode worms, the dominant soil invertebrate group present in more extreme environments, also showed very rapid local increases of one to three orders of magnitude in population density (Convey & Wynn-Williams unpublished data). Such responses by faunal populations are likely to be mediated in part by those of the microbial autotrophic and saprotrophic groups on which they depend for food, as well as by the direct impact of specific environmental variables.

Field "greenhouse" manipulations often lead to changes in growth form of bryophyte and phanerogam vegetation (Smith 1990, Day et al. 1999), which clearly will alter local microclimates and increase the complexity of the three-dimensional habitat structure available to faunal communities. No attempt has yet been made to study the consequences of such changes in the Antarctic.

Impacts of increased UV-B radiation

The deleterious effects of exposure to UV-B radiation on aspects of cell biochemistry are well-known from laboratory studies, as are the range of responses available to organisms (e.g. Vincent & Quesada 1994, Wynn-Williams 1994, Cockell & Knowland 1999). Laboratory studies, including realistic regimes mimicking "real world" conditions, demonstrate deleterious consequences of UV-B exposure (Quesada et al. 1995), but data from field experiments supplementing solar UV-B radiation by fluorescent lamps indicate that there may be few negative effects, at least on higher plant growth, of elevated UV-B arising from realistic levels of ozone depletion (Fiscus & Booker 1995, Allen et al. 1998). Therefore, both wavelength amendment and screening methodologies have been used to study the possible consequences of changes in the UV radiation climate experienced by autotrophic groups (algae, cyanobacteria, bryophytes, phanerogams, lichens) as a result of increased UV-B receipt through the spring ozone hole (Wynn-Williams 1996b, Quesada et al. 1998, Huiskes et al. 1999, Montiel et al. 1999, George et al. in press).

Many terrestrial organisms are not exposed to the peak radiation levels associated with the ozone hole, through remaining under winter snow cover for much of the spring, although they could lose this protection should temperature amelioration lead to earlier spring melt. Maximum receipt of UV-B during the spring ozone hole is comparable to the levels received normally at midsummer. Thus it is to be expected that, if physiologically active, organisms already have appropriate mechanisms to counteract this stress whether exposed to it during spring or summer. This appears to be particularly true of the phanerogams *Deschampsia antarctica* and *Colobanthus quitensis*, which showed no reduction of photosynthetic parameters under realistic ranges of radiation stress (Montiel et al. 1999, Lud et al., in press), although with some evidence of changes in concentrations of UV-B screening pigments in response to exposure (Day et al. 1999, Ruhland & Day 2000, Xiong & Day 2001). Such well-developed stress tolerance perhaps underlies the unusual ecological success of these species, whose distribution extends from the southern Antarctic Peninsula to montane Peru and southern Mexico, respectively.

Consistent patterns have not yet resulted from UV manipulation studies involving bryophytes and cyanobacteria. Reductions in photosynthetic yield and/or protective pigment production after experimental manipulation of radiation climate have been reported in studies of Antarctic cyanobacteria (Quesada et al. 1998, George et al., in press), mosses and liverworts (Montiel et al. 1999, Huiskes et al. 1999), while Newsham (submitted) and Newsham et al. (submitted) have demonstrated increased concentrations of UV-B screening pigments and carotenoids, but no change in photosynthetic yield of two mosses and a liverwort and Lud et al. (in press) found no change in yield in a lichen. More widely, studies of non-polar taxa (e.g. several chapters in Rozema 1999, Paul 2001) have shown changes in utilisation of biochemical pathways and pigment production patterns. In all these examples, some alteration of resource allocation strategies within the organism will be required, while all responses require the organism to be physiologically active. Such responses may have subtle but important impacts elsewhere in food webs. The pigments involved often incorporate polyphenolic elements, a group of chemicals which also influence digestibility to herbivores and detritivores. Both positive and negative impacts of increased pigment production on decomposition have been reported in non-polar studies (Newsham et al. 1999, Rozema 1999, chapters 7 and 8) however, again, this question has not been addressed in the Antarctic. Given the importance of the detritivore pathway in Antarctic terrestrial ecosystems, changes in autotroph pigment production patterns are likely to be an important determinant of responses at other trophic levels. In parallel with thermal manipulation experiments, Antarctic, temperate and Arctic studies have shown that exposure to altered UV-B regimes may lead to morphological changes underlain by switches between biochemical pathways (Day et al. 1999, Gerhke 1999, Rozema 1999, chapter 3, Ruhland & Day 2000), hence again altering habitat structure.

WHERE NEXT?

It is clear from the foregoing that identification of the biological consequences of climate change in the Antarctic is still at a relatively early stage. Although data are available from studies of most of the major taxonomic groups represented, it is rarely possible yet to relate this unequivocally to single or combinations of variables. Much evidence reported so far relates to single species, taxonomic groups or trophic elements of a community. While being suggestive and consistent with the general predictions of consequences of climate change, evidence remains circumstantial in form. These limitations are inevitable in such a recent field. In order to develop the field further, a number of requirements must be met. These include a clear need for long-term field studies, backed by detailed microclimatic data obtained at a scale relevant to the organism and community under study. There should also be a shift in emphasis to allow community and ecosystem level syntheses of the various (currently) isolated aspects of climate change and response studied. It is becoming clear that, while some biological responses are large and rapid, others are likely to be much more subtle under multifactorial influences. Considerably longer periods than the duration of many current studies will be required to identify these consequences, which may nevertheless have fundamental importance to ecosystem structure and function.

CONCLUSIONS

Several climate change processes are occurring within the Antarctic. Temperature amelioration is particularly marked in the region of the Antarctic Peninsula and the sub-Antarctic islands, currently interpreted as regional change processes rather than directly through global warming. Insolation, precipitation and meltwater availability patterns are also changing, although it is not yet possible to model these changes in sufficient predictive detail.

Potential exposure to ultra-violet radiation (UV-B) has altered rapidly over the last two decades due to the presence of the spring Antarctic ozone hole.

Some clear biological changes have been documented, which are consistent with the expected consequences of climate change predictions; these largely relate to local or long-distance colonisation processes, and to rapid expansion of existing populations.

Through geographical isolation, local species and communities are likely to be impacted initially, in advance of any impact of exotic colonising species. The large degrees of physiological stress tolerance and life history flexibility characteristic of current Antarctic terrestrial biota suggests they should face little difficulty in absorbing the levels of change predicted, which are much smaller in comparison and, indeed, are likely to benefit from some aspects of amelioration. Two particular exceptions to this generalisation are highlighted, both of which may act to limit species distributions - (i) exhaustion of local water supplies through changing precipitation patterns and/or local melt rates and (ii) microhabitat temperature increases which exceed the upper lethal limits of existing species.

Potential model or indicator species and communities exist, which are likely to show large and detectable responses to realistic levels of change.

At present most observational evidence is circumstantial, whilst some experimental manipulations are vulnerable to methodological criticism and alternative interpretations.

Large scale scenarios of wholesale ecosystem destruction are exaggerated. Rather, subtle multifactorial responses involving complex resource trade-offs are to be expected. Despite being hard to detect or manipulate experimentally, these may have considerable impacts within ecosystems.

ACKNOWLEDGEMENTS

This contribution has developed through many discussions with colleagues at the British Antarctic Survey and elsewhere. I thank particularly Profs. Andrew Clarke and Lloyd Peck and Dr. Kevin Newsham for helpful comments on the manuscript, along with those of two anonymous referees.

REFERENCES

Adamson H., Whetton P. & Selkirk P.M., 1988, An analysis of air temperature records for Macquarie Island: decadal warming, ENSO cooling and Southern Hemisphere circulation patterns. *Pap. Proc. R. Soc. Tasman.* 122: 107-112.

Addo-Bediako A., Chown S.L. & Gaston K.J., 2000, Thermal tolerance, climatic variability and latitude. *Proc. R. Soc. Lond. B* 267: 739-745.

Allen D.J., Nogués S. & Baker N.R., 1998, Ozone depletion and increased UV-B radiation: is there a real threat to photosynthesis? *J. Exp. Botany* 49: 1775-1788.

Arnold R.J. & Convey P., 1998, The life history of the world's most southerly diving beetle, *Lancetes angusticollis* (Curtis) (Coleoptera: Dytiscidae), on sub-Antarctic South Georgia. *Polar Biol.* 20: 153-160.

Bayley M. & Holmstrup M., 1999, Water vapour absorption in arthropods by accumulation of myoinositol and glucose. *Science* 285: 1909-1911.

Bergstrom D.M. & Chown S.L., 1999, Life at the front: history, ecology and change on southern ocean islands. *Trends Ecol. Evol.* 14: 472-476.

Block W., 1984, Terrestrial Microbiology, Invertebrates and Ecosystems. In: R.M. Laws (ed.) *Antarctic Ecology*. Academic Press, London.

Block W., 1990, Cold tolerance of insects and other arthropods. *Phil. Trans. Roy. Soc. Ser. B* 326: 613-633.

Block W., 1996, Cold or drought - the lesser of two evils for terrestrial arthropods? *Eur. J. Entomol.* 93: 325-339.

Block W. & Convey P. Seasonal and long-term variation in body water content of an Antarctic springtail - a response to climate change? *Polar Biol.*, in press.

Block W. & Harrisson P.M., 1995, Collembolan water relations and environmental change in the maritime Antarctic. *Global Change Biol.* 1: 347-359.

Block W., Burn A.J. & Richard K.J., 1984, An insect introduction to the maritime Antarctic. *Biol. J. Linn. Soc.* 23: 33-39.

Bonner W.N. & Honey M.R., 1987, *Agrotis ipsilon* (Lepidoptera) at South Georgia. *Br. Antarct. Surv. Bull.* 77: 157-161.

Budd W.F. & Simmonds I., 1991, The impact of global warming on the Antarctic mass balance and global sea level. In: G. Weller, C.L. Wilson & B.A.B. Severin (eds.) *Proceedings of the International Conference on the Role of Polar regions in Global Change.*

Cannon R.J.C. & Block W., 1988, Cold tolerance of microarthropods. *Biol. Rev.* 63: 23-77.

Chapuis J.L., Bousses P. & Barnard G., 1994, Alien mammals, impact and management in the French subantarctic islands. *Biol. Conserv.* 67: 97-104.

Chevrier M., Vernon P. & Frenot Y., 1997, Potential effects of two alien insects on a sub-Antarctic wingless fly in the Kerguelen Islands. In: B. Battaglia, J. Valencia & D.W.H. Walton (eds.) *Antarctic Communities: Species Structure and Survival*. Cambridge University Press, Cambridge.

Chown S.L. & Avenant N., 1992, Status of *Plutella xylostella* at Marion Island six years after its colonisation. *S. Afr. J. Antarct. Res.* 22: 37-40.

Chown S.L. & Block W., 1997, Comparative nutritional ecology of grass-feeding in a sub-Antarctic beetle: the impact of introduced species on *Hydromedion sparsutum* from South Georgia. *Oecologia* 111: 216-224.

Chown S.L. & Clarke A., 2000, Stress and the geographic distribution of marine and terrestrial animals. In: K.B. Storey & J. Storey (eds.) *Environmental Stressors and Gene Responses*. Elsevier, Amsterdam.

Chown S.L. & Language K., 1994, Recently established Diptera and Lepidoptera on sub-Antarctic Marion Island. *Afr. Entomol.* 2: 57-76.

Chown S.L. & Smith V.R., 1993, Climate change and the short-term impact of feral house mice at the sub-Antarctic Prince Edward Islands. *Oecologia* 96: 508-516.

Clapperton C.M. & Sugden D.E., 1982, Late quaternary glacial history of George VI Sound area, West Antarctica. *Quat. Res.* 18: 243-267.

Clapperton C.M. & Sugden D.E., 1988, Holocene glacier fluctuations in South America and Antarctica. *Quat. Sci. Revs.* 7: 185-198.

Cockell C.S. & Knowland J., 1999, Ultraviolet radiation screening compounds. *Biol. Rev.* 74: 311-345.

Convey P., 1992, Aspects of the biology of the midge, *Eretmoptera murphyi* Schaeffer, introduced to Signy Island, maritime Antarctic. *Polar Biol.* 12: 653-657.

Convey P., 1994a, Modelling reproductive effort in sub- and maritime Antarctic mosses. *Oecologia* 100: 45-53.

Convey P., 1994b, Growth and survival strategy of the Antarctic mite *Alaskozetes antarcticus*. *Ecography* 17: 97-107.

Convey P., 1996a, Overwintering strategies of terrestrial invertebrates in Antarctica - the significance of flexibility in extremely seasonal environments. *Eur. J. Entomol.* 93: 489-505.

Convey P., 1996b, The influence of environmental characteristics on life history attributes of Antarctic terrestrial biota. *Biol. Rev.* 71: 191-225.

Convey P., 1996c, Reproduction of Antarctic flowering plants. *Antarct. Sci.* 8: 127-134.

Convey P., 1997a, How are the life history strategies of Antarctic terrestrial invertebrates influenced by extreme environmental conditions? *J. Thermal. Biol.* 22: 429-440.

Convey P., 1997b, Environmental change: possible consequences for life histories of Antarctic terrestrial biota. *Kor. J. Polar Res.* 8: 127-144.

Convey P., 1998, Latitudinal variation in allocation to reproduction by the Antarctic oribatid mite, *Alaskozetes antarcticus*. *Appl. Soil Ecol.* 9: 93-99.

Convey P., 2000, How does cold constrain life cycles of terrestrial plants and animals? *Cryo-Lett.* 21: 73-82.

Convey P., 2001, Antarctic Ecosystems. In: S.A. Levin (ed.) *Encyclopedia of Biodiversity*, vol. 1. Academic Press, San Diego.

Convey P. & Block W., 1996, Antarctic Diptera: ecology, physiology and distribution. *Eur. J. Entomol.* 93: 1-13.

Convey P. & Smith R.I.L., 1993, Investment in sexual reproduction by Antarctic mosses. *Oikos* 68: 293-302.

Convey P., Greenslade P., Arnold R.J. & Block W., 1999, Collembola of sub-Antarctic South Georgia. *Polar Biol.* 22: 1-6.

Convey P., Smith R.I.L., Hodgson D.A & Peat H.J., 2000a, The flora of the South Sandwich Islands, with particular reference to the influence of geothermal heating. *J. Biogeogr.* 27: 1279-1295.

Convey P., Smith R.I.L., Peat H.J. & Pugh P.J.A., 2000b, The terrestrial biota of Charcot Island, eastern Bellingshausen Sea, Antarctica an example of extreme isolation. *Antarct. Sci.* 12: 406-413.

Coulson S., Hodkinson I.D., Strathdee A., Bale J.S., Block W., Worland M.R. & Webb N.R. 1993. Simulated climate change: the interaction between vegetation type and microhabitat temperatures at Ny Ålesund, Svalbard. *Polar Biol.* 13: 67-70.

Coulson S.J., Leinaas H.P. Ims R.A. & Søvik G., 2000, Experimental manipulation of the winter surface ice layer: the effects on a High Arctic soil microarthropod community. *Ecography* 23: 299-306.

Crafford J.E., Scholtz C.H. & Chown S.L., 1986, The insects of sub-Antarctic Marion and Prince Edward Islands; with a bibliography of entomology of the Kerguelen biogeographical province. *S. Afr. J. Antarct. Res.* 16: 41-84.

Cullather R.I., Bromwich D.H. & van Woert M.L., 1996, Inter-annual variations in Antarctic precipitation related to El Niño - Southern Oscillation. *J. Geophys. Res.* 101: 19109-19118.

Danks H.V., 1999, Life cycles in polar arthropods - flexible or programmed? *Eur. J. Entomol.* 96: 83-102.

Davey M.C., Pickup J. & Block W., 1992, Temperature variation and its biological significance in fellfield habitats on a maritime Antarctic island. *Antarct. Sci.* 4: 383-388.

Day T.A., Ruhland C.T., Grobe C.W. & Xiong F., 1999, Growth and reproduction of Antarctic vascular plants in response to warming and UV radiation reductions in the field. *Oecologia* 119: 24-35.

Deharveng L., 1981, Collemboles de les Iles Subantarctiques de l'Océan Indien mission J. Travé 1972-1973. *CNFRA* 48: 33-108.

Doake C.S.M. & Vaughan D.G., 1991, Rapid disintegration of the Wordie Ice Shelf in response to atmospheric forcing. *Nature* 350: 328-330.

During, H.J., 1997, Bryophyte diaspore banks. *Adv. Bryol.* 6: 103-134.

During, H.J., 2001, New frontiers in bryology and lichenology: diaspore banks. *Bryologist* 104: 92-97.

Edwards J.A., 1974, Studies in *Colobanthus quitensis* (Kunth) Bartl. and *Deschampsia antarctica* Desv.: VI.: Reproductive performance on Signy Island. *Br. Antarct. Surv. Bull.* 39: 67-86.

Edwards J.A., 1980, An experimental introduction of vascular plants from South Georgia to the maritime Antarctic. *Br. Antarct. Surv. Bull.* 49: 73-80.

Edwards J.A. & Greene D.M., 1973, The survival of Falklands Island transplants at South Georgia and Signy Island, South Orkney Islands. *Br. Antarct. Surv. Bull.* 33 & 34: 33-45.

Ernsting G., 1993, Observations on life cycle and feeding ecology of two recently-introduced predatory beetle species at South Georgia, sub-Antarctic. *Polar Biol.* 13: 423-428.

Ernsting G., Block W., MacAlister H. & Todd C., 1995, The invasion of the carnivorous carabid beetle *Trechisibus antarcticus* on South Georgia (sub-Antarctic) and its effect on the endemic herbivorous beetle *Hydromedion sparsutum*. *Oecologia* 103: 34-42.

Ernsting G., Brandjes G.J., Block W. & Isaaks J.A., 1999, Life-history consequences of predation for a subantarctic beetle: evaluating the contribution of direct and indirect effects. *J. Anim. Ecol.* 68: 741-752.

Farman J.C., Gardiner B.G. & Shanklin J.D., 1985, Large losses of total ozone in Antarctica reveal seasonal ClO_x/No_x interaction. *Nature* 315: 207-210.

Fiscus E.L. & Booker F.L., 1995, Is increased UV-B a threat to crop photosynthesis and productivity? *Photosynthesis Res.* 43: 81-92

Fowbert J.A. & Smith R.I.L. 1994. Rapid population increase in native vascular plants in the Argentine Islands, Antarctic Peninsula. *Arct. Alpine. Res.* 26: 290-296.

Fox A.J. & Cooper A.P.R., 1998, Climate-change indicators from archival aerial photography of the Antarctic Peninsula. *Ann. Glaciol.* 27: 636-642.

Frahm J.P. & Klaus D., 2001, Bryophytes as indicators of recent climate fluctuations in Central Europe. *Lindbergia* 26: 97-104.

Freckman D.W.& Virginia R.A., 1997, Low-diversity Antarctic soil nematode communities: distribution and response to disturbance. *Ecology* 78: 363-369.

Frenot Y., Gloaguen J-C & Trehen P., 1997, Climate change in Kerguelen Islands and colonization of recently deglaciated areas by *Poa kerguelensis* and *P. annua*. In: B. Battaglia, J. Valencia & D.W.H. Walton (eds.) *Antarctic Communities: Species, Structure and Survival*. Cambridge University Press, Cambridge.

Friedmann E.I., 1982, Endolithic microorganisms in the Antarctic cold desert. *Science* 215: 1045-1053.

Friedmann E.I., 1993, *Antarctic Microbiology*. John Wiley & Sons, New York.

Gehrke C., 1999, Impacts of enhanced ultraviolet-B radiation on mosses in a subarctic heath ecosystem. *Ecology* 80: 1844-1851.

George A.L., Murray A.W. & Montiel P.O., Tolerance of Antarctic cyanobacterial mats to enhanced UV radiation. *FEMS Microbiol. Ecol.*, in press.

Gordon J.E. & Timmis R.J., 1992, Glacier fluctuations on South Georgia during the 1970s and early 1980s. *Antarct. Sci.* 4: 215-226.

Greenslade P.J.M., 1983, Adversity selection and the habitat templet. *Amer. Nat.* 122: 352-365.

Greenslade P., 1990, Notes on the biogeography of the free-living terrestrial invertebrate fauna of Macquarie with an annotated checklist. *Pap. Proc. R. Soc. Tasmania* 124: 35-50.

Greenslade P., Farrow R.A. & Smith J.M.B., 1999, Long distance migration of insects to a subantarctic island. *J. Biogeogr.* 26: 1161-1167.

Gremmen N.J.M., 1997, Changes in the vegetation of sub-Antarctic Marion Island resulting from introduced vascular plants. In: B. Battaglia, J. Valencia & D.W.H. Walton (eds.) *Antarctic Communities: Species, Structure and Survival.* Cambridge University Press, Cambridge.

Gremmen N.J.M. & Smith V.R., 1999, New records of alien vascular plants from Marion and Prince Edward Islands, sub-Antarctic. *Polar Biol.* 21: 401-409.

Grime J.P., 1988, The C-S-R model of primary plant strategies - origins, implications and tests. In: R.M. Anderson, B.D. Turner & L.R. Taylor (eds.) *Population Dynamics.* Blackwell, Oxford. pp. 123-139.

Grobe C.W., Ruhland C.T. & Day T.A., 1997, A new population of *Colobanthus quitensis* near Arthur Harbor, Antarctica: correlating recruitment with warmer summer temperatures. *Arct. Alpine Res.* 29: 217-221.

Henry G.H.R. (ed.) 1997, *Global Change Biology* 3 (suppl. 1).

Hodgson D.A. & Johnston N.M., 1997, Inferring seal populations from lake sediments. *Nature* 387: 30-31.

Hodgson D.A., Johnston N.M., Caulkett A.P. & Jones V.J., 1998, Palaeolimnology of Antarctic fur seal *Arctocephalus gazella* populations and implications for Antarctic management. *Biol. Conserv.* 83: 145-154.

Huiskes A., Lud D., Moerdijk-Poortvliet T. & Rozema J., 1999, Impact of UV-B radiation on Antarctic terrestrial vegetation. In: J. Rozema (ed.) *Stratospheric Ozone Depletion: the Effects of Enhanced UV-B Radiation on Terrestrial Ecosystems.* Backuys, Leiden.

Kennedy A.L., 1993, Water as a limiting factor in the Antarctic terrestrial environment: a biogeographical synthesis. *Arct. Alpine Res.* 25: 308-315.

Kennedy A.L., 1994, Simulated climate change: a field manipulation study of polar microarthropod community response to global warming. *Ecography* 17: 131-140.

Kennedy A.L., 1995a, Antarctic terrestrial ecosystem responses to global environmental change. *Ann. Rev. Ecol. Syst.* 26: 683-704.

Kennedy A.L., 1995b, Temperature effects of passive greenhouse apparatus in high-latitude climate change experiments. *Funct. Ecol.* 9: 340-350.

Kennedy A.L., 1995c, Simulated climate change: are passive greenhouses a valid microcosm for testing the biological effects of environmental perturbations? *Global Change Biol.* 1: 29-42.

Kennedy A.L., 1996, Antarctic fellfield response to climate change: a tripartite synthesis of experimental data. *Oecologia* 107: 141-150.

King J.C., 1994, Recent climate variability in the vicinity of the Antarctic Peninsula. *Int. J. Climatol.* 14: 357-369.

King J.C. & Harangozo S.A., 1998, Climate change in the western Antarctic Peninsula since 1945: observations and possible causes. *Ann. Glaciol.* 27: 571-575.

Komarkova V., Scott S. & Toi M., 1985, Summer die-back due to desiccation of *Deschampsia antarctica* Desv. and *Colobanthus quitensis* (Kunth) Bartl. on the largest Stepping Stone Island, Arthur Harbour, Anvers Island, Antarctic Peninsula. *Bull. Ecol. Soc. Amer.* 66: 211.

Leader-Williams N., 1988, *Reindeer on South Georgia, the Ecology of an Introduced Population.* Cambridge University Press, Cambridge.

Longton R.E., 1988, *Biology of Polar Bryophytes and Lichens.* Cambridge University Press, Cambridge.

Lorius C., Jouzel J., Ritz C., Merlivat L. & Barkov N.I., 1985, A 150,000-year climate record from Antarctic ice. *Nature* 316: 591-596.

Lud D., Huiskes A.H.L., Moerdijk T.C.W. & Rozema J., The effects of altered levels of UV-B radiation on an Antarctic grass and lichen. *Plant Ecol.*, in press.

McGraw J.B. & Day T.A., 1997, Size and characteristics of a natural seed bank in Antarctica. *Arct. Alpine Res.* 29: 213-216.

Marshall W.A., 1996, Biological particles over Antarctica. *Nature* 383: 680.

Marshall W.A. & Convey P., 1997, Dispersal of moss propagules in the maritime Antarctic. *Polar Biol.* 18: 376-383.

van der Merwe M., Chown S.L. & Smith V.R., 1997, Thermal tolerance limits in six weevil species (Coleoptera, Curculionidae) from sub-Antarctic Marion Island. *Polar Biol.* 18: 331-336.

Montiel P., Smith A. & Keiller D., 1999, Photosynthetic responses of selected Antarctic plants to solar radiation in the southern maritime Antarctic. *Polar Res.* 18: 229-235.

Müller R., Crutzen P.J., Grooß J.U., Brühl C., Russell J.M., Gernandt H., McKenna D.S. & Tuck A.F., 1997, Severe chemical ozone loss during the Arctic winter of 1995-96. *Nature* 389: 709-712

Newsham K.K. Influence of stratospheric ozone depletion on the Antarctic moss *Andreaea regularis*. Submitted to *Oecologia*.

Newsham K.K., Greenslade P.D., Kennedy V.H. & McLeod A.R., 1999, Elevated UV-B radiation incident on *Quercus robur* leaf canopies enhances decomposition of resulting leaf litter in soil. *Global Change Biol.* 5: 403-409.

Newsham K.K., Hodgson D.A., Murray A.W.A., Peat H.J. & Smith R.I.L. Response of two Antarctic bryophytes to stratospheric ozone depletion. Submitted to *Ecology*.

Norby R.J., Kobayashi K. & Kimball B.A., 2001, Rising CO_2 - future ecosystems. *New Phytol.* 150: 215-221.

Oechel W.C., Cook A.C., Hastings S.J. & Vourlitis G.L., 1997, Effects of CO_2 and climate change on arctic ecosystems. In: S.J. Woodin & M. Marquiss (eds.) *Ecology of Arctic Environments*. Blackwell, Oxford.

Paul N., 2001, Plant responses to UV-B: time to look beyond stratospheric ozone depletion? *New Phytol.* 150: 5-8.

Post A., Adamson E. & Adamson H., 1990, Photoinhibition and recovery of photosynthesis in Antarctic bryophytes under field conditions. In: M. Baltcheffsky (ed.) *Current Research in Photosynthesis* (vol. IV). Kluwer, Dordrecht.

Pudsey C.J. & Evans J., First survey of Antarctic sub-ice shelf sediments reveals mid-Holocene ice shelf retreat. *Geology*, in press.

Pugh P.J.A., 1994, Non-indigenous Acari of Antarctica and the sub-Antarctic islands. *Zool. J. Linn. Soc.* 110: 207-217.

Pugh P.J.A. & Davenport J., 1997, Colonisation vs. disturbance: the effects of sustained ice-scouring on intertidal communities. *J. Exp. Mar. Biol. Ecol.* 210: 1-21.

Quesada A., Mouget J.L. & Vincent W.F., 1995, Growth of Antarctic cyanobacteria under ultraviolet radiation - UVA counteracts UVB inhibition. *J. Phycol.* 31: 242-248.

Quesada A., Goff L. & Karentz D., 1998, Effects of natural UV radiation on Antarctic cyanobacterial mats. *Proc. NIPR Symp. Polar Biol.* 11: 98-111.

Ring R.A. & Danks H.V., 1994, Desiccation and cryoprotection: overlapping adaptations. *Cryo-Lett.* 15: 181-190.

Roberts L., 1989, Does the ozone hole threaten antarctic life? *Science* 244: 288-289.

Rozema J. (ed.) 1999, *Stratospheric Ozone Depletion: the Effects of Enhanced UV-B Radiation on Terrestrial Ecosystems*. Bacchus, Leiden.

Ruhland C.T. & Day T.A., 2000, Effects of ultraviolet-B radiation on leaf elongation, production and phenylpropanoid concentrations of *Deschampsia antarctica* and *Colobanthus quitensis* in Antarctica. *Physiol. Plant* 109: 244-251.

Skvarca P., Rack W., Rott H. & Ibarzábal y Donángelo T., 1998, Evidence of recent climatic warming on the eastern Antarctic Peninsula. *Ann. Glacio.* 27: 628-632.

Smith R.I.L., 1984, Terrestrial Plant Biology of the Sub-Antarctic and Antarctic. In: R.M. Laws (ed.) *Antarctic Ecology*. Academic Press, London.

Smith R.I.L., 1987, The bryophyte propagule bank of Antarctic fellfield soils. *Symp. Biol. Hungarica* 35: 233-245.

Smith R.I.L., 1988a, Recording bryophyte microclimate in remote and severe environments. In: J.M. Glime (ed.) *Methods in Bryology*. Hattori Botanical Laboratory, Nichinan.

Smith R.I.L., 1988b, Destruction of Antarctic terrestrial ecosystems by a rapidly increasing fur seal population. *Biol. Conserv.* 45: 55-72.

Smith R.I.L., 1990, Signy Island as a paradigm of biological and environmental change in Antarctic terrestrial ecosystems. In: K.R. Kerry & G. Hempel (eds.) *Antarctic Ecosystems, Ecological Change and Conservation*. Springer-Verlag, Berlin.

Smith R.I.L., 1991, Exotic sporomorpha as indicators of potential immigrant colonists in Antarctica. *Grana* 30: 313-324.

Smith R.I.L., 1993, The role of bryophyte propagule banks in primary succession: case study of an Antarctic fellfield soil. In: J. Miles & D.W.H. Walton (eds.) *Primary succession on land*. Blackwell, Oxford.

Smith R.I.L., 1994, Vascular plants as indicators of regional warming in Antarctica. *Oecologia* 99: 322-328.

Smith R.I.L., 1996, Introduced plants in Antarctica: potential impacts and conservation issues. *Biol. Conserv.* 76: 135-146.

Smith R.I.L. & Coupar A.M., 1986, The colonization potential of bryophyte propagules in Antarctic fellfield soils. *CNFRA* 58: 189-204.

Smith V.R. & Steenkamp M., 1990, Climatic change and its ecological implications at a sub-Antarctic island. *Oecologia* 85: 14-24.

Sømme L., 1995, *Invertebrates in Hot and Cold Arid Environments*. Springer-Verlag, Berlin.

Southwood T.R.E., 1977, Habitat, the templet for ecological strategies. *J. Anim. Ecol.* 46: 337-365.

Southwood T.R.E., 1988, Tactics, strategies and templets. *Oikos* 52: 3-18.

Strathdee A.T., Bale J.S., Block W.C., Coulson S.J., Hodkinson I.D. & Webb N.R., 1993, Effects of temperature elevation on a field population of *Acyrthosiphon svalbardicum* (Hemiptera: Aphididae) on Spitsbergen. *Oecologia* 96: 457-465.

Strathdee A.T., Bale J.S., Strathdee F.C., Block W.C., Coulson S.J., Webb N.R. & Hodkinson I.D., 1995, Climatic severity and the response to temperature elevation of Arctic aphids. *Global Change Biol.* 1: 23-28.

Sugden D.E. & Clapperton C.M., 1977, The maximum ice extent on island groups in the Scotia Sea, Antarctica. *Quat. Res.* 7: 268-282.

Turner J., Colwell S.R. & Harangozo S., 1997, Variability of precipitation over the coastal western Antarctic Peninsula from synoptic observations. *J. Geophysical Res.* 102: 13999-14007.

Vaughan D.G. & Doake S.M., 1996, Recent atmospheric warming and retreat of ice shelves on the Antarctic Peninsula. *Nature* 379: 328-331.

Vincent W.F. & Quesada A., 1994, Ultraviolet effects on cyanobacteria: implications for Antarctic microbial ecosystems. *Antarct. Res. Ser.* 62: 111-124.

Vogel M., Remmert H. & Smith R.I.L., 1984, Introduced reindeer and their effects on the vegetation and the epigeic invertebrate fauna of South Georgia (sub-Antarctic). *Oecologia* 62: 102-109.

Voytek M.A., 1990, Addressing the biological effects of decreased ozone on the Antarctic environment. *Ambio* 19: 52-61.

Walton D.W.H., 1982, The Signy Island terrestrial reference sites: XV. Microclimate monitoring, 1972-74. *Br. Antarct. Surv. Bull.* 55: 111-126.

Walton D.W.H., 1984, The terrestrial environment. In: R.M. Laws (ed.) *Antarctic Ecology*. Academic Press, London.

Walton D.W.H., Vincent W.F., Timperley M.H., Hawes I. & Howard-Williams C., 1997, Synthesis: polar deserts as indicators of change. In: W.B. Lyons, C. Howard-Williams & I. Hawes (eds.) *Ecosystem Processes in Antarctic Ice-Free Landscapes*. Balkema, Rotterdam.

Webb N.R., Coulson S.J., Hodkinson I.D., Block W., Bale J.S. & Strathdee A.T., 1998, The effects of experimental temperature elevation on populations of cryptostigmatic mites in high Arctic soils. *Pedobiologia* 42: 298-308.

Wharton D.A., 1995, Cold tolerance strategies in nematodes. *Biol. Rev.* 70: 161-185.

Worland M.R., 1996, The relationship between body water content and cold tolerance in the Arctic collembolan *Onychiurus arcticus* (Collembola: Onychiuridae). *Eur. J. Entomol.* 93: 341-348.

Worland M.R. & Convey P., Rapid cold hardening in Antarctic microarthropods. *Funct. Ecol.*, in press.

Worland M.R., Grubor-Lajsic G. & Montiel P.O., 1998, Partial desiccation induced by sub-zero temperatures as a component of the survival strategy of the Arctic collembolan *Onychiurus arcticus* (Tullberg). *J. Insect Physiol.* 44: 211-219.

Wynn-Williams D.D., 1992, Plastic cloches for manipulating natural terrestrial environments. In: D.D. Wynn-Williams (ed.) *BIOTAS Manual of Methods for Antarctic Terrestrial and Freshwater Research*. Scientific Committee on Antarctic Research, Cambridge.

Wynn-Williams D.D., 1993, Microbial processes and the initial stabilisation of fellfield soil. In: J. Miles & D.W.H. Walton (eds.) *Primary Succession on Land*. Blackwell, Oxford.

Wynn-Williams D.D., 1994, Potential effects of ultraviolet radiation on Antarctic primary terrestrial colonizers: cyanobacteria, algae, and cryptogams. *Antarct. Res. Ser.* 62: 243-257.

Wynn-Williams D.D., 1996a, Antarctic microbial diversity: the basis of polar ecosystem processes. *Biodivers. Conserv.* 5: 1271-1293.

Wynn-Williams D.D., 1996b, Response of pioneer soil microalgal colonists to environmental change in Antarctica. *Microb. Ecol.* 31: 177-188.

Xiong F.S. & Day T.A., 2001, Effect of solar ultraviolet-B radiation during springtime ozone depletion on photosynthesis and biomass production of Antarctic vascular plants. *Plant Physiol.* 125: 738-751.

Climate change and ice breeding pinnipeds

BRENDAN P. KELLY
Juneau Center, School of Fisheries and Ocean Sciences, University of Alaska Fairbanks, Juneau, Alaska 99801, U.S.A.

Abstract: Pinniped diversity is greatest in seasonally ice-covered seas where the risk of predation is minimised. In recent decades, the thickness and extent of seasonal ice cover has decreased in the Arctic, and climate models predict that positive feedback from melting ice covers will result in rapid warming in the polar regions. Correlational studies linking arctic marine mammals to climate change are limited by inadequate time series of population counts. Increased understanding of the ecology of individual species is needed as the bases for testable hypotheses. Potential effects of arctic warming on marine mammals have been discussed in terms of decreased areal extent of the ice, but the most immediate effects may result from more subtle changes in the distribution of ice and snow that affect the ecology of individual species.

Ringed seals and walruses are strongly associated with seasonal sea ice and illustrate ecological differences that influence their vulnerability to warming in the Arctic. Earlier snowmelts may prematurely destroy subnivean lairs subjecting ringed seal pups to adverse weather and increased predation. Decreases in the summer extent of arctic sea ice may decrease the Pacific walruses' access to food and increase their exposure to polar bear predation.

INTRODUCTION

The earth's climate is changing in response to natural and anthropogenic influences (Ledley et al. 1999, Crowley 2000, Committee on the Science of Climate Change 2001). Climate models predict amplification of warming in the polar regions due to reduced albedo as ice covers melt (Groisman et al. 1994a, b). The consequences may be especially important for the pinnipeds (seals, sea lions, and walruses), a subfamily of carnivores that is most diverse and numerous at high latitudes. There, many species use sea ice as a substrate for resting, breeding, and molting (Figure 1).

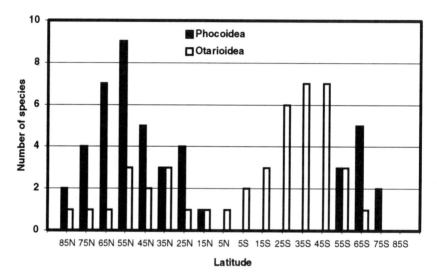

Figure 1. The diversity of breeding species of pinnipeds in the superfamilies Phocoidea and Otarioidea is greatest at latitudes representing the maximal seasonal extent of sea ice. Data from Scheffer (1958), Fay (1982), King (1983). After Kelly (1996).

Populations of pinnipeds and other marine mammals have responded measurably to changes in the physical environment at lower latitudes. For example, in recent decades Steller sea lions (*Eumetopias jubatus*) declined by 75 % in the Gulf of Alaska and Aleutian Islands (Merrick et al. 1987, Loughlin et al. 1992, Hill & DeMaster 1999), Pacific harbor seals (*Phoca vitulina richardsi*) by 85 % in the Gulf of Alaska (Pitcher 1990, Jemison & Kelly 2001), and sea otters (*Enhydra lutris*) by 80 % in the Aleutian Islands (Estes et al. 1998). Considerable on-going research is devoted to the hypothesis that those declines resulted from large-scale changes in the ecosystem, and it is clear that pinnipeds can serve as indicators of environmental change. Pinniped populations at higher latitudes are extremely difficult to track, however, and – with few exceptions – we are unable to say whether or how they are responding to environmental changes. Testa et al. (1991) related the population dynamics of Antarctic pinnipeds to El Niño/Southern Oscillations, but for the most part, we cannot determine if Arctic pinnipeds are responding to climate change. Nonetheless, in the past decade, indigenous hunters of pinnipeds have reported unusual weather and ice conditions affecting the animals and hunting in the Arctic (Noongwook 2000, Pungowiyi 2000, Smith & Harwood 2001). I believe it is important to consider ice-associated pinnipeds in the context of global climate change for at least four reasons: (1) pinniped populations can respond to broad scale environmental changes; (2) pinniped diversity is highest in regions covered by seasonal sea ice; (3) arctic pinnipeds are economically and culturally important to indigenous people; and (4) the extent of sea ice cover may decline rapidly as albedo is reduced. It is important to recognise, however, that likely impacts will depend on the particular ecology of individual species and may involve mechanisms less obvious than reductions in substrate.

I am not aware of any research designed to test hypotheses relating arctic pinniped populations to climate change. A few accounts, however, have suggested links between arctic marine mammal populations and climate change based on ecological studies. Here, I review those accounts and relevant observations from my own work and suggest a framework for further investigation. I begin with background on the relationship between pinnipeds and sea ice.

SEA ICE EXTENT AND PINNIPEDS

Pinnipeds arose from terrestrial carnivores in the late Oligocene of the eastern North Pacific Ocean (Berta 1991), and may have begun to colonise sea ice in the Pleistocene when the shores they bred on became ice bound (Stirling 1975). Modern pinnipeds remain adapted for aquatic locomotion, and they are most vulnerable to predation out of the water where they give birth (Allen 1880). Sea ice provides an extensive substrate on which predation risk is reduced, and there are nine species of pinnipeds that breed on sea ice in the Arctic and four that do so in the Antarctic. The seasonal sea ice of the Arctic provides an average of 7,000,000 km^2 (summer) to 14,000,000 km^2 (winter) of substrate for the pinnipeds that rest, breed, and molt there (Maykut 1985). Two species, ringed seals (*Phoca hispida*) and walruses (*Odobenus rosmarus*), are strongly associated with arctic sea ice with many individuals spending most of the year in contact with the ice.

The importance of sea ice dynamics in the ecology of marine mammals has long been known to indigenous hunters (Shapiro et al. 1979, Noongwook

2000, Pungowiyi 2000) and eventually was appreciated by pioneering marine mammalogists. Vibe (1950, 1967) pointed out that climate, by affecting sea ice conditions, influenced variability in seal and walrus distributions, and Fay (1974) described the relationships between marine mammals in the Bering Sea and particular ice characteristics based on direct observations. With the aid of remotely acquired images of sea ice, Burns et al. (1981) refined and expanded on Fay's descriptions. Mammalogists in Canada and Alaska observed that especially heavy ice conditions in the winter of 1973-1974 coincided with a large reduction in the estimated numbers of ringed seals and bearded seals (*Erignathus barbatus*) in the Beaufort and Chukchi seas (Stirling et al. 1977, 1982, Stirling & Smith 1977, Burns & Eley 1978, Smith & Stirling 1978) and suggested that seal numbers might be regulated by such events, although other causes of the apparent decline could not be ruled out (Kelly 1988).

Recently, Stirling (1997) reviewed the importance of sea ice to marine mammals, and Tynan & DeMaster (1997) speculated on the effects on marine mammals of accelerated warming in the Arctic. They reviewed climate changes affecting the Arctic Ocean and emphasised potential effects of reduced ice cover on pinnipeds, their prey, and polar bears and called for monitoring of sea ice extent and certain "indicator" species including ringed seals.

Reproduction and body condition of one population of ringed seals in the Canadian Arctic was, in fact, monitored in the 1970s and 1990s. One result was the conclusion that the unusually heavy ice conditions in 1974 reduced food availability and led to low rates of ovulation (Harwood et al. 2000, Smith & Harwood 2001). In contrast, in 1998, when the sea ice on which that population gives birth broke up 43 days earlier than the preceding eight-year average, seal prey apparently was abundant, and ovulation rates were high. Survival of pups, however, was low due to premature weaning believed to be the result of the early ice break-up. The early break-up may have enhanced food availability but negatively impacted the population by the interruption of nursing.

Early ice break-up also appears to have negatively impacted a polar bear (*Ursus maritimus*) population, although by a different mechanism (Stirling et al. 1999). Increasingly early ice break-ups in Hudson Bay, Canada between 1981 and 1998 apparently shortened the time bears could hunt seals before going ashore for an annual fast, and the bears' body fat and birth rates declined. Presumably, seals in Hudson Bay enjoyed lower predation rates over the same period.

That early ice break-up may impact ringed seal and polar bear populations by different mechanisms underscores the need for detailed ecological data if we are to make meaningful predictions about the potential effects of warming in the Arctic. Climate change undoubtedly will have multiple negative and positive impacts on the ecology of pinnipeds, and the net effect for each species will be a complicated weighting of diverse effects. The sparseness of population data for arctic marine mammals bodes poorly for our ability to relate population trends to climate or ice data. More detailed information on the ecology of those species, however, can uncover mechanisms by which environmental change might positively or negatively affect individual populations or species. Potential mechanisms, when identified, could provide a basis for developing testable hypotheses.

My co-workers and I have been studying the distribution and demography of Pacific walruses and the behaviour of ringed seals with the aim of improving population monitoring (Fay & Kelly 1989, Fay et al. 1989, 1997, Kelly & Quakenbush 1990, Kelly et al. 1999, 2000, 2001). In the course of those studies, we have made observations that illustrate how the individual nature of each species' use of sea ice can lead to testable hypotheses about the specific effects of climate change.

Ringed seals

Sea ice seasonally covers 5 % of the Northern Hemisphere (Maykut 1985), and that entire area is habitat for several million ringed seals, the most numerous and widespread marine mammal in the Northern Hemisphere (Scheffer 1958, Smith 1987, Kelly 1988). The original human occupation of the Arctic, indeed, depended on the year-round presence of ringed seals, which, in turn, depended on the seals' ability to maintain breathing holes through the ice and to breed in this extreme climate.

Sea ice forms a substrate on which ringed seal pups are born, but the snow on the ice also is critical to pup survival. Ringed seal pups are born primarily in April when winds, combined with temperatures as cold as -30 °C, result in extremely low wind chill temperatures. Weighing about 4 kg at birth, ringed seal pups are the smallest among seals and survive only because they are born in subnivean lairs excavated by their mothers above breathing holes (Hall 1866, Chapskii 1940, McLaren 1958, Smith & Stirling 1975). Lairs typically are excavated in snowdrifts formed next to pressure ridges or other ice deformities. Throughout the winter and spring, the only access to the lair is through the breathing hole below. As temperatures increase in late spring, however, the seals emerge from under the snow to rest where the sun can warm their skin. At that time, they require elevated skin temperatures to promote regeneration of the epidermis (Feltz & Fay 1968, Kelly 1988). The pups are nursed in the lairs where they are concealed for most or all of the first two months of life. As long as the lairs are intact, the pups are completely safe from visual predators such as gulls and ravens

(Kumlien 1879, Lydersen & Smith 1989), but olfactory predators, including polar bears and arctic foxes, detect the pups and prey on them after excavating or collapsing the lairs (Smith 1976, 1980).

I have reviewed evidence that unusually heavy ice and early ice break-ups can be detrimental to ringed seal populations, but warming in the Arctic may reduce pup survival by changing snow conditions before further changes are apparent in the underlying ice cover. In the Beaufort Sea off northern Alaska, we investigated the thermal environment of the subnivean lairs from early in the birthing season through the late spring in 1998-2001 (Kelly et al. 1986, Kelly 1988, Harding et al. 2001, Kelly et al. 2001). We measured air temperature in and outside of lairs and snow temperatures on the ice at 5 cm depth increments. We also monitored seal behaviour telemetrically in 1999, 2000, and 2001.

In April, when most ringed seal pups are born, temperatures inside of lairs were above -9 °C (mean = -5 °C, SD = 1.6) while outside air temperatures were as low as -29 °C (mean = -15 °C, SD = 5.3). The snow cover trapped heat from the seawater below the lair and from the seals' bodies (Figure 2). In May, however, the seals began to emerge from their lairs to bask in the sun by actively excavating through the snow cover. Air temperatures began to converge with or exceeded the temperature in the lairs, and the snow warmed gradually until abruptly becoming wet throughout with the arrival of warm weather systems in late May or early June. Those warming events, marked by the snow pack becoming isothermal at the monitoring site, corresponded with the last of the seals emerging from under the snow. The remaining lairs began to melt and the last radio-tagged seals emerged in early June in 1999, 2000, and 2001.

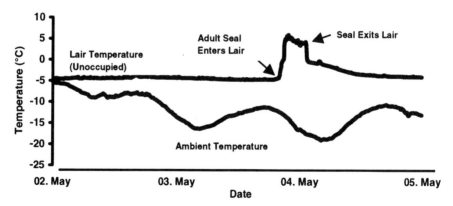

Figure 2. Air temperatures over three days in (upper line) and outside (lower line) a subnivean seal lair. The insulating snow trapped heat from the seawater below the lair. Additional heating took place when a seal (tracked by radio) occupied the lair.

We did not track seals telemetrically in 1998, but we did observe that the snowmelt took place 3 or more weeks earlier than in 1999, 2000, or 2001. That earlier snowmelt meant seals had to emerge from their lairs in 1998 at about the time they were just beginning to emerge in 1999, 2000, and 2001.

The variation in dates of snowmelt we observed in 1998-2001 was consistent with the variability observed between 1975 and 1988 (Robinson et al. 1992). Superimposed on that inter-annual variability, however, is a long-term trend of increasing air temperatures, decreasing depth and areal extent of snow cover, and increasingly earlier snow melts (Jones & Briffa 1992, Brown & Goodison 1996, Brown & Braaten 1998, Serreze et al. 2000).

If snow continues to decrease in depth and to melt earlier, ringed seal pups may be exposed at increasingly earlier ages to predators and freeze-thaw cycles. When lack of snow cover has forced birthing to occur in the open, nearly 100 % of the pups have succumbed to predation (Kumlien 1879, Lydersen et al. 1987, Lydersen & Smith 1989, Smith & Lydersen 1991, Smith et al. 1991). At birth, ringed seal pups lack an insulating blubber layer and depend instead on a woolly natal coat for protection from cold. The natal coat is an effective insulator in air but not when wet. Young pups exposed to freeze-thaw cycles that occur during snowmelt would be prone to hypothermia.

Ringed seal populations might respond to earlier springs by advancing the timing of parturition, although that timing may be constrained by seasonal peaks in marine production (Costa 1991). Of course, early springs might also advance seasonal productivity. In many pinniped species, the seasonal timing of parturition tends to vary little within a population even when there are differences among populations (Boulva 1975, Coulson 1981, Boyd 1991), but Jemison & Kelly (2001) described changes of 1 to 2.5 weeks in the timing of parturition within a harbor seal population. They suggested the changes were in response to decades-long variations in the seasonal availability of food. The degree of phenotypic plasticity in the birthing phenology of ringed seals is unknown.

Walruses

While the small size of ringed seals makes them vulnerable to hypothermia, at least as pups, the large size of walruses (up to 1800 kg) makes them more tolerant of the cold. They are still vulnerable, however, to predation from polar bears, and reduced sea ice extent may increase that vulnerability. Furthermore, the decreased extent of summer sea ice may negatively impact their ability to obtain food.

Walruses are nearly circumpolar in distribution, but they occur only over continental shelves as they feed on benthic invertebrates and cannot effectively feed at depths beyond 90-100 m (Fay 1982, Fay & Burns 1988). After breeding on the winter ice in the Bering Sea, the males retreat to coastal areas of that sea, while the females and their dependent young (up to 3 years old) retreat with the ice in to the Chukchi Sea (Fay 1982). There, they feed intensively in between bouts of resting and suckling their young on the ice.

In 1998, the sea ice in the Beaufort and Chukchi seas retreated unusually far to the north, and by September, it covered 25 % less of the Arctic Ocean than during the minimum for the previous 35 years (Maslanik et al. 1999). Surveying walruses from shipboard that month, we found that substantial portions of the ice edge had receded north of the continental shelf where the water was too deep for walruses to feed. Continued warming in the Arctic might reduce the amount of ice over the continental shelf in summer and fall and, thereby, reduce the amount of forage available to lactating walruses. If walruses respond by concentrating on ice or shorelines near feeding areas, they may increase their risk of predation by polar bears.

CONCLUSIONS AND OUTLOOK

In summary, the reliance of arctic pinnipeds and their major predators, polar bears, on sea ice clearly suggests vulnerability to reduction of the ice cover. Population estimates are crude, however, and we lack time series data with which to seek correlations with ice and weather conditions. Correlations between extreme ice conditions and body condition and reproductive rates of arctic pinnipeds and their predators suggest "fingerprints" from the hand of climate change.

Fingerprints might tie a subject to a location, but criminologists need additional information to establish guilt. Similarly, these biological "fingerprints" are consistent with, but do not confirm, climate change as impacting arctic pinnipeds. How can we determine if the correlations reflect causation? I believe that the answer will lie in ecological understandings and the testing of hypotheses concerning mechanisms by which arctic pinnipeds might be affected by climate change. The ecological knowledge will allow us to ask the right questions. It will be necessary, however, to bear in mind the time scales, ecological or evolutionary, appropriate to specific questions.

On an ecological time scale, we need to know the specific relationships between the animals and sea ice, snow, and other aspects of their habitat (Davis et al. 1998, Martin 2001). It also will be important, on the ecological time scale, to bear in mind that populations are a more appropriate unit of concern than species. For example, decreased summer ice cover might disadvantage some polar bear populations (by decreasing access to ringed seals) while favouring others (by concentrating walruses or other prey). Ringed seals are broadly distributed in all the ice-covered seas and some freshwater lakes of the Northern Hemisphere. Thus, they live in a broad range of biotic and abiotic conditions, and the effects of climate change likely will vary with those conditions.

Responses to climate change on the evolutionary time scale are especially difficult to address (Hoffmann & Blows 1993, Davis & Shaw 2001). The ability of populations to respond adaptively to changing environments depends in part on heritable variation, which may or may not be expressed in the current environment. Typically, adaptive responses are slow and, hence, can only accommodate gradual environmental changes. There are, however, notable exceptions (Berthold et al. 1992, Huey et al. 2000). Comparative studies of species such as ringed seals that occupy a variety of habitats, including relic populations in freshwater lakes, could provide insights into adaptive responses.

ACKNOWLEDGEMENTS

This work was supported by the North Pacific Marine Research Initiative, the U.S. Fish and Wildlife Foundation, and the Coastal Marine Institute of the University of Alaska. Greenpeace supplied ice-breaking ship support for the walrus studies, and ARCO Alaska and BP Exploration provided logistic support for ringed seal field studies. Assistance in data collection was provided by colleagues and students including J. Bengston, K. Blejwas, P. Boveng, M. Cronin, S. Crowley, R. Daniel, L. Dzinich, K. Faloon, J. Garlich-Miller, O. Harding, R. Ingram, G. Jarrell, C. Jay, C. Kelly, K. Kelly, D. McDonald, D. O'Leary, J. Nielsen, J. Pilby, L. Quakenbush, M. Simpkins, P. Singer, B. Taras, D. Wartzok, and J. Womble. My participation in the *"FINGERPRINTS" of CLIMATE CHANGE* conference was made possible by the Centro Stefano Franscini / Swiss Federal Institute of Technology (ETH) Zurich. I am grateful to the institute and the conference organisers for the opportunity.

REFERENCES

Allen J.A., 1880, *History of North American pinnipeds*. A monograph of the walruses, seal lions, sea bears, and seals of North America. U.S. Geol. and Geogr. Surv. Terr., Misc. Publ. 12. U.S. Gov. Print. Off., Washington, D.C. 785pp.

Berta A., 1991, New *Enaliarctos* (Pinnipedimorpha) from the Oligocene and Miocene of Oregon and the Role of "Enaliarctids" in Pinniped Phylogeny. *Smithson. Contrib. Paleobiol.* 69: 1-33.

Berthold P., Helbig A.J., Mohr G. & Querner U., 1992, Rapid microevolution of migratory behaviour in a wild bird species. *Nature* 360: 668-670.

Boulva J., 1975, Temporal variations in birth period and characteristics of newborn harbour seals. *Rapp. P.-v. Réun. Cons. int. Explor. Mer.* 169: 405-408.

Boyd I. L., 1991, Environmental and physiological factors controlling the reproductive cycles of pinnipeds. *Can. J. Zool.* 69: 1135-1148.

Brown R.D. & Braaten R.O., 1998, Spatial and temporal variability of Canadian monthly snow depths, 1946-1995. *Atmosphere-Ocean* 36: 37-54.

Brown R.D. & Goodison B.E., 1996, Interannual variability in reconstructed Canadian monthly snow depths, 1915-1992. *J. Clim.* 9: 1299-1318.

Burns J.J. & Eley T.J., 1978, *The natural history and ecology of the bearded seal (Erignathus barbatus) and the ringed seal (Phoca hispida).* Ann. Rep. RU#230, Outer Continental Shelf Environmental Assessment Program, Juneau, Alaska.

Burns J.J., Shapiro L. & Fay F.H., 1981, Ice as marine mammal habitat in the Bering Sea. In: D.W. Hood & J.A. Calder (eds.) *The eastern Bering Sea shelf: oceanography and resources.* Volume 2. U.S. Dep. Commer., NOAA, Off. Mar. Pollut. Assess., Juneau, AK, 781-797.

Chapskii K.K., 1940, The ringed seal of western seas of the Soviet Arctic (The morphological characteristic, biology and hunting production). *Tr. Vses. Arkt. Inst. (Leningrad)* 145: 1-72. (Transl. from Russian by Fish. Res. Board Can., 1971, Transl. Ser. 1665, 147pp.)

Committee on the Science of Climate Change, 2001, *Climate change science; an analysis of some key questions.* Division on Earth and Life Studies, National Research Council, National Academy Press. Washington, D.C. 29pp.

Costa D.P., 1991, Reproductive and foraging energetics of pinnipeds: implications for life history patterns. In: D. Renouf (ed.) *The behaviour of pinnipeds.* Chapman & Hall, London, 300-344.

Coulson J.C., 1981, A study of the factors influencing the timing of breeding in the grey seal *Halichoerus grypus. J. Zool.* 194: 553-571.

Crowley T.J., 2000, Causes of climate change over the past 1000 years. *Science* 289: 270-277.

Davis A.J., Jenkinson L.S., Lawton J.H., Shorrocks B. & Wood S., 1998, Making mistakes when predicting shifts in species range in response to global warming. *Nature* 391: 783-786.

Davis M.B. & Shaw R.G., 2001, Range shifts and adaptive responses to quarternary climate change. *Science* 292: 673-679.

Estes J.A., Tinker M.T., Williams T.M. & Doak D.F., 1998, Killer whale predation on sea otters linking oceanic and nearshore ecosystems. *Science* 282: 473-476.

Feltz E.T. & Fay F.H., 1966, Thermal requirements *in vitro* of epidermal cells from seals. *Cryobiology* 3: 261-264.

Fay F.H., 1974, The role of ice in the ecology of marine mammals of the Bering Sea. In: D.W. Hood & E.J. Kelley (eds.) *Oceanography of the Bering Sea.* Institute of Marine Science, University of Alaska, Fairbanks, 383-399.

Fay F.H., 1982, Ecology and biology of the Pacific walrus, *Odobenus rosmarus divergens,* Illiger. *North American Fauna* 74.

Fay F.H. & Burns J.J., 1988, Maximal feeding depth of walruses. *Arctic* 41: 239-240.

Fay F.H. & Kelly B.P., 1989, *Development of a method for monitoring the productivity, survivorship, and recruitment of the Pacific walrus population.* Final Report, OCSEAP Study MMS 89-0012. Minerals Management Service, Anchorage, AK, 51pp.

Fay F.H., Kelly B.P. & Sease J.L., 1989, Managing the exploitation of Pacific walruses: a tragedy of delayed response and poor communication. *Mar. Mamm. Sci.* 5: 1-16.

Fay F.H., Eberhardt L.L., Kelly B.P., Burns J.J. & Quakenbush L.T., 1997, Status of the Pacific walrus population, 1950-1989. *Mar. Mamm. Sci.* 13: 537-565.

Groisman P.Y., Karl T.R., Knight R.W. & Stenchinkov G.L., 1994a, Changes of snow cover, temperature and radiative heat balance over the Northern hemishpere. *J. Clim.* 7: 1633-1656.

Groisman P.Y., Karl T.R. & Knight T.W., 1994b, Observed impact of snow cover on the heat balance and the rise of continental spring temperatures. *Science* 263: 198-200.

Hall C.F., 1866, *Arctic researches and life among the Esquimaux: being the narrative of an expedition in search of Sir John Franklin, in the years 1860, 1861, and 1862*. Harper Brothers Publishers, New York.

Harding O.R., Kelly B.P., Quakenbush L.T. & Taras B.D., 2001, Importance of snow for ringed seal pups. In: Abstract Book. *Aquatic Sciences Meeting. American Society of Limnology and Oceanography.* February 12-16, 2001, p. 64.

Harwood L.A., Smith T.G. & Melling H., 2000, Variation in reproduction and body condition of the ringed seal (*Phoca hispida*) in western Prince Albert Sound, NT, Canada, as assessed through a harvest-based sampling program. *Arctic* 53: 422-431.

Hill P.S. & DeMaster D.P., 1999, *Alaska marine mammal stock assessments, 1999*. NOAA Technical Memorandum NMFS-AFSC-110.

Hoffmann A.A. & Blows M.W., 1993, Evolutionary genetics and climate change: will animals adapt to global warming? In: P.M. Kareiva, J.G. Kingsolver & R.B. Huey (eds.) *Biotic interactions and global change.* Sinauer, Sunderland, Massachusetts, 165-178.

Huey R.B., Gilchrist G.W., Carson M.L., Berrigan D. & Serra L., 2000, Rapid evolution of a geographic cline in size in an introduced fly. *Science* 287: 308-309.

Jemison L.A. & Kelly B.P., 2001, Pupping phenology and demography of harbor seals (*Phoca hispida richardsi*) on Tugidak Island, Alaska. *Mar. Mamm. Sci.* 17: 585-600.

Jones P.D. & Briffa K.R., 1992, Global surface air temperature variations during the twentieth century: Part 1, spatial, temporal and seasonal details. *Holocene* 2: 165.

Kelly B.P., 1988, Ringed seal. In: J.W. Lentfer (ed.) *Selected Marine Mammals of Alaska: Species Accounts with Research and Management Recommendations.* Marine Mammal Commission, Washington, D.C., 59-75.

Kelly B.P., 1996, Behavior of ringed seals diving under shore-fast sea ice. Ph.D. Dissertation. Purdue University, West Lafayette, Indiana, 91pp.

Kelly B.P. & Quakenbush L.T., 1990, Spatiotemporal use of lairs by ringed seals (*Phoca hispida*). *Can. J. Zool.* 68: 2503-2512.

Kelly B.P., Quakenbush L.T. & Rose J.R., 1986, *Ringed seal winter ecology and effects of noise disturbance.* Outer Cont. Shelf Environ. Assess. Program, Final Rep. Princ. Invest., NOAA, Anchorage, Alaska 61: 447-536. OCS Study MMS 89-0026; NTIS PB89-234645.

Kelly B.P., Quakenbush L.T. & Taras B.D., 1999, *Correction factor for ringed seal surveys in Northern Alaska.* Annual Report for 1999 to Coastal Marine Institute, University of Alaska Fairbanks, 25pp.

Kelly B.P., Quakenbush L. & Taras B., 2000, Correction factor for ringed seal surveys in northern Alaska. In: V. Alexander (ed.) *University of Alaska Coastal Marine Institute. Annual Report, Federal Fiscal Year 2000.* University of Alaska Fairbanks.

Kelly B.P., Quakenbush L.T., Taras B.D. & Harding O., 2001, *Snow conditions predict optimal survey periods for ringed seals.* Abstract. 14th Biennial Conference on the Biology of Marine Mammals. Vancouver, British Columbia.

King J.E., 1983, *Seals of the world*, 2nd Edition. Comstock Publishing Associates, Ithaca, New York, 240pp.

Kumlien L., 1879, *Contributions to the natural history of Arctic America, made in connection with the Howgate Polar Expedition, 1877-1878.* U.S. national Museums Bulletin 15. 179pp.

Ledley T.S., Sundquist E.T., Schwartz S.E., Hall D.K., Fellows J.D. & Killeen T.L., 1999, Climate change and greenhouse gases. *EOS* 80: 453-473.

Loughlin T.R., Perlov A.S. & Vladimirov V.A., 1992, Range-wide survey and estimation of total number of Steller sea lions in 1989. *Mar. Mamm. Sci.* 8: 220-239.

Lydersen C. & Smith T.G., 1989, Avian predation on ringed seal, *Phoca hispida*, pups. *Polar Biol.* 9: 489-490.

Lydersen C., Jensen P.M. & Lydersen E., 1987, Studies of the ringed seal population in the Van Mijen fjord, Svalbard, in the breeding period 1986. *Norsk Polarinst. Rapportser.* 34: 91-112.

Martin, T.E., 2001, Abiotic vs. biotic influences on habitat selection of coexisting species: climate change impacts? *Ecology* 82:175-188.

Maslanik J.A., Serreze M.C. & Agnew T., 1999, On the record reduction in Western arctic sea ice cover in 1998. *Geophys. Res. Lett.* 26: 1905-1908.

Maykut G.A., 1985, The ice environment. In: R.A. Horner (ed.) *Sea ice biota.* CRC Press, Inc., Boca Raton, Florida, 21-82.

McLaren I.A., 1958, The biology of the ringed seal (*Phoca hispida* Schreber) in the eastern Canadian Arctic. *The Fisheries Research Board of Canada, Bulletin* No. 118, Ottawa.

Merrick R.L., Loughlin T.R. & Calkins D.G., 1987, Decline in abundance of northern sea lions, *Eumetopias jubatus*, in Alaska, 1956-86. *Fishery Bulletin* 85: 351-365.

Noongwook G., 2000, Native observations of local climate change around St. Lawrence Island. In: H.P. Huntington (ed.) *Impacts of changes in sea ice and other environmental parameters in the arctic.* Marine Mammal Commission, Bethesda, Maryland, 21-24.

Pitcher K.W., 1990, Major decline in number of harbor seals, *Phoca vitulina richardsi*, on Tugidak Island, Gulf of Alaska. *Mar. Mamm. Sci.* 6: 121-134.

Pungowiyi C., 2000, Native observations of change in the marine environment of the Bering Strait region. In: H.P. Huntington (ed.) *Impacts of changes in sea ice and other environmental parameters in the arctic.* Marine Mammal Commission, Bethesda, Maryland, 18-20.

Robinson D.A., Serreze M.C., Barry R.G., Scharfen G. & Kukla G., 1992, Large-scale patterns and variability of snowmelt and parameterized surface albedo in the arctic basin. *J. Clim.* 5: 1109-1119.

Scheffer V.B., 1958, *Seals, sea lions and walruses. A review of the pinnipedia.* Stanford University Press, Stanford, California. 179pp.

Serreze M.C., Walsh J.E., Chapin III F.S., Osterkamp T., Dyurgerov M., Romanovsky V., Oechel W.C., Morison J., Zhang T. & Barry R.G., 2000, Observational evidence of recent change in the northern high-latitude environment. *Clim. Change* 46: 159-207.

Shapiro L.H., Metzner R.C. & Toovak K., 1979, *Historical references to ice conditions along the Beaufort Sea coast of Alaska.* Scientific Report, NOAA Contract.

Smith T.G., 1976, Predation of ringed seal pups (*Phoca hispida*) by the arctic fox (*Alopex lagopus*). *Can. J. Zool.* 54: 1610-1616.

Smith T.G., 1980, Polar bear predation of ringed and bearded seals in the land-fast sea ice habitat. *Can. J. Zool.* 58: 2201-2209.

Smith T.G., 1987, The ringed seal, *Phoca hispida*, of the Canadian western Arctic. *Can. Bull. Fish. Aquat. Sci.* 216. Ottawa. 81pp.

Smith T.G. & Harwood L.A., 2001, Observations of neonate ringed seals, *Phoca hispida*, after early break-up of the sea ice in Prince Albert Sound, Northwest Territories, Canada, spring 1998. *Polar Biol.* 24: 215-219.

Smith T.G. & Lydersen C., 1991, Availability of suitable land-fast ice and predation as factors limiting ringed seal populations, *Phoca hispida*, in Svalbard. *Polar Res.* 10: 585-594.

Smith T.G. & Stirling I., 1975, The breeding habitat of the ringed seal (*Phoca hispida*). The birth lair and associated structures. *Can. J. Zool.* 53: 1297-1305.

Smith T.G. & Stirling I., 1978, Variation in the density of ringed seal (*Phoca hispida*) birth lairs in the Amundsen Gulf, Northwest Territories. *Can. J. Zool.* 56: 1066-1071.

Smith T.G., Hammill M.O. & Taugbøl G., 1991, A review of the developmental, behavioural and physiological adaptations of the ringed seal, *Phoca hispida*, to life in the arctic winter. *Arctic* 44: 124-131.

Stirling I., 1975, Factors affecting the evolution of social behavior in the Pinnipedia. In: K. Ronald & A.W. Mansfield (eds.) Biology of the seal. *Rapp. P.-v. Réun. Cons. int. Explor. Mer.* 169: 205-212.

Stirling I., 1997, The importance of polynyas, ice edges, and leads to marine mammals and birds. *J. Mar. Syst.* 10: 9-21.

Stirling I., Archibald W.R. & DeMaster D., 1977, Distribution and abundance of seals in the eastern Beaufort Sea. *J. Fish. Res. Board Can.* 34: 976-988.

Stirling I., Kingsley M. & Calvert W., 1982, The distribution and abundance of seals in the eastern Beaufort Sea, 1974-79. Ottawa: *Canadian Wildlife Service Occasional Paper* 47: 1-23.

Stirling I., Lunn N.J. & Iacozza J., 1999, Long-term trends in the population ecology of polar bears in western Hudson Bay in relation to climate change. *Arctic* 52: 294-306.

Stirling I. & Smith T.G., 1977, Interrelationships of Arctic Ocean mammals in the sea ice habitat. In: *Circumpolar Conference on Northern Ecology*. Section II. Natl. Res. Council. Ottawa, 129-136.

Testa J.W., Oehlert G., Ainley D.G., Bengston J.L., Siniff D.B., Laws R.M. & Rounsevell D., 1991, Temporal variability in Antarctic marine ecosystems: periodic fluctuations in the phocid seals. *Can. J. Fish. Aquat. Sci.* 48: 631-638.

Tynan C.T. & DeMaster D.P., 1997, Observations and predictions of arctic climatic change: potential effects on marine mammals. *Arctic* 50: 308-322.

Vibe C., 1950, The marine mammals and the marine fauna in the Thule District (Northwest Greenland) with observations on ice conditions in 1939-41. *Meddelelser om Grønland* 150: 1-115.

Vibe C., 1967, Arctic animals in relation to climatic fluctuations. The Danish Zoogeographical Investigations in Greenland. *Meddelelser om Grønland* 170. 227pp.

Detection of range shifts: General methodological issues and case studies of butterflies

CAMILLE PARMESAN
Integrative Biology, University of Texas, Austin, Texas, 78712, USA

Abstract: Range shifts are perhaps the least controversial as well as the most easily observed responses expected under global warming scenarios. While changes at single study sites and along single species' range boundaries have been studied in a diversity of taxa, the wealth of historical records for butterflies has allowed changes across entire species' ranges to be analysed. This becomes important in distinguishing local distribution changes from systematic poleward/upward range shifts.

Examination of 58 species in North America and Europe documented poleward shifts of species' ranges in proportions far higher than one would expect by chance. The magnitudes of the observed range shifts (boundary movement from 35 to 200 km) is on the same order as the magnitudes of regional warming (from 0.7 to 0.8 °C, equating to movement of temperature isotherms by 92 to 120 km).

Coupled with general global warming, extreme weather and climate events have been increasing in magnitude and frequency on a global scale. Previous basic research on butterflies has provided detailed information on the mechanistic links between climate and population dynamics, reproductive behaviour, and extinction/colonisation dynamics. A synthesis of these studies indicates that such extreme climate events have driven observed butterfly range shifts.

The sensitivity of butterflies to climate, the temporal and spatial breadth of distributional data (especially for European species), as well as the wealth of basic biological knowledge, allow studies of butterflies to provide an in-depth understanding of current impacts of climate change on wildlife. Further, the effects of non-climatic global change factors have been studied for many species. Species' habitat requirements are often well understood, and butterflies have served as useful organisms for studies of habitat fragmentation and restoration. These complementary fields of study, taken together, poise butterflies for being ideal models for understanding the intersecting effects of modern environmental changes, particularly the hardships imposed by climate change across an increasingly hostile landscape.

PREDICTED BIOLOGICAL RESPONSES TO CLIMATIC WARMING TRENDS

All organisms are influenced by climate and weather events. Physiological and ecological thresholds shape the distributions of species, and the timing of their life-cycles (i.e. periods of growth, reproduction, and dormancy) (Uvarov 1931, MacArthur 1972, Precht et al. 1973, Wieser 1973, Brown et al. 1996, Hoffman & Parsons 1997, Saether 1997, reviewed by Parmesan et al. 2000). Studies of responses to past large-scale climatic warming trends during the Pleistocene ice-ages and early Holocene show that, overwhelmingly, the most common responses were poleward range shifts. A species tended to track long-term climatic change such that it maintained, more or less, a species-specific climatic envelope in which it lived or bred. Typically, a species' range or migratory destination shifted more than 100 km with each 1 °C change in mean annual temperature, moving poleward and upward in altitude during warming trends (Ashworth 1996, Brandon-Jones 1996, Coope 1995, Baroni & Orombelli 1994, Davis & Zabinski 1992, Graham 1992, Goodfriend & Mitterer 1988, Woodward 1987, Barnosky 1986, reviewed by Parmesan et al. 2000, Easterling et al. 2000). Extinctions of entire species and observable evolutionary adapations to temperature change were both rare.

IDENTIFYING INDICATOR SPECIES OR GROUPS

This section discusses some general design and methodological issues inherent in using unplanned correlational data to infer responses of wildlife to recent climate change. Butterflies are discussed as a taxa which may be particularly well-suited to documenting climate change impacts.

General criteria

There are some general traits of study systems which help to make changes in those systems easier to document, easier to interpret and easier to attribute to climate change, (as opposed to any one of many global change or natural factors which might also cause change within natural systems). Ideal target species, communities or systems to look for biotic responses to climate change are those which meet the following criteria (DeGroot et al. 1995, Parmesan 2001):

- Good historical records - either from being a model system in basic research or by having a history of amateur collecting.

- Current data available (from monitoring schemes, long-term research), or easy to gather.
- Short (decadal) or no lag time expected between climate change and response
- Basic research has led to a process-based understanding of the underlying mechanisms by which climate affects the organism or community. This knowledge may come from experimental laboratory or field studies of behaviour and physiology, or from correlational studies between field observations and climatic data.
- Relatively insensitive to other anthropogenic influences so that effects of possible confounding factors are minimised.

Trees provide an example of a category of organisms which meets some, but not all, of these criteria. Tree rings provide long time series of growth rates and data can be gathered over a large geographical area. However, interpretation of differential growth rates can be difficult. For instance, changes in growth could be due to changes in temperature or precipitation, or to other forms of stress, such as insect attack or disease (Briffa et al. 1998, Jacoby & D'Arrigo 1997, Bartholomay et al. 1997, Kullman 1996, Innes 1991). Additionally, tree distributional responses may have a lag time of centuries, and so changes in response to recent climate warming may not yet be detectable (Lavoie & Payette 1996).

Butterflies have been suggested as model indicators of climate change (Dennis 1993, Woiwod 1997, Parmesan 2001) because they broadly meet all of the criteria listed above. Butterflies provide a rare combination of documented climatic sensitivity and sufficiently accurate historical records. Casual observation, extensive field surveys and detailed experiments have established that the physiology, ecology and evolution of many lepidopteran species are dependent on climate (Dennis 1993). Further, habitat requirements are often well-understood. Many of the leading papers in conservation and ecology use butterflies to study effects of habitat fragmentation, land management change (including changes in grazing regime and forestry practices), nitrogen deposition from urban air pollution, and restoration (Hanski 1999, Weiss 1999, Singer & Thomas 1996, Thomas et al. 1996, Thomas 1995, Warren 1995, Thomas 1994, Murphy et al. 1990, Murphy & Weiss 1988, Thomas et al. 1987).

Types of data; their strengths and limitations

In assessing the strengths of studies as indicators of response to climate change, it is helpful to consider where they lie on along axes of time, space and replication (#s of populations, #s of species, etc). To assess changes in species distributions, data over large geographic areas are important, especially areas which represent the boundaries of the species' range or migratory destination. To assess trends through time, frequent (yearly is ideal) observations over many decades are most informative. And to assess the generality of the result, good replication is necessary, with many populations/census sites per species to indicate distributional changes within species, or many species per community to indicate community trends.

There has been a long interest among amateur naturalists in collecting butterflies. The result is an excellent source of data on historical distributions. In Europe, there has been a consistently high density of collectors over the past century, such that decadal and even yearly trends can be documented for many species. Several governments have recently endorsed centralised butterfly monitoring schemes. These utilise a variety of methods. In some countries amateur naturalists record sightings in a systematic, unbiased fashion; in others trained researchers formally census transects. The result is a relatively thorough, unbiased yearly census of butterfly populations for all species in each country. Even though the resolution provided by these techniques may often be insufficient to determine local abundance changes, it is well able to detect changes in local presence/absence of each species. Generally, the data are adequate to determine whether distributional changes have occurred, and whether they are of the magnitude expected from recent global warming.

Responses of mobile vs. sedentary organisms

For very mobile or migratory animals, such as many birds, large mammals, pelagic fish, and some insects, shifts of species' range occur via the process of individuals moving or migration destinations changing. Thus, these movements actually track yearly climatic fluctuations. In contrast, the bulk of organisms, especially plants, are sedentary, living their lives in a single spot either because they have limited mobility or because they lack the behavioural mechanisms which would cause them to disperse from their site of birth. Rather than occurring by individual movements, range changes in sedentary species operate by the much slower process of population extinctions and colonisations.

Ranges of migratory or mobile species can be very sensitive to climate when individuals show an immediate response in their migratory destinations. Such a response is expected for some taxa, (such as birds and lepidoptera), for which a substantial body of basic research documents the importance of weather and climate in shaping the timing and destinations of migrations.

The Monarch butterfly (*Danaus plexippus*) provides a good example. In the Monarch, the overwintering locations, the timing and destinations of northward migration, as well as the return southward, all may be ultimately determined by local climate. Calvert & Brower (1986) showed that the Monarch's overwintering ground in Mexico provides a narrow range of micro-climatic conditions which is cool and moist enough to prevent desiccation and allow individuals to survive the winter on stored reserves, yet warm enough to prevent freezing. This habitat is climatically specialised, with forest cover, slope and aspect interacting to satisfy the specific needs of this species. Malcolm et al. (1987) demonstrated that the northward spring migration of adult *Danaus plexippus*, which progresses via several generations, tracks particular climatic conditions. This suggests that either reproduction and/or larval development is restricted not only by extreme low temperatures, but also by extreme highs, such that placement and timing of breeding along the northward route is driven by these climatic considerations.

As with climatic data itself, to discern pattern in migratory data one then needs long time series in order to distinguish year-to-year variation (noise) from the long-term trends. Lack of detailed yearly data for most migratory species makes it problematic to look for such systematic long-term changes. Even in areas where appropriate data exists for some species, year-to-year variance is often so high that interpretable trends are not possible to discern. For instance, Bretherton (1983) presented an exploration of the data for migrant macro-lepidoptera into Britain. He studied arrival dates and subsequent abundances at this northern-most migratory destination, with some records going back to the 1850s. He noted both a lack of synchrony among species and sporadic changes in abundance that appeared only weakly linked to decadal trends. This mis-match between decadal climatic trends and yearly fluctuations of migrant abundances is nearly impossible to interpret without considerable detailed analysis: it could be due either to high sensitivity of the butterflies to climatic variations more subtle than yearly means, or to relative insensitivity to climate with changes driven by un-explored, non-climatic, factors. Arrival dates appeared even more chaotic (at least with respect to destination climate in the U.K.), leading Bretherton to lament that "forecasting the arrivals of immigrant species is likely to prove a disappointing pastime".

Distributional responses of sedentary species have an inherent lag time stemming from limited dispersal abilities. Neither the numbers of populations nor the geographic location of the range limit may fluctuate strongly between adjacent years. Therefore, the distributions of sedentary species exhibit less fluctuations from year-to-year than do distributions of migratory species. Thus, detectable shifts in species' ranges may take several years, decades or even centuries. In such cases, even when data are not continuous through time, a single year can be taken as representative of the state of the species during the surrounding multi-year period. For most species of any taxa, good yearly census data exists only for the last 20-30 years, i.e. since the advent of formal, centralised monitoring schemes. To look for long-term change it is preferable to have longer time series, on the order of 50-100 years.

Our studies of butterfly distributions have analysed relatively long time series (often more than a century) at the cost of temporally heterogenous data quality: older data are relatively incomplete. For both case studies described here (Parmesan 1996 and Parmesan et al. 1999), only non-migratory species were used, thereby lowering the year-to-year variance in the geographic location of the range limit. This is an important point, because migratory species, by responding to yearly temperature fluctuations, will show rapid shifts in their range limits. In order to pick out an overall trend from the yearly variation, we would require very long time series of records with few gaps. Such continuous data exists for some butterfly species in Great Britain and Finland, but in all other countries there are gaps of 5-10 years in which particular areas were not well-collected or censused. Because of the low dispersal rate of the species in our study, there is little likelihood of rapid, undocumented, change in the intervening years.

Methodological considerations and confounding influences

Studies relating observed changes in natural biota to climatic changes are necessarily correlational. Biologists know that many non-climatic anthropogenic forces affect population dynamics, community stability and species distributions. These largely fall under the main headings of land use change (including habitat loss and habitat fragmentation), hydrological changes, pollution and invasive species. It is not possible to address this question by a standard experimental approach, and so direct cause and effect relationships cannot be established. However, the level of uncertainty can be reduced until it is highly unlikely that any force other than climate change could be the cause of the observed biotic changes. Studies can reduce uncertainty by careful experimental design:

- Design the study to control for major confounding factors
- For confounding factors which remain, directly analyse whether they could explain the biotic changes, and if so, quantify the strength of that relationship.

One key advantage of using butterflies to look for climate change responses is that a lot is known about their habitat requirements. Knowledge of the species' host plants is an essential part of this, but habitat suitability

FOCUS ON RESPONSE TO CLIMATE CHANGE: CASE STUDIES OF BUTTERFLIES

In an attempt to tease apart the many potential influences on distributional change, Parmesan (1996) and Parmesan et al. (1999) designed studies which made intensive effort to exclude changes which were identifiable as due to habitat deterioration, loss or expansion. These two studies focused on current and historical population data from regions which had maintained stable, suitable butterfly habitat over the time period of the records (up to 100 years). Further, suitable habitat extended beyond the known range of the species. For instance, there are species whose ranges reach the distributional limits of their known hosts, and so they would be restricted in any response to climate change by the necessity for their host to first expand its own range. Such species were excluded from this study.

Case one: Edith's Checkerspot

Experimental design and methodology

For *Euphydryas editha* (Edith's Checkerspot) in western North America, habitat assessment was conducted by direct field census of a subset of historical population sites over a five year period (1992-1996). If the site was no longer suitable butterfly habitat (either through overt destruction or through loss of host plants or nectar sources), then it was excluded from the final analysis. If the site still contained the necessary vegetation in sufficient abundance to support a local butterfly population, then it was surveyed for the presence of adults, larvae or eggs. By this means, current population presence could be compared to the historical distribution (compiled from records dating to 1880s).

Results and conclusions

Unfortunately, it was not possible to evaluate whether range expansion was taking place at the cool margins of the species range, because these occur in rugged and remote terrain in Canada where historical records are far from complete. At the highest elevations, *Euphydryas editha* is restricted from further expansion by the range limit of its host plants. The highest known populations are in the Sierra Nevada mountains of California (highest recorded *Euphydryas editha* population was at 3,400 m). The northern-most record of *Euphydryas editha* is from Jasper National Park in Canada. The territory north of this is poorly surveyed due to lack of roads coupled with scarcity of lepidopterists. New populations may have been colonised further north in recent decades, but it would be unlikely for these to have been discovered.

In contrast, there is a clear pattern of population extinctions across the range. Contraction has occurred along the southern range boundary, in Mexico. The most southern known population is currently 160 km further north than it was in 1970. Populations along the southern edge of the range in Baja California, Mexico, were four times as likely to be recorded as currently extinct than populations along the northern range boundary in Canada (80 % extinct vs. 20 % extinct, Figure 1). Populations from lower elevations (sea level to 2,400 m) were more than three times as likely to be extinct than were those at the highest elevations (< 15 % extinct above 2,400 m, Figure 1). These patterns of population extinction had effectively shifted the range of *Euphydryas editha* both northward and upward since the beginning of the century (Parmesan 1996).

The magnitude of the range shift in *Euphydryas editha* matched the observed warming trend over the same region. The mean location of populations shifted 92 km northward and 124 m upward concurrent with mean yearly temperature isotherms shifting 105 km northward and 105 m upward (Karl et al. 1996, Parmesan 1996). Further, the cline in frequency of population extinction had a breakpoint at 2,400 m (fewer extinctions at the highest elevations). This breakpoint coincided with that for snowpack depth and timing of snowmelt, which involved increased depth and later melt-date above 2,400 m, decreased depth and earlier melt below 2,400 m (Johnson 1998). Patterns of surrounding habitat destruction in the vicinity of the target sites did not correlate with the natural extinction patterns (Parmesan, unpublished).

Detection of range shifts using butterflies

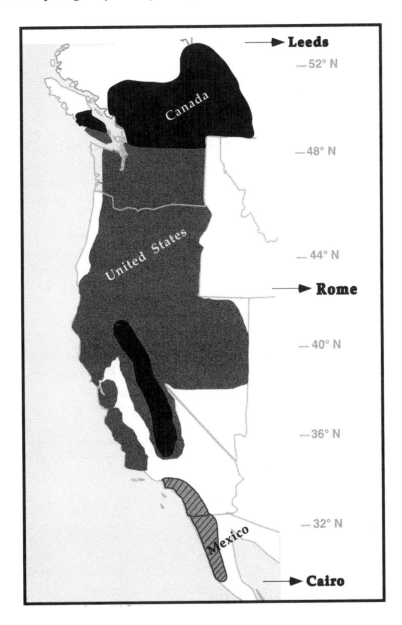

Figure 1. Map of western North America depicting the current status of *Euphydryas editha* populations throughout the species' range. Census conducted from 1992-1996. Historical records are spread out over the decades, dating from 1880-1983. In areas shaded in black, less than 20 % of the populations had gone extinct; in areas shaded in medium grey, about 40 % of the populations had gone extinct; in the area shaded with stripes, nearly 80 % of the populations had gone extinct. Cities in Europe and Africa of equivalent latitude are shown for reference.

Case two: multiple European species

Experimental design and methodology

In Europe, because of better historical coverage and current monitoring schemes, substantial replication was possible, and 57 species of non-migratory butterflies were studied (Parmesan et al. 1999). In this study, the filter for land use changes was generally less direct than with the *Euphydryas editha* study. But the approach remained the same: to isolate climatic impacts, as far as possible, species were eliminated from consideration if they were likely to have been severely affected by other, non-climatic, forces. As with *Euphydryas editha,* the principal direct factor believed to be affecting distributional changes of butterflies was land-use change. Therefore, we instituted criteria designed to restrict our study species to those that had not shown major impacts of land-use change on the position of either the northern or southern boundary.

A species was excluded from the study if:
(1) it is extremely habitat restricted; e.g. requiring such a narrow combination of micro-climate, plant phenology and other characteristics that it is highly localised even within habitats containing its host; or
(2) it is known to be intolerant of even a modest level of human-mediated habitat modification.

By using data from more than one country within a particular latitudinal range, we had replicated datasets for the same species across a single range boundary. But the countries differed in the type and degree of land-use change they had undergone during the study period (since the beginning of the twentieth century). Therefore, for some species we excluded the data from a country if, within that country:
(1) the range boundary lies in an area with so little potential habitat that its distribution could have changed dramatically via relatively small stochastic population extinctions coupled with a poor ability of the species to colonise highly dispersed habitat patches; or
(2) the species was known to have suffered severe habitat loss or degradation.

Application of these criteria caused some species to be excluded from the study in some sections of their range boundaries and included in others. For example, the silver-washed fritillary *(Argynnis paphia)* was excluded from Britain, but retained from Finland. Using these criteria, a few special clusters of species were eliminated from further analysis. For example, in some northern regions, specialists on dry meadows, calcareous grasslands and marshes were excluded due to severe declines in these habitats in recent decades. All *Maculinea* species were excluded due to their extreme specialisation. *Maculinea* are highly habitat-specific and have complex host associations: early instars require a particular plant species and later instars require a particular ant species.

In northern Africa, a seven-year census (late 1980s to early 1990s) of historical butterfly collecting sites (1882-1932) was conducted by Tennent. Even though this area had suffered from large habitat destruction generally, there were many individual sites (often mountainous) which still supported high butterfly diversity and so could be included. As with *Euphydryas editha*, population extinctions at sites with good habitat were then overlaid onto the historical distributions to look for systematic shifts.

There were several large areas of Europe excluded, yet for which good data were available. These were areas in which habitat destruction was so great that it overwhelmed all other factors. Attempts to look for "natural" distribution changes would have been futile. These areas included the Netherlands, Germany, Austria, Denmark and northern France, where prime butterfly habitat (largely traditional hay meadows), has been converted to more intensive agriculture, which subsequent loss of hosts and nectar plants.

These criteria thus focused the study on species in which responses to climate change were least likely to be confounded by other possible factors. They comprised species that were non-migratory, were not extreme specialists and whose ranges were not habitat or host-plant limited. These were the only criteria used to exclude species; prior knowledge of boundary changes was not considered. The result was a replicated dataset of both single boundary and whole range distributions in a set of species relatively uncontaminated by confounding factors. This dataset maximises the opportunity to discern a climate signal in biotic change, while minimising spurious correlation and artefacts.

Results and conclusions

We concluded that significant, systematic northward shifts in both regional distributions and entire ranges had occurred over the course of this century. Of the 57 species for which we had data for at least one range boundary, 66 % had shifted northward at a given boundary (by 35-240 km), and 3 % (2 species) had shifted southward (by less than 50 km). For 35 species for which we had data from both the northern and southern range limits, 63 % had shifted their ranges to the north (by 35-240 km) and only 6 % had shifted to the south (Parmesan et al. 1999).

By studying many species simultaneously we were able to consider each species as a single data point. Among those species that showed range shifts, we asked whether the direction of these changes differed from random. That is, if range shifts were random, we would expect about 50 % of the movements to be towards the north and about 50 % to be towards the south (i.e. no significant difference between the numbers of southward and northward range shifts). The results showed a clear shift northwards, whether we examined data for single boundaries (n = 92 boundaries over 57 species, 44 species moved north at least at one boundary, and 3 moved south at least at one boundary, Binomial test, $p \ll 0.001$) or entire ranges (n = 35 species, 22 shifted north overall and 2 shifted south, Binomial test, $p \ll 0.001$).

FOCUS ON MECHANISTIC LINKS BETWEEN CLIMATE CHANGE AND BIOTIC RESPONSE: A SINGLE-SPECIES APPROACH

Multi-species studies are powerful because they provide replication of a given phenomenon. However, they are often difficult to interpret in any detailed fashion, and mechanistic explanations are typically lacking. Knowledge of precise causal factors underlying large-scale patterns of distributional change is usually unavailable, either because the basic biology is not well-enough understood, or because observations were not made in sufficient detail during relevant periods of time. Though the field is still in its infancy, mechanistic links between climatic change and patterns of response in wildlife are beginning to emerge. There is increasing evidence that short-term extremes of weather and climate events may be driving many of the long-term, large-scale patterns of change being observed in natural systems (Parmesan et al. 2000, Easterling et al. 2000).

A detailed look at one species illustrates the connection between extreme weather events and long-term biological change. The broad patterns of change in both climate and range of *Euphydryas editha* match, giving good support to climatic warming as a candidate for the cause of the range shift, but this tells us little about mechanism. A review of studies at the population level gives us that insight. By good fortune, *Euphydryas editha* has been particularly well-studied from this vantage, with a 40 year history of basic research by more than a dozen researchers. From this, we can draw a population-based, mechanistic understanding of continental-wide distributional changes. In *Euphydryas editha,* single extreme weather events and single extreme climatic years have been shown to drive local population dynamics (including population extinctions and re-colonisations) which in turn drive changes to range limits.

Climate and weather driving population dynamics

This linkage from population-based climate-related studies of *Euphydryas editha* to continental-scale distribution changes has been explored in detail by Parmesan (2001). That material is summarised here. Many extinctions of *Euphydryas editha* have been associated with particular climatic events (Singer & Ehrlich 1979, Ehrlich et al. 1980, Singer & Thomas 1996). The 1975-1977 severe drought over California caused the extinction of 5 out of 21 surveyed populations (Singer & Ehrlich 1979, Ehrlich et al. 1980). Extremely wet years caused opposite responses in two sub-species. Following winters with 50-150 % more precipitation than the average, *Euphydryas editha bayensis* crashed in the vicinity of San Francisco Bay (Dobkin et al. 1987), while *Euphydryas editha quino* exhibited population booms in northern Baja California, Mexico (Murphy & White 1984).

Twenty years of studies at one site in the Sierra Nevada mountains of California have implicated three extreme weather events in carving a pathway to extinction of a whole set of *Euphydryas editha* populations at 2,400 m elevation (Singer & Thomas 1996). The first catastrophe occurred in 1989 when very low winter snow-pack led to an early and unusually synchronous adult emergence in April, almost two months earlier than the usual June flight. So early, in fact, were the adults that flowers were not yet in bloom and most adults died from starvation within a few days of their emergence. Just one year later another relatively light snowpack again caused adults to emerge early. Just when most had emerged, and many were sitting on the ground as mating pairs, a snowstorm that was normal for the season buried the insects under >15 cm of snow. The butterflies, adapted to summertime conditions of warmth and sun, suffered many deaths. Each of these events decreased the population size by an order of magnitude. The finale came but two years later in 1992 "when (unusually low) temperatures of -5 °C on June 16, without insulating snowfall, killed an estimated 97 % of the *Collinsia* (host) plants ... The butterflies had already finished flying and left behind young (caterpillars) that were not killed directly but starved in the absence of hosts," (Singer & Thomas 1996).

Population dynamics driving distribution shifts

Extinctions are more likely to have affected distributional changes than are colonisations by virtue of being more common. While extinctions of populations occurs frequently, colonisation is a rare event. First, colonisation events are rare in time. Across the species' range, Parmesan (1996) found that less than 14 % of empty habitats were re-colonised 30 years (on average) after an extinction event. Secondly, colonisation events are rare in space. In a detailed metapopulation study in the San Francisco Bay area, Harrison (1989) found that colonisation only occurred across very short distances. In a 10 year period following catastophic extinctions of all satellite populations, empty habitat patches which were more than 5 km from the source population were not re-colonised (and colonisation dropped markedly after 3 km). Thus, the observed northward and upward range shift of *Euphydryas editha* during this century appears to have been driven by increased numbers of population extinctions at the southern range boundary and at lower elevations, with a symmetrical tendency towards population survival along the northern range boundary and at the highest elevations (Parmesan 1996). Direct observations of population extinctions implicate influences of extreme weather events on the insect-hostplant relationship. Therefore, infrequent and severe climatic events appear to be driving a gradual range shift in this butterfly species, via short-term responses at the population level.

ARE DESIGN AND METHODOLOGY IMPORTANT?

Distinguishing range movements from local distributional changes

A crucial part of the North American and European butterfly studies was the study of both northern and the southern range limits. By analysing these extreme boundary points, these studies could distinguish actual range shifts from general range expansions and contractions, and from non-systematic changes along different edges (i.e. random local or regional distributional changes). Figure 2 shows a scenario for three hypothetical species exhibiting very different geographical changes across their range. Each species is shown on a generic "continent" in the northern hemisphere. The range of species 'a' has shifted northward (Figure 2a), species 'b' has expanded in all directions (Figure 2b), and species 'c' is inconsistent (parts expanding and parts contracting in an unsystematic fashion). When only a fraction of one edge of the range is studied, as has been typical for most studies, all three species show the same pattern of change: a northward shift of their northern boundary (Figure 2d). Thus, these three very different patterns of change at the scale of the whole range (a continental scale) appear identical when analysed on a local scale.

Does this make a difference? At least in some cases, it does. For example, *Araschnia levana* (the Map butterfly) has shown large population

Detection of range shifts using butterflies

increases and rapid northerly colonisation in the northern parts of its range, in northern France, the Netherlands and Fenno-scandia (Henriksen & Kreutzer 1982, Radigue 1994, van Swaay 1995, Mikkola 1997, N. Ryrholm, Sweden, L. Kaila & J. Kullberg, Finland, 1997, pers. comm.). In fact, the Map butterfly is a new species for Sweden. Since this dramatic expansion has accompanied a period of warm springs, superficially this might appear to be a climate-mediated northward shift, as predicted if it were responding to global warming. But a closer look shows that it has expanded equally rapidly in the south. The first sighting of *Araschnia levana* in Spain was in 1962 in the Pyrenees. Since then, *Araschnia levana* has steadily expanded into Catalonia, and is now common nearly to Barcelona (Viader 1993, C. Stefanescu, pers. comm.). Thus, for unknown reasons, *Araschnia levana* is expanding its range in all directions (as in Figure 2b). The southward extension is unlikely to have been climate-mediated, and its existence cautions us against interpreting expansions at northern range boundaries as a northward shift of an entire species' range.

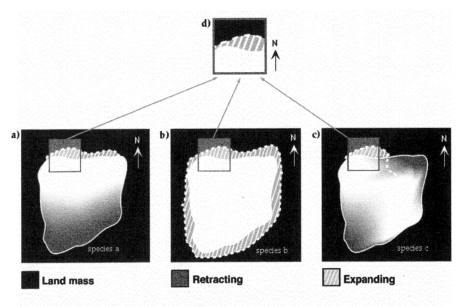

Figure 2. Hypothetical scenarios for three imaginary species. The black squares represent generic continental land masses, each containing a species whose range spans the bulk of the continent (e.g. a species that covers most of western Europe, from Spain to Great Britain).
a) Species 'a' shown disappearing from its southern range edge and expanding along its northern range edge. b) Species 'b' shown expanding along all range edges, in all directions. c) Species 'c' shown disappearing from parts of its southern edge as well as parts of its northern edge, and expanding along only one portion of its northern edge. d) A blow-up of the northwestern corner of the range for all three species, showing only the expansion along that part of the northern range edge.

A second example comes from *Carterocephalous palaemon* (the Chequered Skipper). Over the past 30 years, *Carterocephalous palaemon* has gone extinct in southern Britain and southern Finland. In northern parts of both these countries, populations have remained stable over the same time. To observers based in Britain and Finland, this species appears to have undergone a substantial northward range shift. However, examination of the southern boundary of the species, in Spain, shows that it has remained stable in recent decades (since the 1950s, Parmesan et al. 1999).

Thus, the Chequered Skipper is very variable in its status throughout its range, similar to the hypothetical species 'c' in Figure 2c. Species-wide interpretation of the regional patterns in northern countries would have led to a false conclusion of a poleward range shift when, in fact, the range of *Carterocephalous palaemon* has remained stable over this century. Such a false interpretation was never suggested, for the biology of *Carterocephalous palaemon* is well-understood. Regional losses of habitat in the more disturbed, southern parts of Britain and Finland are the clear culprits for loss of this very sensitive butterfly (Thomas & Lewington 1991, J. Kullberg & L. Kaila, Finland, 1997, pers. comm.). For many other parts of the world, however, such detailed biological knowledge is lacking and the raw changes in distribution may be the only available information.

Comparisons: alternate methodologies

In Parmesan et al. (1999), when the British butterfly dataset was analysed prior to any "filter" (i.e. including even those species most severely affected by habitat loss), we found that 47 % of all 38 nonmigratory butterflies had shifted north at the northern boundary and only 8 % had shifted south. The ratio of northward to southward movement of the northern boundary is still significantly different from the 50:50 ratio expected by random ($P < 0.001$). Thus, we are again lead to the same conclusion that systematic northward expansions indicate a general response to regional warming. Close examination of the British dataset reveals that northward movement has, indeed, occurred in spite of general habitat-related decline for some species in recent years.

Parting comments

Parmesan (1996) and Parmesan et al. (1999) provide continental-scale evidence of poleward shifts of whole species' ranges. Few other databases exist at such a large scale for other taxa. The notable exceptions are among birds of North America and Europe, where histories of amateur watching and modern organised "bird counts" provide parallel datasets to those for butterflies. In a study of a subset of birds of the western USA and southwestern Canada, Johnson (1994) documented more complex sets of response than seen in butterflies. Rather than simply shifting northward, many species appeared to be expanding into the Great Basin from all directions, possibly in response to increased precipitation into this very arid region in recent decades.

However, such complexity appears the exception. In reviewing a suite of smaller-scale (local or regional) studies, (on butterflies, as well as on other insects, vertebrates and plants), a concordance emerges. Besides being on a different scale, most studies do not attempt to purge their datasets of confounding influences. Yet, taken together, the results provide similar indications of poleward shifting (reviewed by Hughes 2000, IPCC Report 2001). In other words, the overall picture is of similarity in recent observed changes among studies of very different levels of detail, design and scale. It appears that rather unexpectedly strong responses to the current warming trends are swamping other, potentially counteractive, global change forces. Thus, the more visible aspects of human impacts on species' distributions, such as habitat loss, are not completely masking the impacts of climate change.

REFERENCES

Ashworth A. C., 1996, The response of arctic Carabidae (Coleoptera) to climate change based on the fossil record of the Quaternary period. *Annales Zoologici Fennici* 33: 125-131.

Barnosky A. D., 1986, "Big game" extinction caused by late Pleistocene climatic change: Irish elk (*Megaloceros giganteus*) in Ireland. *Quaternary Research* 25: 128-135.

Baroni C. & Orombelli G., 1994, Abandoned penguin rookeries as Holocene paleoclimatic indicators in Antarctica. *Geology* 22: 23-26.

Bartholomay G A, Eckert R.T. & Smith K.T., 1997, Reductions in tree-ring widths of white pine following ozone exposure at Acadia National Park, Maine, U.S.A. *Can. J. Forest Res.* 27(3): 361-368.

Brandon-Jones D., 1996, The Asian Colobinae (Mammalia: Cercopithecidae) as indicators of Quaternary climate change. *Biol. J. Linnean Soc.* 59: 327-350.

Bretherton R.F., 1983, The incidence of migrant Lepidoptera in the British Isles. In: J. Heath & A.M. Emmet (eds.) *Moths and Butterflies of Great Britain and Ireland*, Harley Books, Colchester, United Kingdom, pp. 9-34.

Briffa K R, Schweingruber F.H., Jones P.D., Osborn T.J., Harris I.C., Shiyatov S.G., Vaganov E.A. & Grudd H., 1998, Trees tell of past climates: But are they speaking less clearly today? *Philos. Trans. R. Soc. London [Biol.]* 353(365): 65-73.

Brown J.H., Stevens G.C. & Kaufman D.M., 1996, The geographic range: size, shape, boundaries and internal structure. *Ann. Rev. Ecol. Syst.* 27: 597-623.

Calvert W.H. & Brower L.P., 1986, The location of monarch butterfly danaus-plexippus l. Overwintering coloniesin mexico in relation to topography and climate. *Journal of the Lepidopterists' Society* 40(3): 164-187.

Coope G. R., 1995, Insect faunas in ice age environments: why so little extinction? In: J.H. Lawton & R.M. May (eds.) *Extinction Rates*, Oxford University Press, Oxford, pp. 55-74

Davis M. B. & Zabinski C., 1992, Changes in geographical range resulting from greenhouse warming: effects on biodiversity in forests. In: R.L. Peters & T.E. Lovejoy (eds.) *Global Warming and Biological Diversity*, Yale University Press, New Haven. ch 22

DeGroot R.S., Ketner P. & Ovaa, A.H., 1995, Selection and use of bio-indicators to assess the possible effects of climate change in Europe. *J. Biogeography* 22: 935-943.

Dennis R.L.H., 1993, *Butterflies and Climate Change*. Manchester University Press, Manchester, 203pp.

Dobkin D.S., Olivieri I. & Ehrlich P.R., 1987, Rainfall and the interaction of microclimate with larval resources in the population dynamics of checkerspot butterflies (*Euphydryas editha*) inhabiting serpentine grassland. *Oecologia* 71: 161-166.

Easterling D.R., Meehl G.A., Parmesan C., Chagnon S., Karl T. & Mearns L., 2000, Climate extremes: observations, modelling, and impacts. *Science* 289: 2068-2074.

Ehrlich P.R., Murphy D.D., Singer M.C., Sherwood C.B., White R.R. & Brown I.L., 1980, Extinction, reduction, stability and increase: the responses of checkerspot butterfly (*Euphydryas editha*) populations to the California drought. *Oecologia* 46: 101-105.

Goodfriend G. A. & Mitterer R. M., 1988, Late Quaternary land snails from the north coast of Jamaica: Local extinctions and climatic change. *Palaeogeography Palaeoclimatology Palaeoecology* 63: 293-312.

Graham R.W., 1992, Late Pleistocene faunal changes as a guide to understanding effects of greenhouse warming on the mammalian fauna of North America. In: R.L. Peters & T.E. Lovejoy (eds.) *Global Warming and Biological Diversity*, Yale University Press, New Haven, Connecticut, USA, pp. 76-87.

Hanski I., 1999, *Metapopulation Ecology*, Oxford University Press, Oxford.

Harrison, S. 1989. Long-distance dispersal and colonization in the bay checkerspot butterfly, *Euphydryas editha bayensis*. *Ecology* 70: 1236-1243.

Henriksen H.J. & Kreutzer I.B., 1982, *The Butterflies of Scandinavia in Nature*, Skandinavisk Bogforlag, Denmark, 210 pp.

Hoffman A.A. & Parsons P.A., 1997, *Extreme Environmental Change and Evolution*, Cambridge University Press, Cambridge.

Hughes L., 2000, Biological consequences of global warming: is the signal already apparent? *Trends Ecol. Evol.* 15: 56-61.

Innes J.L., 1991, High-altitude and high-latitude tree growth in relation to past, present and future global climate change. *The Holocene* 1,2: 168-173.

Intergovernmental Panel on Climate Change (IPCC), 2001, Third Assessment Report (Working Group II): *Impacts, Adaptations and Vulnerabilities*, Cambridge Univ. Press

Jacoby G.C. & D'Arrigo R.D., 1997, Tree rings, carbon dioxide, and climatic change. *Proc. Natl. Acad. Sci. USA* 94(16): 8350-8353.

Johnson N.K., 1994, Pioneering and natural expansion of breeding distributions in western North American birds. In: J.R. Jehl, Jr. & N.K. Johnson (eds.) *A Century of Avifaunal Change in Western North America*, Cooper Ornithological Society, Camarillo, California, pp. 27-44.

Johnson T., 1998, *Snowpack accumulation trends in California*. M.S. thesis, Bren School of Environmental Sciences, University of California at Santa Barbara, Santa Barbara, California.

Karl T.R., Knight R.W., Easterling D.R. & Quayle R.G., 1996, Indices of climate change for the United States. *Bull. Amer. Meteorological Soc.* 77: 279-292.

Kullman L., 1986, Recent tree-limit history of *Picea abies* in the southern Swedish Scandes. *Can. J. Forest Res.* 16: 761-771.

Lavoie C. & Payette S., 1996, The long-term stability of the boreal forest limit in subarctic Quebec. *Ecology* 77(4): 1226-1233.
MacArthur R. M., 1972, *Geographical Ecology*, Harper and Row, New York.
Malcolm S.B., Cockrell B.J. & Brower L.P., 1987, Monarch butterfly voltinism effects of temperature constraints at different latitudes. *Oikos* 49(1): 77-82.
Marttila O., Haahtela T., Aarnio H. & Ojalainen P., 1990, *Suomen Päiväperhoset*. Kirjayhtymä, Helsinki.
Mikkola K., 1997, Population trends of Finnish Lepidoptera during 1961-1996. *Entomologica Fennica* 3: 121-143.
Murphy D.D. & White R.R., 1984, Rainfall, resources, and dispersal in southern populations of *Euphydryas editha* (Lepidoptera: Nymphalidae). *Pan-Pacific Entomology* 60: 350-354.
Murphy D.D., Freas K.E. & Weiss S.B., 1990, An environment-metapopulation approach to population viability analysis for a threatened invertebrate. *Conserv. Biol.* 4: 41-51.
Murphy D.D. & Weiss S.B., 1988, A long-term monitoring plan for a threatened butterfly. *Conserv. Biol.* 2:367-374.
Parmesan C., 2001: Butterflies as Bio-Indicators for Climate Change Impacts. In: C.L Boggs, W.B.Watt & P.R., Ehrlich (eds.) *Evolution and Ecology Taking Flight: Butterflies as Model Systems*, University of Chicago Press.
Parmesan C., 1996, Climate and species' range. *Nature* 382: 765-766.
Parmesan C., Ryrholm N., Stefanescu C., Hill J.K., Thomas C.D., Descimon H., Huntley B., Kaila L., Kullberg J., Tammaru T., Tennent W.J., Thomas J.A. & Warren M., 1999, Poleward shifts of butterfly species ranges associated with regional warming. *Nature* 399: 579-583.
Parmesan C., Root T.L. & Willig M., 2000, Impacts of extreme weather and climate on terrestrial biota. *Bull. Amer. Meteorological Soc.* 81: 443-450.
Precht H., Christophersen J., Hensel H. & Larcher W., 1973, *Temperature and Life*. Springer-Verlag, New York.
Radigue F., 1994, Une invasion pacifique: la Carte géographique (*Araschnia levana* L.) dans l'Orne (1976-1992). *Alexanor* 18: 359-367.
Saether B-E, 1997, Environmental stochasticity and population dynamics of large herbivores: a search for mechanisms. *Trends Ecol. Evol.* 12: 143-149.
Singer M.C. & Ehrlich P.R., 1979, Population dynamics of the checkerspot butterfly *Euphydryas editha*. *Fortschr. Zool.* 25: 53-60.
Singer M.C. & Thomas C.D., 1996, Evolutionary responses of a butterfly metapopulation to human and climate-caused environmental variation. *Amer. Nat.* 148: 9-39.
Thomas C.D., 1994, Extinction, colonization, and metapopulations: Environmental tracking by a rare species. *Conserv. Biol.* 8: 373-378.
Thomas C.D., 1995, Ecology and conservation of butterfly metapopulations in the fragmented British landscape. In: A.S. Pullin (ed.) *Ecology and Conservation of Butterflies*. Chapman and Hall, London, pp. 46-64.
Thomas J.A. & Lewington R., 1991, *The Butterflies of Britain and Ireland*. Dorling Kindersley Lim., London, 224pp.
Thomas C.D., Ng D., Singer M.C., Mallet J.L.B., Parmesan C. & Billington H.L., 1987, Incorporation of a european weed into the diet of a north american herbivore. *Evolution* 41: 892-901
Thomas C.D., Singer M.C. & Boughton D., 1996, Catastrophic extinction of population sources in a butterfly metapopulation. *Amer. Nat.* 148: 957-975.
Uvarov B.P., 1931, Insects and climate. *Royal Entom. Soc. of London* 79: 174-186.

van Swaay C.A.M., 1995, Measuring changes in butterfly abundance in The Netherlands. In: A.S. Pullin (ed.) *Ecology and Conservation of Butterflies*. Chapman and Hall, London, pp. 230-247.

Viader J., 1993, Papallones de Catalunya: *Araschnia levana* (Linnaeus, 1758). *Bullt. Soc. Catalonian Lepidoptera* 71: 49-62.

Warren M.S., 1995, Managing local microclimates for the high brown fritillary, *Argynnis adippe*. In: A.S. Pullin (ed.) *Ecology and Conservation of Butterflies*. Chapman & Hall, London, pp. 198-210.

Weiss S. B., 1999, Cars, cows, and butterflies: Nitrogen deposition and management of nutrient-poor grasslands for a threatened species. *Conserv. Biol.* 13: 1476-1486.

Weiss S.B., Murphy D.D. & White R.R., 1988, Sun, slope, and buttterflies: topopgraphic determinants of habitat quality for *Euphydras editha*. *Ecology* 69: 1486-1496.

Weiser W. (ed.), 1973, *Effects of Temperature on Ectothermic Organisms*. Springer-Verlag, New York.

Woiwod I.P., 1997, Detecting the effects of climate change on Lepidoptera. *J. Insect Conserv.* 1: 149-158.

Woodward F.I., 1987, *Climate and Plant Distribution*, Cambridge University Press, Cambridge.

Climate and recent range changes in butterflies

JANE K. HILL[#,‡], CHRIS D. THOMAS[*] & BRIAN HUNTLEY[‡]
[#] *Department of Biology, PO Box 373, University of York. York YO10 5YW. UK*
[*] *Centre for Biodiversity & Conservation, School of Biology, Univ. of Leeds, Leeds LS2 9JT. UK*
[‡] *Environmental Research Centre, School of Biological Sciences, University of Durham, Durham DH1 3LE. United Kingdom*

Abstract: In order to make realistic predictions of species' responses to future climate change we need to understand the relative importance of biotic versus abiotic factors in limiting species distributions. We focus on British butterflies, a group of species for which there are good current and historical distribution records. We review our previous studies investigating the relative importance of climate and habitat availability in limiting butterfly distributions. Our studies have used a combination of modelling and analysis of distribution records to investigate factors determining limits to species' distributions in Europe, and to investigate recent range expansions of butterflies in Britain. Climates in Europe have warmed during the 20th century and many northern areas are improving for butterflies in terms of climate suitability. However, the widespread loss and fragmentation of natural habitats means that many climatically suitable areas are beyond the reach of dispersing adults and so species are unable to keep track of climate changes. In the future, many species may have the potential to occupy many northerly regions that are currently unsuitable. However, most of these newly available areas are remote from current distributions and many species are unlikely to be able to keep track of rapidly warming future climates.

BACKGROUND

Over the past 200+ years, many European butterflies species have undergone marked changes in their distributions (Heath et al. 1984, Emmet & Heath 1990, Asher et al. 2001). Approximately 60 % of British butterflies had more extensive distributions during the 19th century (Emmet & Heath 1990). For many species, recent declines are due to the widespread loss and fragmentation of natural and semi-natural habitats (Pollard & Yates 1993), but for other species, these distribution changes are related to climate (Pollard 1979, Parmesan et al. 1999, Hill et al. 1999, Roy et al. 2001). In the UK for example, several species including *Pararge aegeria* and *Pyronia tithonus* (Satyrinae), *Thymelicus sylvestris* (Hesperidae) and *Polygonia c-album* (Nymphalidae) occurred throughout most of Britain during the 19th century, probably occurring as far north as central Scotland (Thomson 1980). However, these species underwent marked contractions of their ranges towards the mid to end of the 19th century at a time when the climate appears to have been cooler (Hulme & Barrow 1997). By the 1930s, *Pyronia tithonus* had disappeared from Scotland and *Pararge aegeria* was essentially restricted to southwest England and Wales, but with a few refuge populations in northern England and southwest Scotland (Emmet & Heath 1990). *Polygonia c-album* became restricted to southwest England (Pratt 1986-87) and *Thymelicus sylvestris* also had a more restricted distribution in southern Britain (Thomson 1980). From the 1940s onwards, however, the climate in the UK has been warming (Hulme & Barrow 1997), and at least 12 species of British butterflies are currently expanding their breeding ranges (Pollard & Eversham 1995, Parmesan et al. 1999, Asher et al. 2001), including *Pararge aegeria, Pyronia tithonus, Thymelicus sylvestris* and *Polygonia c-album*.

As with all insects, butterflies would be expected to be particularly responsive to climate changes due to their poikilothermic nature and their high reproductive rates (Uvarov 1931, Turner et al. 1987); many aspects of butterfly growth and survival increase at warmer temperatures (Dennis 1993, Dennis & Shreeve 1991, Roy & Sparks 2000). However, butterflies will not be able to respond to climate warming and shift their distributions polewards if these newly-available, climatically-suitable habitats do not contain areas of suitable breeding habitat, or if this habitat is too remote to be colonised by dispersing adults. Thus, understanding interactions between climate change and habitat availability will be important for determining why some species have responded to recent climate warming while others have not, and will be important for ecologists and conservationists in the future for predicting which species may respond and keep track of climate changes in the future, and which may not. In this chapter, we review our previous studies investigating the relative importance of climate and habitat availability in determining current range limits in butterflies. We then show how our climate models can be used to predict potential distributions of butterflies in the future, by using output from a climate change scenario for the period

2070-99. Finally, we discuss interactive effects of climate and habitat availability in determining recent range expansions of butterflies in Britain.

RECENT RANGE CHANGES IN BRITISH BUTTERFLIES

Due to the charismatic nature of butterflies, a lot of information now exists on many aspects of their ecology. In addition, information on butterfly distributions has been gathered over a long period of time in Europe and it is now possible to use these data to investigate patterns of range changes over several decades. Such long-term data sets are necessary for investigating species' responses to climate change, and the British butterfly data set (data collated by Butterfly Conservation and the Biological Records Centre, CEH – Monks Wood) is probably one of the best for this type of study. We have investigated 20th century range changes in butterflies in Britain over a time when there have been several periods of intensive recording prior to the publication of national distribution atlases (Heath et al. 1984, Emmet & Heath 1990, Asher et al. 2001).

For those species which are expanding, there is evidence that rates of expansion are consistent with climate changes over the same period. For example, three species of Satyrinae have shifted their range margins northwards in Britain by 107-178 km since 1940 (Hill et al. 2001b), during a period when the climate isotherms have shifted northwards by an average of 120 km (Watson et al. 1998). Other butterfly species across Europe show range shifts of similar magnitude (35-240 km; Parmesan et al. 1999). Additional evidence for range limits being climate driven is that UK range expansions temporarily contracted or halted during the relatively cooler climates of the 1950s and 1960s, before resuming expansion from the 1970s onwards (Jackson 1980). Figure 1 gives an indication of the degree to which some species have expanded their distributions in Britain over the past 60 years, and shows recent range expansion in Britain of *Thymelicus sylvestris* (small skipper, Hesperidae). It is possible to estimate the rate of range expansion from the slope of a plot of the relationship between area occupied against time (Lensink 1997, Figure 2). This produces an estimate of range expansion rate of approximately 3 km yr^{-1} for *Thymelicus sylvestris*, similar to estimates for other species (Hill et al. in press, 2001b). However, estimates taken from data at relatively coarse spatial scales (such as a 10 km grid resolution used here) are likely to produce overestimates; estimates taken from finer scale distribution data for other species with broadly similar dispersal abilities produce more realistic estimates in the region of 0.5-1 km yr^{-1} (Hill et al. 2001b).

Figure 1. Recent range changes in the butterfly *Thymelicus sylvestris* (Hesperidae). Black squares show records (10 km grid) from 1940-1989. Grey squares show 1990s range expansion.

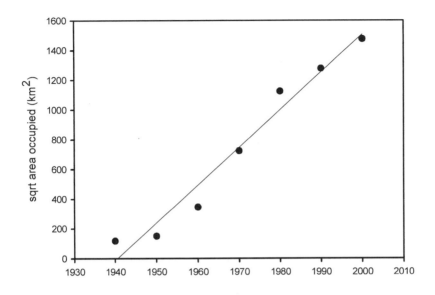

Figure 2. Changes in the area occupied by *Thymelicus sylvestris* ($\sqrt{}$area km^2 of number of 10 km grid cells with records) in Britain each decade since 1930.

MODELLING THE ROLE OF CLIMATE IN LIMITING RANGE MARGINS OF BUTTERFLIES AT CONTINENTAL SCALES IN EUROPE

We have used 'climate response surface' models to investigate the role of climate in limiting the distributions of butterfly species in Europe. These models use locally weighted regression techniques to fit species' distributions to a number of climate variables. The methods used to generate these response surfaces are explained in detail elsewhere (Beerling et al. 1995, Huntley et al. 1995, Hill et al. 1999), but we will briefly outline the methods here. Current European distributions of butterfly species are obtained from distribution atlases (Tolman 1997, Asher et al. 2001) and converted to butterfly presence/absence on a 50 km UTM grid extending from the Azores east to 30° E longitude and from the Mediterranean Sea north to Svalbard (total of 2648 grid squares). Records from outside this area are likely to be very incomplete and unreliable for this type of analysis and so were not included in the models. We do not attempt to simulate either current or future distributions for these areas even though many species have Palaearctic distributions which extend eastwards into Asia. We have computed three bioclimatic variables that reflect principal limitations on butterfly growth and survival (Hill et al. 1999, Hill et al. in press, 2001b). These three bioclimate variables are: (1) annual temperature sum above 5 °C (GDD5; developmental threshold for larvae); (2) coldest month mean temperature (MTCO; related to overwintering survival); (3) moisture availability (AET/PET; related to host plant quality and expressed as an estimate of the ratio of actual to potential evapotranspiration - Huntley et al. 1995). Our previous studies have shown that this combination of variables consistently produces the best-fitting models (Hill et al. 1999), but we are currently investigating different combinations of several other bioclimate variables in addition to those used here, and it is possible that other variables (e.g. sunshine) may also be important. We have computed values for the three bioclimate variables listed above for the mean elevation of each grid square for the climate normal period of 1931-60 (Leemans & Cramer 1991). We then fit climate response surfaces describing European distributions of butterfly species in terms of these three variables (Hill et al. 1999, Hill et al. in press, 2001b) and then use the models to simulate current distributions of species for the current climate. The goodness-of-fit between observed and simulated distributions can be tested using the kappa statistic (range 0-1; 0 = no fit (i.e. random), 1 = perfect fit) which measures the proportion of grid squares which are correctly assigned or are mismatches (Monserud & Leemans 1992).

To date, the goodness-of-fit of the climate models has generally been good across all species we have investigated (Hill et al. in press, 2001b, Hill & Telfer in press). Figure 3 illustrates the goodness-of-fit between observed and simulated distributions for the satyrid butterfly *Melanargia galathea* (marbled white). This species reaches northern and southern range margins within the study region, although isolated populations also occur outside the study region in the Atlas Mountains, Morocco (Tolman 1997). The kappa goodness-of-fit value between observed and simulated distributions of *Melanargia galathea* was typically high showing that climate is important in limiting its distribution at a continental scale (kappa = 0.77 at a threshold probability of butterfly occurrence of 0.50, 1289 grid cells with records vs. 1275 grid cells simulated occupied; Hill & Telfer, in press). Comparison of observed and simulated distributions showed that the fit was generally good throughout *Melanargia galathea*'s range, particularly at range margins; the few mismatches may be due to inaccuracies of the coarse-grained distribution maps rather than poor performance of the climate models. The results for *Melanargia galathea* are typical of many southerly-distributed European species and distributions are generally not simulated in highly topographically diverse areas, such as the Alps (Hill et al. 1999, Hill et al., in press, 2001b). This is probably because the bioclimate variables were calculated for the mean elevation of grid squares but the butterflies are occurring in specific warm microhabitats below the mean elevation of the grid cell (Hill et al. 1999). Nonetheless, we have shown in our previous studies that the goodness-of-fit between observed and simulated distributions is generally very good for a wide range of butterfly species, showing that climate is important in determining limits to butterfly distributions at continental scales.

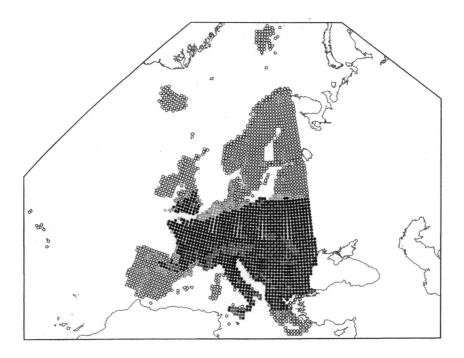

Figure 3. Goodness-of-fit between current observed and simulated distribution of the satyrine butterfly *Melanargia galathea*. Black and hollow circles show correctly assigned 50 km grid cells (black = currently recorded and simulated present; hollow = not recorded and simulated absent), light and dark grey circles show mismatches (light grey = simulated present but not recorded; dark grey = recorded present but simulated absent).

MODELLING THE ROLE OF CLIMATE AND HABITAT AVAILABILITY IN LIMITING RANGE MARGINS OF BUTTERFLIES AT REGIONAL SCALES IN BRITAIN

At smaller spatial scales, it is likely that factors other than climate are important in limiting species' distributions and we have investigated the dual roles of habitat availability and climate in limiting butterfly distributions in Britain (Hill et al. 1999, in press, 2001b). The satyrid butterfly *Pararge aegeria* (speckled wood) is currently expanding its distribution at its northern range margin in Britain and Scandinavia; its range margin has shifted northwards by 107 km in Britain (Hill et al. 2001b). *Pararge aegeria* had a far more extensive distribution in Britain during the 19th century; although its distribution is currently expanding, it has not yet re-colonised all areas it occupied historically (Hill et al. 1999). *Pararge aegeria* is essentially restricted to woodland habitats, particularly at its range margins, and 20th century loss and fragmentation of these habitats may be preventing it re-colonising areas that it occupied historically. In order to investigate this, we obtained bioclimate data and butterfly distribution data (Asher et al. 2001) for Britain at a finer spatial resolution (10 km grid cells). The distribution of woodland habitat in the UK was determined from remotely-sensed satellite Landcover data, by combining coniferous and deciduous landcover types (Hill et al. 1999).

We fitted a climate response surface which described the European distribution of *Pararge aegeria* in terms of three bioclimate variables (MTCO, GDD5 and AET/PET) using the techniques described above (model kappa = 0.80 at a threshold probability of butterfly occurrence of 0.45; 2096 simulated occurrences vs. 2064 observed occurrences; Hill et al. 1999). We then used this climate model to simulate butterfly distributions in Britain at a 10 km grid resolution for the current climate This allowed us to quantify the extent of climatically suitable areas in Britain for *Pararge aegeria*. We then used logistic regression to model *Pararge aegeria*'s distribution in Britain in relation to climate suitability (represented by the probability of occurrence predicted by the response surface) and woodland cover (from the Landcover dataset; Hill et al. 1999). This model showed that butterfly presence was positively related to both climate suitability and woodland availability (logistic regression predicted 78 % of butterfly presence/absence correctly; Hill et al. 1999), but there were areas in northern Britain where the model predicted *Pararge aegeria* to occur but where it does not. In many cases, these were areas that the butterfly occupied during the 19th century, but which it has not yet re-colonised (Hill et al. 1999). These results show that even after taking account of habitat availability, *Pararge aegeria* is lagging behind current climate.

The logistic regression model takes account of the percentage cover of woodland within a 10 km grid cell, but the spatial distribution of woodland, as well as the area of woodland per se is likely to affect *Pararge aegeria*'s distribution and our results indicate that finer-scale habitat fragmentation is likely to be affecting colonisation rates. We have investigated this in more detail by studying rates of range expansion of *Pararge aegeria* in two areas in Britain which differ in their availability of woodland (Hill et al. 2001). We have investigated range expansion in a 5000 km^2 area in northern England and a 17000 km^2 area in Scotland. *Pararge aegeria* occurred in only a few locations in these areas during the 1970s, but distributions have been expanding during the past 30 years. We determined rates of range expansion from a time series of distribution maps and showed that expansion rates were approximately 40 % slower in northern England where there was 20 % less woodland (Hill et al. 2001a). In comparison with other UK butterflies, *Pararge aegeria* is fairly mobile and does not have particularly specialised habitat requirements, and yet its rate of range expansion is directly related to the availability of habitat. By contrast, the majority of British butterflies are far more sedentary with more specialised habitat requirements; thus the lack of breeding habitat and/or the difficulty sedentary species have in reaching these habitats may explain why the majority of British butterflies have not responded to recent climate warming (Pollard & Yates 1993, Asher et al. 2001).

PREDICTING FUTURE EUROPEAN DISTRIBUTIONS

We have used our response surface models to predict potential future distributions of butterflies in Europe. In order to do this, we obtained output from a climate change scenario for the period 2070-99 (UK Hadley Centre; HADCM2) using methods described in Hill et al. (1999). We computed values for the three bioclimate variables (MTCO, GDD5 and AET/PET) for the period 2070-99 and used the climate response surfaces to predict potential future butterfly distributions for 2070-99. The HADCM2 climate scenario takes into account the positive forcing effects of greenhouse gases as well as the negative effects of sulphate aerosols (Mitchell et al. 1995) and can be considered as a relatively conservative simulation for the magnitude of climate change by the end of the 21st century. In the future, our studies predict butterfly species to shift their distributions northwards; many northern regions that are currently beyond range margins are predicted to become climatically suitable, but many areas in the south of species ranges will become unsuitable (Hill et al. 1999, 2001b). However, many northern areas that will become suitable are remote from current distributions so that even the most mobile species are unlikely to keep track of changes. Thus species may disappear from southerly areas as they become unsuitable, but may be unable to colonise new areas in the north. This is likely to result in greatly reduced distributions compared with current distributions. In addition, northerly species (e.g. *Coenonympha tullia* – large heath, *Erebia aethiops* – scotch argus, *Erebia epiphron* – mountain ringlet) which reach their southern range margins in northern Euope, will have little opportunity to shift northwards (or to higher altitudes) and are likely to have very reduced distributions in future (Hill & Telfer, in press).

CONCLUSIONS

Our studies show that climate is important in limiting butterfly distributions at continental scales, but that habitat availability, in addition to climate, limits distributions at smaller spatial scales in Britain. Approximately 12/57 British butterfly species are currently expanding their ranges but even relatively mobile species whose habitat requirements are not particularly specialised, such as *Pararge aegeria*, show rates of range expansion that are directly related to the availability of breeding habitat. Unlike *Pararge aegeria*, the majority of British butterflies have far more specialist habitat requirements and are more sedentary. In general, these species have not expanded their distributions during a period of climate warming (Asher et al. 2001), almost certainly because of lack of suitable habitat. Our studies indicate that even those species which are currently expanding (e.g. *Pararge aegeria*) may lag behind climate in areas where there is little suitable breeding habitat. Thus, interactions between climate and habitat availability, together with species mobility, will be important factors in predicting future responses of species to climate change. There is evidence that evolutionary changes at range margins may increase species' ability to respond to climate changes (Thomas et al. 2001), although Quaternary responses indicate that migration, rather than evolution in situ, is the most likely response of species to climate change (Coope 1978). Our data suggest that for butterflies, expanding distributions in response to climate changes is likely to be the exception rather than the rule.

ACKNOWLEDGEMENTS

We thank the huge number of volunteers whose records have contributed to the Biological Records Centre and Butterfly Conservation data sets. We also thank Wolfgang Cramer (PIK Potsdam) for providing the spline surfaces used to interpolate both the present climate and the HADCM2 anomalies. The output from the HADCM2 transient simulation was supplied by the Climate Impacts LINK Project (Department of the Environment Contract EPG 1/1/16) on behalf of the Hadley Centre and the UK Meteorological Office. David Viner (CRU UEA) kindly facilitated access to these data. This study was funded by NERC grants GR9/3016 and GR3/12542.

REFERENCES

Asher J., Warren M., Fox R., Harding, P., Jeffcoate, G. & Jeffcoate S., 2001, *The Millennium Atlas of Butterflies in Britain and Ireland*. Oxford University Press, UK.

Beerling D. J., Huntley B. & Bailey J.P., 1995, Climate and the distribution *of Fallopia japonica*: use of an introduced species to test the predictive capacity of response surfaces. *J. Veg. Sci.* 6: 269-282.

Coope G.R., 1978, Constancy of insect species versus inconstancy of Quaternary environments. In: L.A. Mound & N. Waloff (eds.) *Diversity of Insect Faunas*. Blackwell, Oxford, pp. 176-187.

Dennis R.L.H., 1993, *Butterflies and Climate Change*. Manchester University Press, UK.

Dennis R.L.H. & Shreeve T.G., 1991, Climatic change and the British butterfly fauna: opportunities and constraints. *Biol. Conserv.* 55: 1-16.

Emmet A.M. & Heath J., 1990, *The Butterflies of Great Britain and Ireland*. Harley Books, Colchester, UK.

Heath J., Pollard E. & Thomas J.A., 1984, Atlas of Butterflies in Britain and Ireland. Viking, London.

Hill J.K., Thomas C.D. & Huntley B., 1999, Climate and habitat availability determine 20th century changes in a butterfly's range margins. *Proc. R. Soc. Lond. B*, 266: 1197-1206.

Hill J.K., Collingham Y.C., Thomas C.D., Blakeley D.S., Fox R., Moss D. & Huntley B., 2001a, Impacts of landscape structure on butterfly range expansion. *Ecology Letters*.

Hill J.K., Thomas C.D., Fox R., Moss D. & Huntley B., 2001b, Analysing and modelling range changes in UK butterflies. In: I. Woiwod & D. Reynolds (eds.) *Insect Movement: mechanisms and consequences*. Royal Entomological Society 20th symposium, CABI, London, 415-441.

Hill J.K. & Telfer M.G., in press, Modelling the role of climate in limiting species' distributions. In: T.G. Shreeve , A.S. Pullin & J. Settele (eds.) *Ecology of European Butterflies*, Chapman & Hall.

Hill J.K., Thomas C.D. & Huntley B., in press, Modelling present and potential future ranges of European butterflies using climate response surfaces. In: W. Watt, C. Boggs & P. Erhlich (eds.) *Proceedings of the 3rd International Butterfly Ecology & Evolution Symposium*, Yale University Press.

Hulme M. & Barrow E., 1997 (eds.) *Climates of the British Isles; present, past and future*. Routledge, UK.

Huntley B., Berry P.M., Cramer W. & McDonald A., 1995, Modelling present and potential future ranges of some European higher plants using climate response surfaces. *J. Biogeog.* 22: 967-1001.

Jackson S.M., 1980, Changes since 1900 in the distribution of butterflies in Yorkshire and elsewhere in the north of England. *Entomol. Rec.* 105: 139-142.

Leemans R. & Cramer W., 1991, *Research Report RR-91-18*, International Institute for Applied Systems Analysis (IIASA). Laxenburg, Austria.

Lensink R., 1997, Range expansion of raptors in Britain and the Netherlands since the 1960s: testing an individual-based diffusion model. *J. Anim. Ecol.* 66: 811-826.

Mitchell J.F.B., Johns T.C., Gregory J.M. & Tett S., 1995, Climate response to increasing levels of greenhouse gases and sulphate aerosols. *Nature* 376: 501-504

Monserud R.A. & Leemans R., 1992, Comparing global vegetation maps with the Kappa statistic. *Ecol. Model.* 62: 275-293.

Parmesan C., Ryrholm N., Stefanescu C., Hill J.K., Thomas C.D., Descimon H., Huntley B., Kaila L., Kullberg J., Tammaru T., Tennant J., Thomas J.A. & Warren M., 1999, Polewards shifts in geographical ranges of butterfly species associated with regional warming. *Nature* 399: 579-583.

Pollard E., 1979, Population ecology and change in range of the White Admiral butterfly *Ladoga camilla* L. in England. *Ecol. Entomol.* 4: 61-74.

Pollard E. & Eversham B.C., 1995, Butterfly monitoring 2 - interpreting the changes. In: A.S. Pullin (ed.) *Ecology and Conservation of Butterflies.* Chapman & Hall, London, pp. 23-36.

Pollard E. & Yates T.J., 1993, *Monitoring Butterflies for Ecology and Conservation.* Chapman & Hall, London.

Pratt C., 1986-87, A history and investigation into the fluctuations of *Polygonia c-album* L. *Entomol. Rec. J. Var.* 98: 197-203, 244-250, 99: 21-27.

Roy D.B., Rothery P., Moss D., Pollard E. & Thomas J.A., in press, Butterfly numbers and weather: predicting historical trends in abundance and the future effects of climate change. *J. Anim. Ecol.* 70: 201-217.

Roy D.B. & Sparks T.H., 2000, Phenology of British butterflies and climate change. *Global Change Biol.* 6: 407-416.

Thomas C.D., Bodsworth E.J., Wilson R.J., Simmons A.D., Davies Z.G., Musche M. & Conradt L., 2001, Ecological and evolutionary processes at expanding range margins. *Nature* 411: 577-581.

Thomson G., 1980, *The butterflies of Scotland.* Croom-Helm, London.

Tolman T., 1997, *Butterflies of Britain and Europe.* HarperCollins, London.

Turner J.R.G., Gatehouse C.M. & Corey C.A., 1987, Does solar energy control organic diversity? Butterflies, moths and the British climate. *Oikos* 48: 195-205.

Uvarov B.P., 1931, Insects and climate. Trans. *Entomol. Soc. Lond.* 79: 1-247.

Watson R.T., Zinyowera M.C. & Moss R.H. (eds.), 1998, *The Regional Impacts of Climate Change; an assessment of vulnerability.* IPCC, Cambridge University Press.

Expansion of Mediterranean Odonata in Germany and Europe – consequences of climatic changes

JÜRGEN OTT
L.U.P.O. GmbH, Friedhofstrasse 28, D-67705 Trippstadt, Germany

Abstract: Whereas a few years ago a clear northward expansion was shown only for the dragonfly *Crocothemis erythraea*, a Mediterranean element of the German dragonfly fauna, now for a lot of dragonflies a comparable situation is very obvious. In this paper an overview of recent expansion of many dragonfly species in Germany and Europe will be given, as well as some information on the biological and ecological consequences. Beside this clear trend of expansion towards the north, the increase of population sizes and the colonisation of biotopes in higher altitudes, also several biological and behavioural adaptations could be registered, which will be shown in detail. Consequences and scenaria for the future of several dragonfly species and for the aquatic systems as a whole are pointed out.

INTRODUCTION

During the last decades meteorologists stated an overall rise of temperature (Schönwiese et al. 1993) and presently an average global warming of 1-3,5 °C over the next century is expected. For Germany we have now a comprehensive documentation of this climatic change (Rapp & Schönwiese 1994, 1995, 1996, Rapp 1994), while the concise account of actual climatic data by Bissolli (1999) covers the whole of Europe. Climatic changes clearly modify the fauna (e.g. Peters & Lovejoy 1992, Gates 1993), but in Germany this influence is analysed only by very few detailed and reliable studies (BfN 1995). In particular, the invertebrate fauna has been largely neglected in this respect. Special public attention was paid to the phenomenon of climatic change only when its consequences for local agriculture and forestry, and its role in the expansion of certain diseases and in causing floods and other catastrophes became obvious (Lozan et al. 1998, Glogger 1998).

A noticeable expansion of some Mediterranean Odonata species occurred in Germany and Europe, beginning around 1980. Because the northward spread of these species is especially well documented, it seems worthwhile trying to correlate it with detailed climatic data now at hand.

MATERIAL AND METHOD

The present study is an updated summary of a compilation of probably all relevant publications on this topic, including the so called "grey literature" (Ott 2000). It includes also many hitherto unpublished records by specialists and colleagues who were contacted directly. Occurrence and spread of *Crocothemis erythraea* in Rhineland-Palatinate was investigated mainly by the author through a period of about 20 years.

Climatic data were obtained from the "Staatliche Lehr- und Forschungsanstalt" in Neustadt/Weinstrasse, a research station for viniculture (SLFA 1960-1996).

RESULTS

The example of *Crocothemis erythraea*

The ecology of the species

This species inhabits stagnant waters of various types (deep as well as shallow pools and lakes, eutrophic water bodies, ditches, backwaters etc., vide Robert 1959, Jurzitza 1988, Askew 1988). In Germany this species exhibits a particular preference for secondary waters like gravel and sandpits (Ott 1988, Sternberg 1989). *Crocothemis erythraea* has repeatedly been characterised as a "migrating species" by several authors (e.g. Dumont 1967, d'Aguilar et al. 1985) but no proof for this characterisation has ever been observed. Single specimens are met with sometimes several 100 km apart from autochthonous populations, and also new water bodies are rather quickly colonised. On the other hand, mark-recapture experiments in a suitable habitat revealed a considerable connection of the individuals to this particular habitat (Ott 1988).

The univoltine larva is generally considered thermophilic (Schanowski & Buchwald 1987: "sommerwarme Gewässer", Sternberg 1989), but this assumption is not experimentally verified. Definitely the larva tolerates very severe winter conditions with ice covers of 40 cm thickness on the water (Ott 1988, Trockur & Didion 1994, pers. obs.). Even after the very severe winter 1995/96 emergence could be observed by the author at several water bodies in Rhineland-Palatinate at gravel pits near Ludwigshafen. However, this severe winter extinguished a larval population in a shallow water near Rodenbach/Kaiserslautern, where the water froze down to the bottom.

Distribution and recent status of *Crocothemis erythraea* in Germany

Recently the distribution of the species has been documented in detail by the author (Ott 1996, 2000), and the actual situation is presented in the following table, including also actual records and observations from the year 2000 by comparing the recent status of *Crocothemis erythraea* with that documented by Lohmann (1980) (Table 1).

It could be summarised, that the species expanded its range during the past two decades in Germany over several hundreds of kilometres to the north and about 400 metres in altitude, and the populations increased all newly settled localities in sizes and numbers.

Spreading of *Crocothemis erythraea* into other European countries

In France the species is widely present in the southern regions. It was present in some northern regions already before 1900, but, especially north of the river Loire, the number of records has fluctuated considerably over the years depending on climatic conditions (Dommanget in litteris 1995). The French national Odonata recording programme had already revealed an increase of records during the past decades, especially in the northern and northeastern regions (Dommanget 1987, 1994). Brugiere (1999) regards *Crocothemis erythraea* as an "expanding species". Nowadays it is fairly common in the Ile-de-France (Arnaboldi & Dommanget 1996). Also in the northernmost department, Pas-de-Calais, the number of records is increasing continuously. This is also true for Normandy, where *Crocothemis erythraea* was not present in earlier times (Lecompte 1999).

Table 1. Comparison of recent distribution of *Crocothemis erythraea* in Germany with data obtained before 1980.

Province	Situation up to 1980 (LOHMANN 1980)	Situation up to 2000 (OTT 1996, 2000, actual data added)
Bavaria	lost	several new records between 1992 and 2000, reproducing at several sites obvious tendency to spread, reproduction up to an altitude of 550 meters a.s.l.
Baden-Württemberg	some local populations, reproducing	increasing population number in the Rhine valley, breeding, expansion also in other parts the federal state and up to 420 meters a.s.l.
Rhineland-Palatinate	single records, not breeding	reproducing populations in the Rhine valley and Palatinate (up to 250 meters a.s.l.), rising number of populations and density, expansion into other parts of the federal state
Saarland	no records	breeding populations and spreading
Hessen	no records	breeding populations in the Rhine valley up to around Frankfurt, single records up to Giessen in middle-Hessen
Northrhine-Westfalia	no records	single records from 1977-2000, breeding near Cologne, probably reproduction also farther north near Dortmund (1994/95), several records 1994 and 1995 in the Senne/Eastern Westphalia very probably reproduction, several individuals in the alluvial zone of the river Ems (Münsterland) in 2000
Saxonia	no records	first record in 1997
Brandenburg	only one historical record (Bollow 1919)	no actual records (but probably present)

Also in Switzerland the number of records is increasing. The species is now present in the provinces of St. Gallen, Genf and Zürich (Kiauta & Kiauta 1984, Meier 1989). Successful settlements are also reported by Vonwil & Osterwalder (1994) and Hoess (in litteris 2000).

In Austria some old records exist (e.g. Landmann 1983). Nowadays the number of records is continuously increasing. New records from 1989, even some with confirmed larval development, are reported by Schweiger-Chvala 1990). New artificial water bodies are immediately colonised (Chovanec 1994), which fact indicates a certain overall presence of the species.

The first observation in England was made on August 8, 1995, in Cornwall (Jones 1996). The observed individual probably originated from France or Spain. Earlier records are known from the Channel Islands (Silsby & Silsby 1988). Several subsequent British records are documented and discussed by Parr (2000, 2001 in litteris) and in 2001 the fourth British record was made.

In Poland a few records were reported 1922 and 1968, and in 1989 a juvenile female was observed in the eastern Carpathian mountains (Czekaj 1994). More individuals were registered in 1998 and 1999 (Bernard 2000, Kalkman & Dijkstra 2000) and an established population has been observed in the Bialowiecza region.

In the Netherlands only a single individual was observed in 1959 (Geijskes & van Tol 1983), but in 1994 a small population was noticed in Vlaanderen (Wasscher 1995). Gubbels et al. (1995) report records from Limburg where the species appeared along with *Orthetrum brunneum* 1995 for the first time. Meanwhile several well established populations are known (Verbeek in prep.).

More and more single records demonstrated an expansion into Belgium (e.g. Goffart 1984, De Knijf & Anselin 1996). Today more than 100 established populations are known in Belgium. In several regions *Crocothemis erythraea* is the prevailing anisopterous species (teste De Knijf 2000). Its category in the Belgian "Red List" based on IUCN-criteria is not yet fixed.

After first observations in Luxembourg in the 80s, the species now definitely has there a growing number of reproducing populations (Gerend & Proess 1994, Proess 1996).

Table 2 summarises the distribution data of *Crocothemis erythraea* in several European countries (for all references see Ott 2000).

Table 2. Expansion of *Crocothemis erythraea* in Europe (see Ott 2000).

State	Observations
Switzerland	increasing number of records, reproducing populations near Martigny and Fanel in 1999
Austria	increasing number of records, reproducing populations
Poland	single record, three new records in 1999, probably one population
Czech Republic	rediscovery in Bohemia after more than 100 years
Belgium	several records with distinct spreading
France	increasing number of records and spreading, 1995 several records in Pas-de-Calais since several years remarkable expansion in Dept. Yvelines and the whole Ile-de-France
The Netherlands	new records in Limburg (1995) and reproducing population in Vlaanderen, meanwhile several small but stable populations e.g. also along the Dutch coastline
Luxembourg	first records in the 80s, increasing populations in the 90s
Great Britain	first record on the mainland 1995, additional records in 1997, 1998 and 2000

Expansion of further Mediterranean Odonata into Germany

A summary of records through the last 10 years as shown in Table 3 clearly reveals northward expansion of certain other Mediterranean species and a growing number of records in already occupied regions in Germany. Compared with *Crocothemis erythraea*, some of the species exhibit a stronger tendency to short term invasions (e.g. *Sympetrum fonscolombii, Hemianax ephippiger, Aeshna affinis*), but also in these species a growing number of records, even with reproduction, is to be noted in Germany. Some inconspicuous Mediterranean species may have been overlooked in earlier years (e.g. *Erythromma viridulum, Cercion lindenii*), but also in these species a spreading is now obvious. It would be very noteworthy if these species should colonise habitats at higher altitudes. This has already been observed in *Orthetrum brunneum*, a species new for the district of Koblenz where it settles in habitats even at altitudes around 300 metres above sea level.

The severe winter 1995/96 did not harm larvae of *Aeshna affinis*, which survived in Sachsen-Anhalt (Müller 1996) and in Niedersachsen (Drees et al. 1996). It is also noteworthy that nowadays in *Sympetrum fonscolombii* a second generation is sometimes produced as is normally the case in southern countries. In earlier years this species was a mere immigrant (Lempert 1987).

Expansion of further Mediterranean Odonata into various European countries

Analysis of recent publications as listed in Table 4 again clearly shows a general northward spreading in several southerly species as well as a rising number of reproducing populations in the north.

Individuals of *Hemianax ephippiger*, a typical "migrating" species, have been observed very far from their core area in earlier times, but the recent massive recording of *Hemianax ephippiger* in Middle Europe and even the Ukraine opens a new dimension. Also *Aeshna affinis, Crocothemis erythraea* and some of the less vagrant Zygoptera (e.g. *Cercion lindenii, Erythromma viridulum*) are expanding widely in Europe. In Great Britain northward expansion is especially obvious in *Anax imperator* and *Libellula depressa*, when comparing the data of Heath (1978) and Merritt et al. (1996). The so far northernmost European record of the rapidly expanding *Anax imperator* was made in Estonia (Ellwanger & Zirpel 1996), and new records in Denmark in 1994 and 1995 are documented by Nielsen (1996).

Table 3. Expansion of other Mediterranean Odonata in Germany (selected notes, for references see Ott 2000).

Species	Behaviour / records
Aeshna affinis	rediscovery 1994 in Thuringia; exuviae found in Sachsen-Anhalt first record for Niedersachsen in 1994, several new records in 1995 rediscovery in Brandenburg 1994 after 40 years; numerous records 1995 along the river Elbe, further records in other parts of Sachsen-Anhalt; first record in Schleswig-Holstein 1994; first observed reproduction in Rheinland-Pfalz; survival of a severe winter in Niedersachsen, further reproduction with emergence in the Drömling in 1998; first record in Niederlausitz 1996; reproduction observed in many habitats in Sachsen
Hemianax ephippiger	juvenile specimen found in Baden-Württemberg; record in Rheinland-Palatinate; massive invasion in Bayern, strong invasion also 1995, further records in 1996, reproduction/emergence in autumn 1998 in Bavaria
Anax parthenope	colonising coal pits in Saxony between1993 and 1997; distinct increase – also with clear autochthony – since 1990 along the river Spree (Brandenburg)
Orthetrum brunneum	expansion in Rheinland-Pfalz; expansion in Nordrhein-Westfalen; first record in Saxony; first records 1994 in Schleswig-Holstein; second record in Niedersachsen; first record in the district of Koblenz in in 1995 at 300 meters above sea level
Sympetrum fonscolombei	expansion and sometimes invasions with breeding populations at several sites in eastern Westphalia; breeding and emergence of a second generation in 1991 (North Hessen/Niedersachsen) and in 1995 in Bavaria; second generation in eastern Saxony along with *Ischnura pumilio;* colonisation of Helgoland with northernmost reproduction
Sympetrum meridionale	emergence in northern Bavaria; emergence in the Schwäbische Alb (Baden-Württemberg)
Cercion lindenii	expansion in Northrhine-Westphalia; expansion in Bavaria; first record in Lower Saxony and expansion in Lower Saxony; expansion in Hessen; expansion in the Harz mountains; first record from Thuringia in 1997
Erythromma viridulum	general spreading; new records in Bremen; expansion in Northrhine-Westphalia; conspicuous expansion in Schleswig-Holstein; northernmost population with reproduction on Helgoland; first record from the island of Rügen, even farther north

Table 4. Actual records and the distribution of Mediterranean Odonata in some European countries (selected notes, for all references see Ott 2000).

Species	Observations/records
Aeshna affinis	several single records in Poland; since 1986 records from the Reusstal (Switzerland); rediscovery after 30 years (1993) in Luxembourg, then increasing records in 1995; records from several sites in the province of Limburg (1995), first records since 1995; invasion at 17 sites in Vlandern/Belgium between 1994 and 1996
Hemianax ephippiger	mass emergence in the Reusstal (Switzerland) in 1989, also records in 1994/95; several records from April to May 1989 near Vienna; also mating immigration to France up to Lyon; reproduction in the Ukraine; several new records in Great Britain; repeated invasion to the Isle of Man in 1998; record near Kiev/Ukraine 1995; several records at 7 sites in Western Poland, with mating; several new records in Hungary with larvae/exuviae and adults from 1992 to 1996; further records 1996 in Bavaria, Brandenburg, Switzerland and The Netherlands; first record from The Netherlands 1996; first record from Norway 1995; records from Öland/Sweden; first record from Latvija; emergence in the French Dept. Jura
Anax parthenope	first record from Great Britain and further records in 1998, first proof of autochthony at two sites in 1999; several observations in Switzerland
Anax imperator	first record in Northern Poland; remarkable increase of records in Great Britain; reproduction in Estonia; first record from Denmark, possibly reproduction
Crocothemis erythraea	see this paper
Orthetrum brunneum	records from Limburg (1995), some 93 years after the last record from NL; breeding population in The Netherlands
Sympetrum fonscolombei	record of a single female from Poland; record of larvae near Zalom/Poland; influx and spreading in Great Britain 1996; heavy influx 1996 into The Netherlands with reproduction; increasing records in Tyrolia; dispersal in Cornwall 1998; reproduction in the Midlands/GB; increasing observations in Switzerland in 1999, especially in flooded areas; invasions at the Dutch coastline
Sympetrum flaveolum	heaviest invasion for years to Great Britain in 1995, several records of reproduction
Libellula depressa	rediscovery in Norway; expansion in the UK; repeated invasion to Finland
Trithemis annulata	expansion in Spain and first records in France reproduction in the French Pyrenees; 1999 again autochthony proven in southern France
Cercion lindenii	first record 1992 from Poland, large established population, records from 11 sites; new records from 30 sites in Poland; expansion in Belgium
Lestes barbarus	rediscovery in Belgium 1994, even mating successful; settlement / reproduction in Belgium increasing records from Luxembourg; 1995 records in the French Dept. Pas-de-Calais; rediscovery after decades in 1995 in the region of Meinweg / Groote Peel in The Netherlands; invasions at the Dutch coast; expansion in Switzerland

Alterations within regional species groups – the example Palatinate (Southern Germany)

Itzerott (1965) analysed the anisopterous fauna of the Palatinate, comparing for the Vorderpfalz (vineyard region along the eastern mountains and the Rhine valley) on the one hand and for the Westpfalz (Western Palatinate, including Northern Palatinate and the forest regions of the Pfälzer Wald) on the other hand the respective numbers of Mediterranean and Eurosiberian species (sensu St. Quentin 1960). He found less species in the Westpfalz than in the climatically more favourable Vorderpfalz. In the Westpfalz he registered twice as much Eurosiberian species as in the Vorderpfalz where, on the other hand, he counted three times as much Mediterranean species than in the Westpfalz. Itzerott (1965) gave the following interpretation: "The impact of climatic conditions upon Anisoptera in the Palatinate is obvious, and we are within the mark in saying that the Palatinate even odonatologically is divided into the warm Vorderpfalz and the colder Western Palatinate. Clearly the Vorderpfalz 16 more favourable for the sun-dependent Odonata and as a consequence in the Rhine valley the Mediterranean species are dominant compared with the Western Palatinate (mean yearly temperature +9 °C, mean temperature of July 17-18 °C)."

Thirty years later it seemed worthwhile to check if Itzerott's statement is still true. The data in Table 5 are based upon the literature, beginning with Niehuis (1984), and upon the author's own research in the last two decades (for the description of all species of the two regions see Ott 2000). As shown in Table 5, today the proportion of Mediterranean and Eurosiberian Anisoptera in the two compared regions of Palatinate is almost even: also the faunal difference described by Itzerott (1965) is no longer present after 30 years, as nearly the same species are occurring in the two regions (Ott 2000). This could be only interpreted as an effect of the climatic change in the whole area (see next chapter).

Table 5. Changes in the composition of the dragonfly fauna in two regions in Palatinate / Germany within 30 years.

Year	Area	Number of species	Species of Mediterranean origin (%)	Species of Eurosiberian origin (%)
1965	Vorderpfalz	29	55	45
	Westpfalz	16	31	69
1995	Vorderpfalz	33	52	48
	Westpfalz	30	47	53

Similar observations were made in a 14-year-monitoring-study also in the Reusstal in Switzerland (Vonwil in Vonwil & Osterwalder 1994, Vonwil in litteris) and by Handke (2000) for the Niedervieland near Bremen.

Finally it can be shown, that the expansion of southern Odonata species is not a mere European phenomenon: for a compilation of the first American and Japanese data see Ott (2000) and a continuous rise of the number of Afrotropical species in Algeria has been noted also (teste Samraoui 2000).

CHANGE OF CLIMATE - THE TRIGGERING FACTOR?

When thermophilic and poikilothermic species like Odonata expand their area, we may in general assume climatic causes, because temperature is one of the main ecological factors which influences biology, ecology, behaviour and even the community composition and the distribution of Odonata over broad geographical ranges (e.g. Corbet 1999).

Higher temperatures for example lead to a faster larval development: whereas in northern Sweden *Coenagrion hastulatum* has a larval stage of 3-4 years, its duration in southern Sweden is only 1-2 years (Norling 1984). On the other hand at too high temperatures larvae show clear behavioural responses and burrow in the cooler mud; together with the reduced content of dissolved oxygen too high temperatures could easily – e.g. in temporary waters – increase larval mortality (Corbet 1999). In adults the takeoff temperature differs in the so called "fliers" (endotherms – see May 1991) and the "perchers" (ectotherms) and so the perchers have to attain the takeoff temperature by passive heat gain and tend to fly only when thoracic temperature exceeds the ambient temperature by about 7 °C. May (1991) postulated – assuming that the ability to cope with extreme or variable temperature is a significant selective force on Odonata – that in warm and relatively equable tropics and subtropics Zygoptera and percher Anisoptera have an advantage, whereas in colder temperatures fliers would be more dominant and he also gave some first good indications. The same effect was shown by Samways (1989) along an altitudinal gradient in South Africa: in higher and colder regions the aeshnids were dominating, in lower and warmer the libellulids. Flying season and daily phenology are also clearly influenced by the temperature, so as spacing behaviour of adults (Lutz & Pittman 1970, Corbet 1999). The increase of temperature thus could easily influence the community structure and increase competition (e.g. through common emergence of earlier separated species), and also the effects on water quality and water vegetation must be considered.

But still relatively little is known about the effect of certain climatic factors (e.g. summarised temperature, duration of daily sunshine, length of day, length of summers etc.) on larvae or adults of a given species nor about the temperature range preferred by these species.

Only very few analyses of the effect of climate on dispersal of animals or plants exist, though there is a strong need of knowledge in this respect (Rapp & Schönwiese 1994, 1995). Undoubtedly a long term change of climate is in progress in Germany and Middle Europe (Schönwiese et al. 1993, Rapp 1994, Rapp & Schönwiese 1996), marked by a rise of mean temperature of 0,5 to 1,5 °C through the last 100 years, especially during the last 20 years. In Frankfurt/Main the yearly mean temperature 1992 was 0,77 °C higher than in 1857. Even the duration of spring and summer in Germany is extended, with actual duration depending on the region. Also in the Palatinate the warming up through the last 30 years is substantial (SLFA 1960-1996). While the mean daily temperature in June is ascending at two stations of the German Weather Service (Kaiserslautern and Neustadt), it is descending at the two other stations in the Palatinate (Mannheim and Karlsruhe). On the other hand, all four stations registered a substantial rise (up to nearly 20 °C) of daily mean temperature in May, July, August (see Ott 2000). Also the yearly mean temperatures are ascending at all four stations, especially in Kaiserslautern (slightly over 1 °C). For the whole of Rhineland-Palatinate the yearly mean temperature rose by 1,4 °C from 1961 to 1990, a period which does not even include the extraordinary hot summers of the very last years.

Serious effects of higher temperature upon vines have been noted by the vintagers in the Palatinate (Ott 2000): In 1995 flowering of vines (around middle of June) and vintage (around middle of August) took place up to 6 days earlier than in 1963 (data SLFA 1960-1996). Regarding all this, we may assume an effect especially upon the adults of Odonata. The assumed general effect of higher mean temperature in summer possibly is enforced by certain meso- and microclimate factors: sand- and gravel-pits are often especially hot habitats. In the Palatinate the higher rise of mean temperature in Neustadt and Kaiserslautern compared with other sites may be responsible for the alteration of local Odonata communities as described above for the Westpfalz. Similar changes in the phenology of the growing season were recently published for the whole of Germany by Menzel (2000).

PROBLEMS AND CONSEQUENCES

Immigrating species and "Red Lists"

In earlier Red Lists *Crocothemis erythraea* was classified as an "immigrant species" (Germany: Clausnitzer et al. 1988, Rheinland-Pfalz: Itzerott et al. 1985). Then, after increasing its numbers and populations, it was in the Red List of Rheinland-Pfalz (Eislöffel et al. 1993) classified as "A3 - species" (= threatened), in the Bavarian Red List (Kuhn 1992) as "A4" (= possibly threatened because of rareness), in the Red List of Baden-Württemberg (Borsutzki et al. 1993) as "A2 - species" (= seriously threatened).

The Belgian Red List has a previous categorisation of *Crocothemis erythraea* as "IUCN - insufficiently known" (De Knijf & Anselin 1996). The Red List of Luxembourg (Gerend & Proess 1994) regards the species as "Kategorie 3", although there is definitely no recession in records.

To me it seems more adequate to categorise this species as "Neozoon" (sensu AN 1996), even though in the case of flying insects the criterion "new to a particular region not earlier than 1492 by direct or indirect human influence" is definitely not applicable. In any case a categorisation as "A - species" seems not to be reasonable. This categorisation would urge us to support this thermophilic species with adequate measures, even to promote higher temperatures. This could not be accepted, acknowledging the negative impact of rising temperature upon other threatened species particularly adapted to lower temperature, like the strongly threatened mooreland species (*Aeshna caerulea* and *Aeshna subarctica*, *Somatochlora arctica* etc.). Consequently, *Crocothemis erythraea* was not at all included in the newest Red List for Germany (Ott & Piper 1998), as this would counteract the sense of the Red List to protect the indigenous fauna.

Expansion of Mediterranean species – an indication of general alteration in species composition and biotope quality?

Biological diversity and nature conservancy – consequences for the coenosis

Climatic changes and correlated dynamics in flora and fauna exist from the beginning and will exist forever (McElroy 1994, Coope 1994, Küster 1996). Variation of inhabited area – even when caused by climatic changes –

is principally a natural phenomenon and a more diversified fauna in a particular area is not at all to be regarded as negative from the standpoint of nature conservancy. Furthermore, *Crocothemis erythraea* and other Odonata immigrating from the Mediterranean are by no means aggressive species imposing a threat on the autochthonous Odonata. We do not know anything about the synecology of larval *Crocothemis erythraea*, but for the adults we may say that they settled here in an unoccupied niche distinctly separated from other Odonata (Ott 1988, 1995b, Sternberg 1989). A relegation of *Sympetrum*-species as mentioned by Krach (1996) was never observed at the many sites investigated in Rhineland-Palatinate by the author and this is especially well demonstrated at the "Schleusenloch", the Odonata fauna of which is continuously observed since 1980. Here, for two decades, *Sympetrum striolatum, Sympetrum vulgatum* and – for several years – *Sympetrum sanguineum* have coexisted with *Crocothemis erythraea*. Fluctuations in population density in Odonata at this water body are only due to its fish population (Ott 1995b).

Mass invasion of one or more species along with a long term change of abiotic conditions, e.g. temperature and precipitation, may cause considerable alterations in the floral and faunal composition of whole coenoses. Four general reactions of organisms under such new conditions have been formulated by Graves & Reavey (1996):

- changing of inhabited area
- a priori tolerance against new conditions
- adaptation to new conditions by microevolution
- dying out.

Predictions concerning microevolution and reaction of species to changing conditions (Holt 1990) may only be given on a very weak basis, if at all.

As far as Odonata are concerned we have to consider as determining factors an earlier drying out of shallow waters, diminishing flow of springs and small brooks, besides alteration of remineralisation of biomass in peat bogs, a general deterioration of habitat conditions (e.g. eutrophication and temperature increase in dark mooreland waters).

Possible effects upon Odonata resulting from a rise of temperature at our northern latitude are listed in Table 6, and most of them have already been observed and published in the last few years (see also Ott 2000).

Table 6. Rise of temperature: biological effects on Odonata.

Effect	already observed / published
emergence earlier in the season	yes
breeding of a second generation per year	yes
more rapid larval development, earlier emergence	yes
more northerly breeding, also breeding at higher altitudes	yes
overall alteration of phenology:	yes
earlier beginning of flight period and longer duration in thermophilic species – shortening of flight period in other species	
shifting in the diurnal activity	–
alteration of type of preferred biotope (e.g. of habitats in general, preferred territorial structures, hunting sites)	–
changes in the composition of the fauna	yes
increasing competition / exclusion of the former fauna	–
more prominent tendency for expansion	yes

The last years brought observations in Germany on the breeding of a second yearly generation in some Odonata species like *Sympetrum fonscolombei, Ischnura elegans* and *Ischnura pumilio*, and maybe also for *Anax parthenope* and *Enallagma cyathigerum* (details see Ott 2000); univoltine development of *Aeshna cyanea* (usually regarded as semivoltine) in a small pond in Germany was documented by Jödicke (1999). This aspect, as well as possible changes of duration of flight period (see Jödicke 1998), should be given more attention in future investigations.

Probably we have to consider also an alteration of general fitness and fertility in these species. We may assume that unspecialised Odonata species will endure new conditions more easily and that they will recolonise the habitats more quickly („*winners*"). Probably these generalists simply by their dense population mass will then impede and restrain other species. Rearrangement within the food chain may also be suspected.

An overall negative effect of climate change, combined with still unknown synergetic effects from other ecological conditions like acidification, eutrophication, and invasions by exotic species, may be assumed for the other species group („*losers*" – see Table 7).

As we know, larvae and adults of peat bog Odonata are especially sensitive to varying habitat temperature (Sternberg 1993, 1995), so it may be assumed that climatic changes will affect the coenosis. After its settlement in two habitats near Bremen, *Erythromma viridulum* ousted *Ceriagrion tenellum* there (teste Burckhart 1999).

Table 7. „winners" and „losers" in Odonata of climatic changes.

„winners"	„losers"
species with preference for higher temperature lowland species widespread / common species species of eutrophic waters ubiquitous / euryoecious species good flyers fast larval development r-strategists species with aggressive larvae examples: *Libellela depressa, Anax imperator, Ischnura elegans*	species with preference for lower temperature mountain species locally distributed / rare species species of oligotrophic waters stenoecious species bad flyers slow / long larval development K-strategists species with sensitive larvae examples: *Somatochora arctica* and *Somatochora alpestris, Aeshna subarctica, Leucorrhinia albifrons*

A rise of temperature combined with earlier drying up of the waters presumably is the main cause for regression of *Coenagrion armatum* in Poland (Buczynski 2000). We have to take into account also a synergetic effect of warming up and continuating eutrophication (Ellenberg 1989). Again peat bogs with their particular coenosis will be hit most. Species of temporarily drying up waters would suffer because larval development would be interrupted by regular earlier drying up of the habitat. A good example is *Lestes dryas* (Ohliger 1990, see also the remarks of Vonwil & Osterwalder 1994).

Should climate zones show an even heavier drift, then several species would be threatened throughout the whole country because of lack of suitable habitats. This is shown in Figure 1: a rise of temperature causes northward shift of the distribution area of a species or pushes it up to higher altitudes. This northward shift of the home area of genuine southern species would automatically restrict the home area of northern species which would be pushed even farther north, thus disappearing from whole regions. This might become true e.g. in Great Britain and Scandinavia for some peat bog species with their small distribution areas or for those restricted to the northernmost regions (Merritt et al. 1996, Aguilar et al. 1985, see also Figure 2). We know such massive shifts of distribution area following changes of mean temperature already from Carabid beetles (Coope 1994).

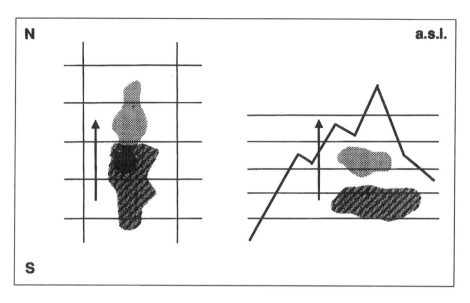

Figure 1. Fictitious changes in the distribution (old area = lines, new area = dots; left: northward expansion - right: expansion in higher altitudes).

Bioindication and Biomonitoring

The relevancy of Odonata as indicator species has already been pointed out (Schmidt 1983), and some long term investigations have demonstrated the sensitivity of Odonata to changing biotope conditions (Donath 1988, Ott 1995a, b, Corbet 1999), such as impacts on the food web through intensive fish stocking, lowering of the water table, loss or changes of the water vegetation, acidification or climatic changes. Especially *Crocothemis erythraea* appears as an ideal indicator species. It reacts rather quickly to climate change and it is, by its marked colonisation, easily detected. Exophytic egg deposition makes it independent of vegetation which itself reacts very slowly to changing biotope conditions. As a strong flyer it is not hindered in its dispersal by any barriers. It inhabits secondary waters and thus may be studied without the need to enter biotopes highly sensitive to e.g. mechanical destruction as would be the case with peat bogs.

Attention should be paid to alterations of proportions between Mediterranean and Eurosiberian species (sensu St. Quentin 1960) in selected areas, as well as to relationships between ecologically similar species, e.g. the Eurosiberian *Sympetrum vulgatum* and the Mediterranean *Sympetrum strioliatum*.

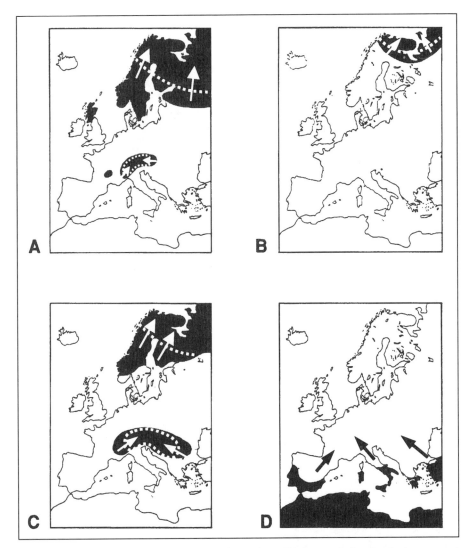

Figure 2. Distribution of four European species and possible respectively already documented consequences (A: *Aeshna caerulea*, B: *Somatochlora sahlbergi*, C: *Somatochlora alpestris* - restriction of the distribution range, white arrows; D: *Trithemis annulata* - expansion of the range, black arrows) – distribution after Aguilar et al. 1985.

The density of investigators appears to be sufficiently high (Ott 1994) to collect as many data as we need to attain a reliable basis for accurately monitoring changes.

It is obvious that mapping only vegetation or biotope types will not provide us with sufficient data to monitor a development caused by climate change. Continuous mapping of the fauna at least at selected waters has to be added to achieve a comprehensive view of the expected ecological alterations.

Scenario "Further regional and global climatic change" and strategic aspects

The rise of temperature we are confronted with may no longer be regarded as a genuine natural phenomenon independent of man-made causes (Rapp 1994, Rapp & Schönwiese 1995, Lozan et al. 1998, Glogger 1998). Some of the man-made factors (gases, emissions etc.) have been identified and it is of interest to discuss their future effect. Regarding the still accelerating rise of energy metabolism and correlated emissions we are confronted with a continuation of certain trends, such as temperature rise. New data issued by the Umweltbundesamt and the Statistisches Bundesamt (UBA/SBA 1998) reveal a not noteworthy reduction in yearly production of carbon dioxide from 1,091 million tons in 1980 to 1,014 million tons in 1990 and 910 million tons in 1996 in Germany, while on a world scale the yearly production increases constantly, as well as the atmospheric concentration of carbon dioxide (UNDP et al. 2000).

A similar situation is to be noted with other gases (see DB 1994). Accordingly we have to expect a further man-made rise of temperature with all its consequences for fauna and flora. Synergistic effects with other general impacts, though still insufficiently known in their quality, must be expected for the Odonata (Ott 1995b). Lakes in the Black Forest have already a mean pH of 3,7, which considerably reduces development of Amphibia and has an effect even on Odonata, reducing their species diversity (Böhmer & Rahmann 1992). We know that in spite of the reduction of acid emissions the acidification of waters is proceeding, because a more effective filtering of smoke emissions also reduced the amount of neutralising basic particles. A recent investigation at alpine lakes demonstrated particular hitherto unknown synergistic effects of local climate upon the chemistry of these lakes (Sommaruga-Wögrath et al. 1997). Continuous rise of temperature caused extended phases without snow and ice provoking in consequence increased erosion around these lakes. This was responsible for increasing acidification and an increment of biological production and phytoplankton. Interactions of climate change with eutrophication by atmospheric nitrogen deposition are not yet understood (Jefferies & Maron 1997). Probably nitrogen deposition is responsible for increased carbon accumulation in plants observed since the 50s, a process compensated by eutrophication and acidification.

We have no strategies yet to counterbalance environmental changes under the aspects of biotope and species conservation, e.g. by construction of new biotopes or corridors enabling species to disperse. Moreover, we have to doubt whether even large reserves will guarantee long term conservation of their fauna (see the problems in the Biosphärenresrevat Schorfheide-Chorin,

SABD 1995). The general reaction of communities and the landscapes to loss or gain of species still cannot be described. Strategies for conservation of specialised species of peat bogs or alpine regions or, in general, species depending on biotopes with a very long development, are probably not realistic because conditions at such water bodies can simply not be kept artificially constant at a particular level. Then traditional concepts of nature conservancy (species protection, biotope protection etc.) will not make sense any more. Protection of biological diversity, the core of nature conservancy measures, will be possible in future only by protection of complex geobiocenoses (Müller-Motzfeld 1996). But even this strategy may become doubtful under altered climate conditions because it will be senseless to sustain e.g. peat bogs as a landscape for glacial relic species, if these species do not find appropriate conditions there. It may well be far from certain if it will then be possible to create new habitats in time in an intensively exploited landscape.

ACKNOWLEDGEMENTS

Many thanks to Rainer Rudolph for his help in preparing the English version.

REFERENCES

Aguilar J. d`, Dommanget J.-L. & Prèchac R., 1985, *Guide des libellules d`Europe et d`Afrique du Nord*. Delachaux & Niestlé, Paris.
AN 1996, Arbeitsgruppe Neozoa, „Stuttgarter Thesen" zur Neozoen-Thematik. In: H. Gebhardt, R. Kinzelbach & S. Schmidt-Fischer (eds.) *Gebietsfremde Tierarten*. ecomed-Verlag, Landsberg.
Arnaboldi F. & Dommanget J.-L., 1996, Les Odonates du massif forestier de Rambouillet (Département des Yvelines). *Martinia* 12(4): 87-108.
Askew R.R., 1988, *The dragonflies of Europe*. Harley Books, Colchester.
Bernard R., 2000, An interesting record of *Crocothemis erythraea* (Brullé) in midwestern Poland (Anisoptera: Libellulidae). *Notul. Odonatol.* 5(5): 65-66.
BfN (Bundesamt für Naturschutz), 1995, *Klimaänderungen und Naturschutz - Klimabedingte Vegetations- und Faunenänderungen und Konsequenzen für den praktischen Naturschutz*. Angewandte Landschaftsökologie, Heft 4, Bonn-Bad Godesberg.
Bissolli P., 1999, Temperatur-Rekonstruktion für die letzten 600 Jahre. *Naturwiss. Rundschau* 52(12): 491-492.
Böhmer J. & Rahmann H., 1992, *Gewässerversauerung*. ecomed Verlag, Landsberg am Lech.
Bollow C., 1919, *Crocothemis erythraea* BRULLÉ in der Mark. *D. Ent. Z.* 191.
Borsutzki H., Buchwald R., Höppner B., Schanowki A., 1993, *9. Sammelbericht (1993) über Libellenvorkommen (Odonata) in Baden-Württemberg*. Stand Februar 1993. Hrsg.: Schutzgemeinschaft Libellen in Baden-Württemberg.

Brugiere D., 1999, Pré-inventaire des Odonates du département de la Loire. *Martinia* 15(2): 47-53.

Buczynski P., 2000, On the occurence of *Coenagrion armatum* (Charpentier, 1840) in Poland (Odonata: Coenagrionidae). *Opusc. Zool. Flumin.* 179: 1-10.

Chovanec A., 1994, Man-made wetlands in urban recreational areas - a habitat for endangered species? *Landscape and Urban Planning* 29: 43-54.

Clausnitzer H.J., Pretscher P. & Schmidt E., 1984, Rote Liste der Libellen (Odonata). In: J. Blab, E. Nowak, W. Trautmannn & H. Sukopp (eds.) *Rote Liste der gefährdeten Tiere und Pflanzen in der Bundesrepublik Deutschland.* Kilda, Greven.

Coope G.R., 1994, The response of insect faunas to glacial-interglacial climatic fluctuations. *Phil. Trans. Royal Soc. London B* 344: 19-26.

Corbet P.S., 1999, *Dragonflies – Behaviour and Ecology of Odonata.* Harley Books, Colchester.

Czekaj A., 1994, New records of *Crocothemis erythraea* (Brullé) and *Tarnetrum fonscolombii* (Sel.) from Poland (Anisoptera, Libellulidae). *Notul. Odonatol.* 4 (3): 53.

DB, 1994, (Deutscher Bundestag), Drucksache 12/8600, *Schlußbericht der Enquete-Kommission "Schutz der Erdatmosphäre".* 746pp.

De Knijf G. & Anselin A., 1996, Een gedocumenteerde Rode lijst van de libellen van Vlanandern. - *Mededelingen van het Instituut voor Natuurbehound* 4: 1-90.

Dommanget J.-L., 1987, Étude faunistique et bibliographique des odonates de France. *Inventaires de faune et de flore*, Vol. 36, Paris.

Dommanget J.-L., 1994, *Atlas préliminaire des Odonates de France.* Collection Patrimoines Naturels, Vol. 16, Paris.

Donath H., 1988, Bestandsänderungen in der Libellenfauna von Ober- und Unterspreewald innerhalb von drei Jahrzehnten. *Natur. Landsch. Bez. Cottbus* 10: 59-63.

Drees C., Eggers T.O., Jökel I., Kühne B. & Zeiss C., 1996, Entwicklungserfolg von *Aeshna affinis* Vander Linden nach einem strengen Winter in Norddeutschland (Anisoptera: Aeshnidae). *Libellula* 15(3/4): 203-206.

Dumont H., 1967, A possible scheme of the migration of *Crocothemis erythraea* BRULLÉ - populations from the Camargue (Odonata. Libellulidae). *Biol. Jaarboek* 35: 222-227.

Eislöffel F., Niehuis M. & Weitzel M., 1993, *Rote Liste der bestandsgefährdeten Libellen (Odonata) in Rheinland-Pfalz*, Mainz.

Ellenberg H., 1989, Eutrophierung - das gravierendste Problem im Naturschutz ?. *NNA-Ber.* 2/1: 4-13.

Ellwanger G. & Zirpel S., 1996, Entwicklungsnachweis von *Anax imperator* Leach in einem Hochmoor in Estland (Anisoptera: Aeshnidae). *Libellula* 14(1/2): 41-48.

Gates D.M., 1993, *Climate Change and its Biological Consequences.* Sinauer Associates, Sunderland, Massachusetts.

Geijskes D.C. & Van Tol J., 1983, *De Libellen van Nederland (Odonata).* - Koniklike Nederlandse Natuurhistorische Vereniging, Hoogwond (N.H.), Zutphen.

Gerend R. & Proess, R., 1994, Nachweis neuerer und interessanter Libellen aus Luxemburg nebst einer provisorischen Fassung der Roten Liste der einheimischen Odonaten (Insecta, Odonata). *Bull. Soc. Nat. Luxemb.* 95: 299-314.

Glogger B., 1998, *Heisszeit - Klimaänderungen und Naturkatastrophen in der Schweiz.* Vdf-Verlag, Zürich.

Goffart P., 1984, Observations de *Crocothemis erythraea* et *Anax parthenope* en Belgique durant l'été 1983. *Gomphus* 1(3): 1-3.

Graves J. & Reavey D., 1996, *Global Environmental Change - Plants, Animals & Communities.* Longman, Essex.

Gubbels R., Hermans J. & Krekels R., 1995, De Zuidelijke Ooverlibel na 93 jaar weer in Nederland. *Natuurhist. Maandbl.* 12: 284-291.
Handke K., 2000, Veränderungen in der Insektenwelt der Bremer Flussmarschen 1982 - 1999. Zeichen eines Klimawandels? *NNA-Ber.* 2: 37-54.
Heath J., 1978, *Provisional atlas of the insects of the British Isles*, Part 7, Odonata. Huntingdon.
Holt R.D., 1990, The Microevolutionary Consequences of Climate Change. *Trends Ecol. Evolt.* 5(9): 311-315.
Itzerott H., 1965, Die Verbreitung und Herkunft der Pfälzer Großlibellen. *Mitt. Poll.* 12: 164-168.
Itzerott H., Niehuis M. & Weitzel M., 1985, *Rote Liste der bestandsgefährdeten Libellen (Odonata) in Rheinland-Pfalz.* Mainz.
Jefferies R.J. & Maron J.L., 1997, The embarrassment of richness: atmospheric deposition of nitrogen and community and ecosystem processes. *Trends Ecol. Evolut.* 12(2): 74-78
Jödicke R., 1998, Herbstphänologie mitteleuropäischer Odonaten. 2. Beobachtungen am Niederrhein, Deutschland. *Opusc. Zool. Flumin.* 159: 1-20.
Jödicke R., 1999, Nachweis einjähriger Entwicklung bei *Aeshna cyanea* (Müller) (Anisoptera: Aeshnidae). *Libellula* 18(3/4): 169-174.
Jones S.P., 1996, The first British record of the Scarlet Dragonfly *Crocothemis erythraea* (Brullé). *J. Brit. Dragonfly Soc.* 12(1): 11-12.
Jurzitza G., 1988, *Welche Libelle ist das ?* Die Arten Mittel- und Südeuropas. Kosmos, Stuttgart.
Kalkman V. & Dijkstra K.D., 2000, The dragonflies of Bialowieza area, Poland and Belarus (Odonata). *Opusc. Zool. Flumin.* 185: 1-19.
Kiauta B. & Kiauta M., 1984, *Crocothemis erythraea* (Brullé) recorded from the Canton St. Gallen, Switzerland (Anisoptera: Libellulidae). *Notul. Odonatol.* 2(4): 65-66.
Krach J. E., 1996, Bemerkenswerte Libellenfunde aus dem Schuttermoos. *Globulus* 3: 23-34.
Küster H., 1996, Auswirkungen von Klimaschwankungen und menschlicher Landschaftsnutzung auf die Arealverschiebung von Pflanzen und die Ausbildung mitteleuropäischer Wälder. *Forstwiss. Cbl.* 115: 301-320.
Kuhn K., 1992, Rote Liste gefährdeter Libellen (Odonata) Bayerns. *Schriftenr. Bayer. Landesamt für Umweltschutz* 111: 76-79.
Landmann A., 1983, Zum Vorkommen und Status der Feuerlibelle *(Crocothemis erythraea* Brullé, 1832) in Österreich (Insecta: Odonata, Libellulidae). *Ber. Naturwiss.-Med. Ver. Innsbruck* 70: 105-110.
Lecompte T., 1999, Les Odonates du marais Vernier (Département de l'Eure). *Martinia* 15(1): 15-22.
Lemoine G., 1996, Libellules rare dans le Nord. *Le Nord* (Journal du Conseil général du Nord) p. 10.
Lempert J., 1987, Das Vorkommen von *Sympetrum fonscolombii* in der Bundesrepublik Deutschland. *Libellula* 6(1/2): 59-69.
Lohmann H., 1980, Faunenliste der Libellen *(Odonata)* der Bundesrepublik Deutschland und Westberlins. *Soc. Int. Odonat. Rap. Comm.* No. 1. Utrecht.
Lozán J., Graßl H. & Hupfer P., 1998, *Warnsignal Klima - Wissenschaftliche Fakten.* Verlag Wissenschaftliche Auswertungen, Hamburg.
Lutz P.E. & Pittman A.R., 1970, Some ecological factors influencing a community of adult Odonata. *Ecology* 51: 279-284.
May M., 1991, Thermal adaptions of dragonflies, revisited. *Adv. Odonatol.* 5: 71-88.
Merritt R., Moore N.W. & Eversham B.C., 1996, *Artlas of the dragonflies of Britain and Ireland.* ITE research publication no. 9. London.

McElroy M.B., 1994, Climate of the earth: an overview. *Envir. Poll.* 83: 3-21.

Meier C., 1989, Die Libellen der Kantone Zürich und Schaffhausen. *Neujahrsblatt Naturforsch. Ges. Schaffhausen* 41, 124pp.

Menzel A., 2000, Trends in phenological phases in Europe between 1951 and 1996. *Int. J. Biometeorol.* 44: 76-81.

Müller J., 1996, Fortschreibung der Roten Listen, dargestellt am Beispiel der Kenntnis- und Bestandsentwicklung der Libellenfauna Sachsen-Anhalt. *Berichte LfU Sachsen-Anhalt* 21: 66-70.

Müller-Motzfeld G., 1996, Vielfalt ohne Ende - Die Biodiversitäts-Diskussion aus der Sicht der Entomologie. In: NABU (ed.) *Biologische Vielfalt in Deutschland.* Dokumentation der NABU-Fachtagung, Potsdam.

Niehuis M., 1984, Verbreitung und Vorkommen der Libellen (*Insecta: Odonata*) im Regierungsbezirk Rheinhessen-Pfalz und im Nahetal. *Naturschutz und Ornithologie in Rheinland-Pfalz* 3(1): 1-203.

Nielsen O.F., 1996, *Anax imperator* fundet igen i Danmark. *Nord. Odonat. Soc. Newsl.* 2(1): 23.

Norling U., 1984, Life history patterns in the northern expansion of dragonflies. *Adv. Odonatol.* 2: 127-156.

Ohliger S., 1990, Die Glänzende Binsenjungfer (*Lestes dryas*), eine Charakterart periodisch austrocknender Flachsümpfe. *Mitt. Poll.* 77: 371-383.

Ott J., 1988, Beiträge zur Biologie und zum Status von *Crocothemis erythraea* (Brullé, 1832). *Libellula* 7(1/2): 1-25.

Ott J., 1994, Zum Stand des Libellenschutzes in Deutschland - Ergebnisse einer bundesweiten Umfrage. *Libellula* 12(3/4): 119-138.

Ott J., 1995a, Do dragonflies have a chance to survive in industrialised countries? In: P.S. Corbet et al. (eds.) *Proceedings International Symposium on Conservation of Dragonflies.* Kushiro, Japan.

Ott J., 1995b, Die Beeinträchtigung von Sand- und Kiesgruben durch intensive Angelnutzung – Auswirkungen auf die Libellenfauna und planerische Lösungsansätze. *Limnol. aktuell* 7: 155-170.

Ott J., 1996, Zeigt die Ausbreitung der Feuerlibelle *Crocothemis erythraea* BRULLÉ in Deutschland eine Klimaveränderung an? *Naturschutz und Landschaftsplanung* 2: 53-61.

Ott J., 2000, Die Ausbreitung mediterraner Libellenarten in Deutschland und Europa. *NNA-Ber.* 2: 13-35.

Ott J. & Piper W., 1998, Rote Liste Libellen (Odonata). In: BfN (ed.) *Rote Liste gefährdeter Tiere Deutschlands.* Landwirtschaftsverlag, Bonn-Bad Godesberg.

Parr A.J., 2000, Migrant and dispersive dragonflies in Britain during 1999. *J. Br. Dragonfly Soc.* 16(2): 52-58.

Peters R.L. & Lovejoy T.E. (eds.), 1992, *Global Warming and Biological Diversity.* Yale Univ. Press, New Haven & London.

Proess R., 1996, Überblick über die Libellenfauna der stehenden Gewässer Luxemburgs. *Bull. Soc. Nat. Luxemb.* 97: 163-80.

Rapp J., 1994, Klimatrends in Deutschland und Europa. *Natur und Museum* 12: 434-439.

Rapp J. & Schönwiese C.-D., 1994, "Thermische Jahreszeiten" als anschauliche Charakteristik klimatischer Trends. *Meteorol. Z. N.F.* 3(2): 91-94.

Rapp J. & Schönwiese C.-D., 1995, Trendanalyse der räumlich-jahreszeitlichen Niederschlags- und Temperaturstruktur in Deutschland 1891-1990 und 1960-1990. *Ann. Meteorol.* 31: 33-34.

Rapp J. & Schönwiese C.-D., 1996, *Atlas der Niederschlags- und Temperaturtrends in Deutschland 1891-1990.* Frankfurter Geowissenschaftliche Arbeiten, Serie B - Meteorologie und Geophysik, Vol. 5,

Robert P.-A., 1959, *Die Libellen.* Kümmerly und Frey, Bern.

SABD, 1995, (Ständige Arbeitsgruppe der Biosphärenreservate in Deutschland, ed.) *Biosphärenreservate in Deutschland - Leitlinien für Schutz, Pflege und Entwicklung.* Springer, Berlin.

Samways M., 1989, Taxon turnover in Odonata across a 3000 m altitudinal gradient in southern Africa. *Odonatologica* 18: 263-74.

Schanowski A. & Buchwald R., 1987, 4. *Sammelbericht (1987) über Libellenvorkommen (Odonata) in Baden-Württemberg.* Herausgegeben von der Schutzgemeinschaft Libellen Baden-Württemberg.

Schmidt E., 1983, Odonaten als Bioindikatoren für mitteleuropäische Feuchtgebiete. *Verh. DZG* 1983: 131-136.

Schönwiese C.-D., Rapp J., Fuchs T. & Denhard M., 1993, *Klimatrend-Atlas Europa 1891 - 1990.* Berichte des Zentrums für Umweltforschung Nr. 20., ZUF-Verlag, Frankfurt.

Schweiger-Chwala E., 1990, *Hemianax ephippiger* (BURMEISTER 1839) und *Crocothemis erythraea* (BRULLÉ 1832) (Odonata) in der Oberen Lobau in Wien, Österreich. *Lauterbornia* 4: 31-34.

Silsby J.D. & Silsby, R.I., 1988, Dragonflies in Jersey. *J. Brit. Dragonfly Soc.* 4(2): 31-36.

SLFA (Staatliche Lehr- und Forschungsanstalt für Weinbau), 1960-1996, *data sets.* Neustadt-Mußbach.

Sommaruga-Wögrath S., Koinig K.A, Schmidt R., Sommaruga R., Tessadri R. & Psenner R., 1997, Temperature effects on the acidity of remote alpine lakes. *Nature* 387: 64-67.

Sternberg K., 1989, Beobachtungen an der Feuerlibelle (*Crocothemis erythraea*) bei Freiburg im Breisgau. *Veröff. Nat. Landschaftspfl. Bad.-Württ.* 64/65: 237-254.

Sternberg K., 1993, Bedeutung der Temperatur für die (Hoch-)Moorbindung der Moorlibellen (Odonata: Anisoptera). *Mitt. dtsch. Ges. Allg. Angew. Ent.* 8: 521-527.

Sternberg K., 1995, Influence of oviposition rate and temperature upon embryonic development in *Somatochlora alpestris* and *S. arctica* (Odonata: Corduliidae). *J. Zool. Lond.* 235: 163-174.

St. Quentin D., 1960, Die Odonatenfauna Europas, ihre Zusammensetzung und Herkunft. *Zool. Jb. Syst. Ökol. Geograph. Tiere* 87(4/5): 301-316.

Trockur B. & Didion A., 1994, Bemerkenswerte Libellenfunde für das Saarland aus den Jahren 1988 bis 1993. *Faun.-flor. Notizen Saarland.* 26(2): 329-344.

UBA/SB, 1998, (Umweltbundesamt/Statistisches Bundesamt), *Umweltdaten Deutschland 1998*, Berlin.

UNDP/UNEP/WB/WIR, 2000, (United Nations Development Programme, United Nations Enironment Programme, World Bank, World Resources Institute), *World Resources 2000-2001*, Elsevier Science, Amsterdam.

Vonwil G. & Osterwalder R., 1994, *Kontrollprogramm NLS Libellenfauna Reusstal 1988-1992.* Grundlagen und Berichte zum Naturschutz, Vol. 7. 82 p.

Wasscher M., 1995, Het Libellenjaar 1994. *Contactbl. Nederl. Libellenond.* 23: 10-13.

Phytophenological trends in different seasons, regions and altitudes in Switzerland

CLAUDIO DEFILA & BERNARD CLOT
MeteoSwiss, Krähbühlstrasse 58, CH - 8044 Zürich, Switzerland

Abstract: Phenological observations have been made in Switzerland since 1951. Trends for 896 phenological time series have been calculated with data from 1951 to 1998. A tendency towards earlier appearance dates in spring and later appearance dates in autumn could be made out. This results are in accordance with those from the airborne pollen monitoring network. It must be noted that the different phenophases and plant species react differently to various environmental influences.

INTRODUCTION

The aim of the science of phenology is to temporally register the annually recurrent growth and development of plants, as well as to study the influences thereon. The phenophases – such as foliation, beginning of flowering, full bloom, ripening of fruit, leaf colouring and leaf fall – are observed and the relevant occurrence times noted. The Swiss phenological observation network, that covers all the regions and altitudes, was founded in 1951 (Primault 1955) and initially consisted of 70 observation posts. 37 plant species and 70 phenophases were observed. A real-time phenological observation network was introduced in 1986. 40 selected stations report 17 phenophases spread over the entire vegetation period immediately on their appearance. Based on this information, up-to-date bulletins can be composed which are published on the Internet. The phenological observation programme was slightly modified in 1996. Today the 160 observation posts register 69 phenophases of 26 different plant species. The observation programme focuses mainly on wild growing plants.

In recent times the phenological data has been focussed upon in connection with the possibility of a global climatic change (Menzel 1997).

Airborne pollen season can also be considered in the same way as a phenophase, because it reflects the maturity of the flowers. Airborne pollen concentrations result mostly from a combination of the different development stages of the single plants in the surroundings, even if medium to long range pollen dispersal can also affect the data. The national pollen monitoring network was founded in 1993, however continuous data are available since 1969 (Basle) and 1979 (Neuchâtel). As pollen analysis requires a lot of time, the number of observation posts (14) is much smaller.

To what extent phenological series can be used for vegetation monitoring is to be discussed. Beside being influenced by the length of the day, the phenological appearance dates are mainly induced by meteorological conditions. In spring the rising temperature is an important factor (Defila 1991). A possible man-induced climatic global change would lead to expect a global warming. Higher temperatures in winter and spring induce earlier appearance dates of phenological phenomena. Thus, a warming should become evident in the trends of appearance dates of phenological data. Based on some examples, the shift in the phenological appearance dates shall be studied and discussed.

DATA SETS AND METHODS

For the study of phenological trends in Switzerland, data are available from the national phenological observation network. The almost fifty-years observations series are well suited for trend analysis. It needs to be noted that not all the observation posts were operational in 1951, and that many posts of them have meanwhile been abandoned. There are also numerous gaps where the observations of entire cycles or certain phenophases within a year are missing. A further problem arises with the changing of observing personnel. Due to a certain subjectivity in phenological observation, a change in observers can lead to a break in the series. In spite of all these difficulties, 896 phenological time series, carried out at 68 observation stations in different regions and altitudes of Switzerland and concerning 19 different phenophases, could be examined for trend analysis. Only time series of a minimal duration of 20 years have been evaluated.

Pollen data from Neuchâtel 1979-1999 are available from the national pollen monitoring network. These data are obtained with an objective method: volumetric traps are used (Hirst 1952).

A linear regression model was used for the trend analysis. The significance was determined with an F-test with the error limits at $p < 0.05$.

RESULTS AND DISCUSSION

From the 896 tested phenological time series, 269 (30%) show a significant trend. This result can be compared to the one of the International Phenological Gardens (IPG), where about a quarter significant trends for Europe have been found (Menzel & Fabian 1999). Out of the 269 significant analysis in Switzerland, 98 (36,4 % or 10,9 % of all examined time series) show a positive trend (to later appearances) and 171 (63,6 % or 19,1 % of all time series) a negative trend (to earlier appearances). On Table 1 all the phenophases in question together with the number of the positive and the negative trends are listed as well as the total number of the significant trends and the number of the examined time series. As the number of the examined time series – depending on the phenophase – is quite different, the percentage of the significant trends has been calculated. At almost all the phenophases positive and negative trends are occurring. Thus there are almost no phenophases on all the stations which show only a tendency to earlier or later appearance dates. The shares of the positive or negative trends however are different. Without taking into consideration the vintage – as it is greatly influenced by man - the biggest share of the negative trends (early appearance date) is found at the blossoming of the hazel. This phenophase occurs at an early stage of the year (partly already in January). As the occurrence of the phenophases depend greatly on the temperature of the air (Defila 1991), this result is an indication of the mild winter temperatures of the recent years. The biggest share of the positive trends (late appearance date) is found at the colouring of the leaves of the horse-chestnut. In Figure 1 and 2 an example of linear trend at a phenological spring, respectively autumn, phase is shown.

Due to these results the question arises whether a tendency towards a prolongation of the vegetation period can be detected. In Figure 3 all the phenological spring (510 phenophases), summer (252 phenophases) and autumn (134 phenophases) phases are put together (see also Table 1). In spring and summer a distinct predominance of the negative trends (early appearance date) can be pointed out whereas the positive trends (late appearance date) predominate only slightly. Taking into consideration all stations and phenophases showing a significant trend, the average is an early appearance date in spring of -11,6 days and a late appearance date in autumn of +1,7 days. This results in a prolongation of the vegetation period of 13,3 days within 50 years (1951-2000). At the International Phenological Gardens a prolongation of the vegetation period of 10,8 days for Europe has been observed (Menzel & Fabian 1999), which converted in a linear way means for Switzerland a prolongation of 6,4 days for 30 years.

Table 1. Studied phenophases with number and percentage of significant trends and indication of the season.

phenophases	positive trends	negative trends	total sign. trends	total examined
hazel (*Corylus avellana*) full bloom (spring)	2 (3,6%)	18 (32,7%)	20 (36,4%)	55
wood anemone (*Anemone nemorosa*) full bloom (spring)	2 (4,3%)	14 (30,4%)	16 (34,8%)	46
horse chestnut (*Aesculus hippocastanum*) foliation (spring)	3 (6,1%)	6 (12,2%)	9 (18,4%)	49
larch (*Larix decidua*) leaf appearance (spring)	5 (7,7%)	16 (24,6%)	21 (32,3%)	65
hazel (*Corylus avellana*) foliation (spring)	5 (8,8%)	7 (12,3%)	12 (21,1%)	57
beech (*Fagus sylvatica*) foliation (spring)	6 (10,3%)	4 (6,9%)	10 (17,2%)	58
spruce (*Picea abies*) leaf appearance (spring)	10 (18,5%)	9 (16,7%)	19 (35,2%)	54
dandelion (*Taraxacum* sp.) full bloom (spring)	2 (3,0%)	15 (22,4%)	17 (25,4%)	67
daisy (*Leucanthemum vulgare*) full bloom (spring)	9 (15,3%)	8 (13,6%)	17 (28,8%)	59
elder (*Sambucus nigra*) full bloom (summer)	4 (7,4%)	13 (24,1%)	17 (31.5%)	54
linden (*Tilia platyphyllos*) full bloom (summer)	2 (5,4%)	10 (26,3%)	12 (31,6%)	38
linden (*Tilia cordata*) full bloom (summer)	0 (0%)	6 (19,4%)	6 (19,4%)	31
grapevine (*Vitis vinifera*) full bloom (summer)	1 (9,1%)	2 (18,2%)	3 (27,3%)	11
autumn crocus (*Colchicum autumnale*) full bloom (autumn)	11 (27,5%)	6 (15,0%)	17 (42,5%)	40
horse chestnut (*Aesculus hippocastanum*) leaf colouring (autumn)	4 (8,3%)	13 (27,1%)	17 (35,4%)	48
beech (*Fagus sylvatica*) leaf colouring (autumn)	9 (16,7%)	10 (18,5%)	19 (35,2%)	54
horse chestnut (*Aesculus hippocastanum*) leaf fall (autumn)	14 (30,4%)	4 (8,7%)	18 (39,1%)	46
beech (*Fagus sylvatica*) leaf fall (autumn)	9 (17,3%)	6 (11,5%)	15 (28,8%)	52
vintage (*Vitis vinifera*) (autumn)	0 (0%)	4 (33,3%)	4 (33,3%)	12

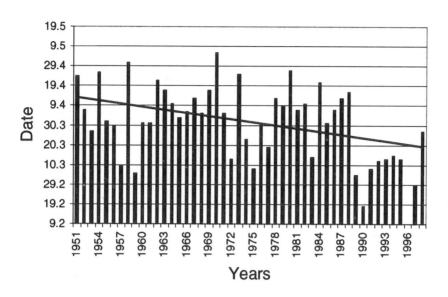

Figure 1. Wild anemone (*Anemone nemorosa*) full bloom in Ennetbuehl (altitude 900 m a.s.l.) F = 8,8, p = 0.005.

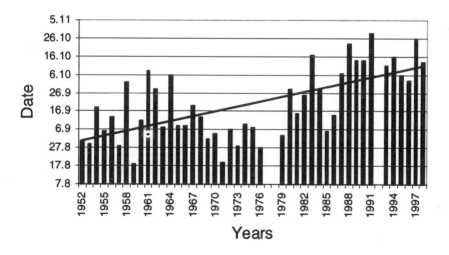

Figure 2. Beech (*Fagus sylvatica*) leaf colouring in Wiliberg (altitude 650 m a.s.l.) F = 28,3, p = 3.7E-6.

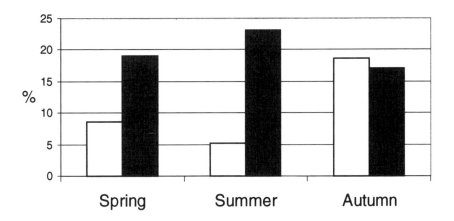

Figure 3. Percent of significant positive (white bars) and negative (black bars) trends over the 3 considered seasons.

Trends in the starting and end dates of the pollen season in Neuchâtel have been analysed for 19 pollen types: *Alnus, Salix, Fraxinus, Carpinus, Betula, Fagus, Platanus, Picea, Quercus, Pinus,* Poaceae, *Urtica, Ambrosia, Taxus* and Cupressaceae, *Plantago, Corylus, Castanea, Artemisia, Populus.* From the 38 series examined, 24 (63 %) show a significant negative trend. Not any time series revealed a positive trend. 12 series presented a negative trend with an average early starting date of 21 days, 12 series presented a negative trend with an average early end date of 30 days. In Figure 4 the example of the starting and end dates of airborne birch pollen season is shown: these results are in accordance with those from Emberlin et al. (1997), who revealed a strong trend to an earlier appearance date of birch pollen in England. It must be stressed that 22 (92 %) of the significant pollen series occur during the spring or early summer and only 2 (8 %) during autumn. From the 14 series showing no significant trend, 7 (18 % of the total) occur during the autumn. These results, obtained with a different approach, fully confirm those revealed by the phenological observations, showing a strong tendency towards earlier appearance dates in spring and a weaker situation in autumn.

As the phenological observation stations in Switzerland are situated at different altitudes (from 300 to 1800 m a.s.l.), it has been examined if the altitude can have an influence on the difference between positive and negative trends. The observation stations have been divided into slopes of 200 m each, whereas the last category includes all altitudes over 1300 m. (300-500 m include the study of 173 phenophases, 500-700 m : 208, 700-900 m : 201, 900-1100 m : 156, 1100-1300 m : 115, above 1300 m : 43). In

Figure 5 the difference between the positive and the negative trends are displayed in relation to the slopes of altitude.

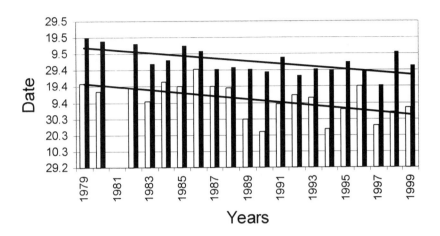

Figure 4. Starting date (white bars, F = 8,1, p = 0.011) and end date (black bars, F = 13,46, p = 0.002) of birch (*Betula sp.*) pollen season in Neuchâtel.

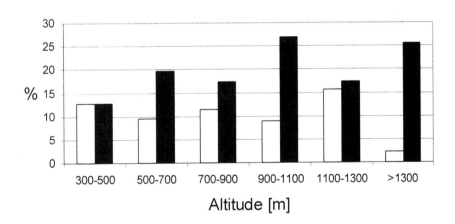

Figure 5. Percent of significant positive (white bars) and negative (black bars) trends in different classes of altitude.

In the lowest slope of altitude there is no difference between the positive and the negative trends. In the next three categories (until 1100 m a.s.l.) the negative trends are predominant. At the category 1100 until 1300 m a.s.l. the differences are not considerable whereas at the highest category the negative trends are predominating. The last result must not be overvalued, as there are only a few stations in this category. As to global warming it can be estimated that – above all – plants in higher regions react strongly, as the temperature is considered as a limiting factor for the growing and the development of the plants. There is a tendency that the negative trends increase in relation to the altitude above sea level, but this increase is not clear.

Switzerland can be divided into different regions of climate due to its topographical character. In terms of plant phenology the division into the following seven regions have worked out well: Jura (12 stations - 167 phenophases), Swiss Plateau (18 stations - 276 phenophases), northern slope of the Alps (17 stations - 250 phenophases), Rhone Valley (6 stations - 45 phenophases), Rhine Valley (7 stations - 81 phenophases), southern side of the Alps (7 stations - 72 phenophases), Engadine (1 station - 5 phenophases). In Figure 6 the differences between the positive and the negative trends for the above mentioned seven regions are displayed.

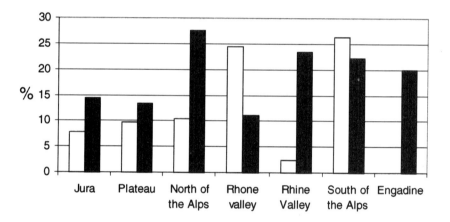

Figure 6. Percent of significant positive (white bars) and negative (black bars) trends in different regions of Switzerland.

The differences from region to region are considerable. The regions of the Rhone Valley and the southern side of the Alps, where the positive trends predominate, and the region of the Rhine Valley, where the negative trends predominate heavily, are remarkable. In the region of Engadine there are only negative trends: this result however should not be overvalued, as in this region only one station is considered. The fact is that the two regions with a predominant share of positive trends (Valais and the southern side of the Alps) are located in a warmer zone than the other regions. In these regions the temperature is not necessarily a limiting factor for the growth and the development of the plants. The regions with a bigger share of negative trends are located in the higher situated regions (northern slope of the Alps, Rheinbünden and Engadine), which reflects the dependence on altitudes.

CONCLUSIONS

A strong trend towards earlier appearance dates of the phenophases in spring during the recent decades could be proved. At the phenological autumn phases the tendency towards later appearance dates in the recent decades is less remarkable. Within 50 years, a prolongation of the vegetation period of 13,3 days has been established. This prolongation can be traced back in the first place to the earlier appearance dates of the phenological occurrences in spring. As above all the phenological spring phases depend to a large extend on the temperature, the thesis has been reinforced that a climate change has an effect on the phenological appearance dates. This work makes clear that not all the phenophases react to changes in the environment to the same extend. The location of the observations can also have an influence on the results of a trend analysis of phenological time series, as the local climate and the microclimate contribute to the phenological appearance dates. Depending on the slopes of altitudes and regions the results of the trend analyses turn out differently. A uniform trend towards an earlier or a later vegetation development could not be found, however the existing results represent a tendency. The phenology is particularly useful for the monitoring of the vegetation in connection with the subject of a global or regional climate change. Thorough examinations have yet to stabilise many of the results or tendencies in question.

REFERENCES

Defila C., 1991, Pflanzenphänologie der Schweiz. Diss. Uni Zürich, *Veröff. Schweiz. Meteorologischen Anstalt*, Nr. 50: 1-235.

Emberlin J., Mullins J., Corden J., Millington W., Brooke M., Savage M. & Jones S., 1997, The trend to earlier birch pollen seasons in the U.K.: a biotic response to changes in weather conditions? *Grana* 36: 29-33.

Hirst J.M., 1952, An automatic volumetric spore trap. *Ann. Appl. Biol.* 39: 257-265.

Menzel A., 1997, Phänologie von Waldbäumen unter sich ändernden Klimabedingungen – Auswertungen der Beobachtungen in den Int. Gärten und Möglichkeiten der Modellierung von Phänodaten. *Forstliche Forschungsberichte*, München, 186: 1-147.

Menzel A. & Fabian P., 1999, Growing season extended in Europe. *Nature*, 397: 659.

Primault B., 1955, Cinq ans d'observations phénologiques systématiques en Suisse. *Ann. Schweiz. Meteorol. Anst.* 92: 7/4 - 7/5.

Plant Phenological Changes

ANNETTE MENZEL & NICOLE ESTRELLA
Department of Ecology, Technical University Munich, Am Hochanger 13, D-85354 Freising, Germany

Abstract: Various indications for shifts in plant and animal phenology due to climate change have been observed. First, this article presents a review of plant phenological changes observed in the last few years for Europe and North America. In the second part, the analysis of phenological seasons in Germany of more than 4 decades (1951-1996) is reported as an example for bio-monitoring by plant phenology. This study has the major advantages of a dense geographical coverage due to data of the phenological network of the German Weather Service and a good seasonal coverage due to 16 phenophases. The results of different methods of trend analysis, and then the spatial and annual variability of these trends are discussed.

INTRODUCTION

Phenology is the science of seasonal plant and animal activity driven by environmental factors. Leaf unfolding, flowering, fruit ripening, leaf colouring or leaf fall of plants are observed as well as arrival dates of migrant birds, dates of egg laying of birds or spawning of fish and the timing of first appearance of butterflies. Spring phases of plants in high and mid latitudes are mainly triggered by air temperature during winter and spring. Thus, climate change may change the onset of phenological phases and have impacts on their (re-)production, distribution and fitness.

But phenological changes in plants and animals as well as abiotic indices such as ice cover (Magnuson et al. 2000) are also discussed as "fingerprints" of climate change. Changes of spring phases can be often linked to temperature changes; however, the most important factors influencing autumn phases are not as evident (Defila 1991, Mahrer 1985).

REVIEW OF REPORTED CHANGES IN PLANT PHENOLOGICAL PHASES

Various indications for shifts in plant and animal phenology have already been observed in the boreal and temperate zones of the northern hemisphere. Table 1 summarises the results of the latest plant phenological studies.

In general, only a few locations show records covering a century, while most of the data is only from the last 4 to 5 decades. There are a lot of results revealing an advance of spring phases in Europe and North America in the last 5 to 9 decades. The mean advance for the last 5 decades ranges between -0.12 to -0.40 days/year, but the latter figure is for an urban time series incorporating an additional heat island effect. Wide geographic coverage is available for very few species and phases only (e.g. Schwartz & Reiter 2000, Cayan et al. 2001). Other studies analysing a lot of different spring phases only include few sites/stations (e.g. Bradley et al. 1999, Ahas 1999, Jaagus & Ahas 2000). Only data analyses of the International Phenological Garden network (Menzel & Fabian 1999, Menzel 2000, Chmielewski & Rötzer 2000) comprise several phases on a European scale, but contrary to all other studies, the plants observed in this network consists of clones of trees and shrubs. Thus, comparing results of these different studies should be done with care, as the underlying time periods and species differ, and strong regional differences are described.

However, on average, spring has advanced by up to -0.3 days/year over the last 4 decades in Europe. But, as several studies report, these changes are mean values and not all underlying time series possess significant trends. There are differences between species (e.g. Bradley et al. 1999, Menzel 2000) as well as strong differences between sites (e.g. Schwartz & Reiter 2000, Menzel 1998, Defila 2000).

Analyses of the annual amplitude of CO_2 records also reveal an earlier onset of spring by -0.2 to -0.28 days/year for the last 2 to 3 decades (Keeling et al. 1996). The NDVI (Normalized Difference Vegetation Index) derived from AVHRR/NOAA satellite data also gives evidence for an advance of spring in the period 1982/3-1989/90 (Myneni et al. 1997).

Only a few studies deal with autumn phases whose shifts seem to be less pronounced and smaller (delay by +0.09 to +0.16 days/year for the International Phenological Gardens in Europe, Menzel & Fabian 1999, Menzel 2000, Chmielewski & Rötzer 2000). Here, the onset of autumn

Table 1. Changes in plant phenological phases in Europe (review).

Spring	Period	Linear trend	Author
Flowering of locust tree in Hungary	1851-1994	from -3 to -8 days (total)	Walkovszky 1998
Plant flowering in Estonia	1952-1996	from -0.14 to -0.29 days/year	Ahas 1999
Trees ("Int. Phenological Gardens" IPG Europe)	1959-1993	-0.20 days/year	Menzel & Fabian 1999
Trees (North & Central Europe IPG)	1959-1993	-0.31 days/ year	Menzel & Fabian 1999
Trees (Germany IPG)	1959-1993	-0.31 days/year	Menzel & Fabian 1999
55 diff. phases (one site in Wisconsin, USA)	1936-1998	On average -0.12 days/year	Bradley et al. 1999
Trees (IPG Europe)	1959-1996	-0.21 days/year	Menzel 2000
Trees (IPG Europe, start of growing season)	1969-1998	-0.27 days/year	Chmielewski & Rötzer 2000
Diff. Phases (68 stations in Switzerland)	1951-1998	-0.19 days/year	Defila 2000
Flowering of 4 species in 10 European cities	1951-1995	from -0.12 (apple) to -0.40 (forsythia urban) d/y	Rötzer et al. 2000
Plant flowering (4 species, one site in Estonia)	1919-1995	from -0.05 to -0.17 days/year	Jaagus & Ahas 2000
Grapevines in Bordeaux	1952-1997	tendency toward earlier events in the last 2 decades	Jones & Davis 2000
Flowering of *Populus tremuloides* (Edmonton, Canada)	1900-1997	-0.26 days/year	Beaubien & Freeland 2000
Spring flowering index (Edmonton, Canada)	1936-1996	-0.13 days/year	Beaubien & Freeland 2000
Spring index first leaf / first bloom date (USA)	1959-1993	-0.18 days/year / -0.14 days/year	Schwartz & Reiter 2000
Blooming of lilac and honeysuckle (western US)	1957(68)-94	-0.20 days/year / -0.38 days/year	Cayan et al. 2001
CO_2-record (Mauna Loa)	1970-1994	-0.28 days/year	Keeling et al. 1996
CO_2-record (Pt. Barrow)	1975-1994	-0.20 days/ year	Keeling et al. 1996
AVHRR NDVI data 45°- 70° latitude north	1982/3-1989/90	-8 +/- 3 days (total), -1.0 days/ year	Myneni et al. 1997
Thermal seasons in Germany	1949-1985	up to -0.38 days/year (max)	Rapp & Schönwiese 1994
Autumn	**Period**	**Linear trend**	**Author**
Trees (IPG Europe)	1959-1993	+0.16 days/ year	Menzel & Fabian 1999
Trees (IPG Europe)	1959-1996	+0.15 days/year	Menzel 2000
Trees (IPG Europe, end of growing season)	1969-1998	+0.09 days/year	Chmielewski & Rötzer 2000
Diff. phases (68 stations in Switzerland)	1951-1998	+0.03 days/year	Defila 2000
AVHRR NDVI data 45°-70° latitude north	1982/3-1989/90	+4 +/- 2 days (total), +0.5 days/year	Myneni et al. 1997
Thermal seasons in Germany	1949-1985	up to +0.32 days/year (max)	Rapp & Schönwiese 1994
Vegetation period	**Period**	**Linear trend**	**Author**
Trees in Europe (IPG)	1959-1993	+0.36 days/ year	Menzel & Fabian 1999
Trees (IPG Europe, length of growing season)	1969-1998	+0.35 days/year	Chmielewski & Rötzer 2000
Betula pendula / *Fagus sylvatica* (Germany)	1951-1996	+0.17 days/year / +0.11 days/year	Menzel 1998

derived from temperature records (Rapp & Schönwiese 1994) and from satellite data (Myneni et al. 1997) shows higher delays in the last decades (see Table 1). In total, a lengthening of the growing season is detected by plant phenological observations, but this is more due to an earlier onset of spring phases than to changes in autumnal phases (Menzel 1998, 2000, Defila 2000).

PHENOLOGICAL CHANGES IN GERMANY (1951-1996)

Since 1951, the German Weather Service has been running an extremely dense phenological network with ~1600 stations today. Out of this huge observational programme, 16 phases of wild plants were selected in order to cover the whole growing season (see Table 2). The length of the growing season was determined directly as the time span between leaf colouring and leaf unfolding of four deciduous tree species. After intensive data quality checks two different methods, linear trend analyses and comparison of averages of subintervals, were applied in order to determine shifts in these 20 phenological phases in the last 46 years (Menzel et al. 2001).

Linear trend analysis

For all phenological phases linear regressions have been calculated between the dates on onset (dependent) and the year (independent variables), if at least 20 years of observations were available. A positive trend or correlation indicates that the dates of onset are delayed over the years, whereas a negative trend describes advancing onset. The significance of the linear regressions was determined by the F-test, even though that requires a normal Gaussian distribution of the data. However, trends which are significant following the F-test ($p<0.05$) are often significant after the trend noise ratio and the Mann-Kendall-trend test also (Menzel 1997, Estrella 1999); therefore, the F-test can be generally used.

The spatial distribution of these linear trends is shown on maps (Figure 1). The trend of every station is characterised by its sign (black = negative trend or earlier onset, grey = positive trend) and its significance (not significant or $p>0.05$, significant or $p<0.05$ and highly significant or $p<0.01$) indicated by the size of the dot. Eight examples of spatial distribution of trends in Germany of all records 20+ are given by Menzel et al. 2001. Another 8 new examples can be found in Figure 1: Flowering of *Forsythia suspensa*, key indicator of the early spring, has almost exclusively negative trends (earlier onset). Leaf unfolding of *Betula pendula* has predominantly

Plant phenological changes

Table 2. Mean trends and mean differences between the subintervals 1951-1973 / 1974-1996 and 1951-1980 / 1967-1996 respectively (20+ / 30+ = records of 20 / 30 years and more, sig. = only significant records p<0.05, n = number, mean = mean difference [days], sd = standard deviation [days]) (Menzel et al. 2001).

	Mean trends [day / year]				Difference 1951-1973 and 1974-1996 mean			Difference 1951-1980 and 1967-1996 mean		
Phenophase (F flowering, LU leaf unfolding, M May shoot, FR fruit ripening LC leaf colouring, GS growing season)	20+	20+ (sig.)	30+	30+ (sig.)	n	mean	sd	n	mean	sd
G. nivalis F	-0,16	-0,48	-0,18	-0,48	235	6,26	5,28	227	2,46	3,46
F. suspensa F	-0,24	-0,53	-0,23	-0,48	121	6,03	3,47	115	3,04	2,36
B. pendula LU	-0,14	-0,37	-0,13	-0,34	179	3,02	3,17	172	1,50	2,05
A. hippocastanum LU	-0,16	-0,38	-0,15	-0,33	218	3,08	3,95	209	1,75	2,46
P. avium F	-0,05	-0,18	-0,03	-0,10	166	1,02	2,56	157	0,31	1,63
F. sylvatica LU	-0,08	-0,23	-0,06	-0,17	156	0,09	3,56	150	0,08	2,41
M. domestica F	-0,05	-0,21	-0,03	-0,08	178	0,78	2,97	175	0,49	1,94
Q. robur LU	-0,12	-0,31	-0,09	-0,23	127	0,87	3,24	117	0,85	2,28
P. abies M	-0,13	-0,31	-0,09	-0,22	94	1,42	4,01	90	0,93	2,73
T. platyphyllos F	-0,09	-0,24	-0,10	-0,24	78	1,34	4,46	71	1,29	2,66
S. nigra FR	-0,12	-0,28	-0,13	-0,30	128	2,78	6,00	121	2,53	3,82
A. hippocastanum FR	-0,02	-0,06	-0,03	-0,06	145	0,65	5,09	137	0,52	3,29
A. hippocastanum LC	-0,00	-0,01	-0,01	-0,01	169	-0,33	5,94	163	-0,03	3,80
B. pendula LC	0,03	0,07	0,04	0,10	167	-2,16	7,16	159	-1,42	4,74
F. sylvatica LC	0,03	0,06	0,03	0,07	175	-2,30	6,19	168	-1,18	3,96
Q. robur LC	0,09	0,23	0,10	0,23	119	-3,74	5,61	112	-2,22	3,49
B. pendula GS	0,18	0,44	0,18	0,39	60	-4,80	7,57	58	-2,92	5,30
F. sylvatica GS	0,11	0,33	0,10	0,29	58	-2,92	6,64	55	-1,53	3,91
A. hippocastanum GS	0,17	0,44	0,14	0,31	116	-3,56	7,11	114	-1,89	4,44
Q. robur GS	0,22	0,49	0,18	0,39	72	-4,61	5,94	66	-3,29	3,90

negative trends – only few positive trends appear to be concentrated in the low mountain range of central Germany. In contrast, a late spring phase, such as leaf unfolding of *Fagus sylvatica,* shows several positive trends (later onset) across Germany, but negative trends are clearly predominant. Flowering of *Malus domestica* has a higher percentage of insignificant trends. The mid-summer flowering phase of *Tilia platyphyllos* has both negative and some positive trends. The late autumn phases of leaf colouring of *Betula pendula* and of *Fagus sylvatica* have predominantly positive trends with occasional negative trends equally distributed over Germany. In total, a

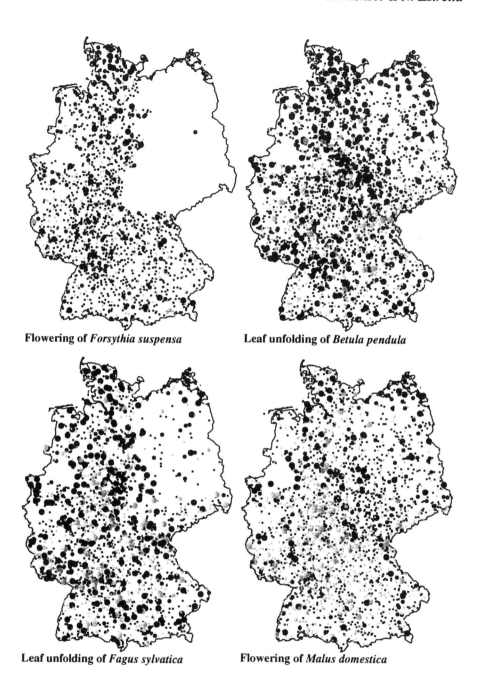

Figure 1. Linear trend and its significance according to F-test of all records with 20 years of observations and more. Black dot: negative trend, grey dot: positive trend, small sized dot: $p>0.05$, medium: $p<0.05$, great: $p<0.01$).

Plant phenological changes

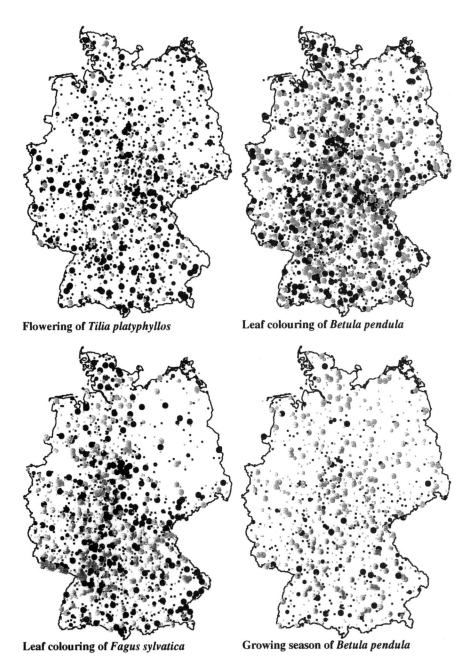

Flowering of *Tilia platyphyllos* **Leaf colouring of** *Betula pendula*

Leaf colouring of *Fagus sylvatica* **Growing season of** *Betula pendula*

Figure 1 (continued). Linear trend and its significance according to F-test of all records with 20 years of observations and more. Black dot: negative trend, grey dot: positive trend, small sized dot: $p>0.05$, medium: $p<0.05$, great: $p<0.01$).

clear lengthening of the growing season is detected. But, generally neither the sign of the trend nor the significance level seems to be dependent on geography. Menzel et. al 2001 showed that there is no meaningful and relevant correlation of trends with geographical parameters (longitude, latitude and altitude).

Comparison of mean values of different subintervals

Trends determined by linear regressions strongly depend on the time period analysed (Rapp & Schönwiese 1995, Menzel et al. 2001). Therefore, these results were verified by a second method by comparing phenological phases averaged over different subintervals. When dividing the whole period in two subintervals of equal length (1951-1973 and 1974-1996), the second subintervals comprises not only the years 1989 and 1990, but also 1974 and 1976, which were characterised by early plant development. Using two overlapping 30 year periods (1951-1980 and 1967-1996), the differences are less pronounced than for first comparison. Positive differences suggests an earlier, while negative differences a later onset in the second period. As mean values strongly depend on the number of observations (Schnelle 1955, Gornik 1994), only records with more than 20 observations in the two periods of 23 years or more than 27 observations in the two periods of 30 years were used. This precondition reduced the number of records considerably, especially in the eastern parts of Germany.

This second method confirmed the results of the linear trend analysis (see Table 2): Early spring phases are up to 6 days earlier in the second subinterval of 1974-1996, leaf unfolding of deciduous trees (*Aesculus hippocastanum, Betula pendula, Quercus robur*) are 1 to 3 days earlier, whereas the mean onset of leaf unfolding of *Fagus sylvatica* and flowering of *Malus domestica* does not change. Fruit ripening of *Sambucus nigra* is earlier, of *Aesculus hippocastanum* unchanged in the second interval. Autumn phases are generally later, up to 4 days (leaf colouring of *Quercus robur*). The growing season is longer by 3 (*Fagus sylvatica*) to 5 days (*Betula pendula*) in the second half of the period analysed.

In addition to these results (Menzel et al. 2001), histograms of all 20 phenophases showing the frequency distribution of these differences of mean values (1951-1973 and 1974-1996) (Figure 2) and maps showing the spatial distribution of these differences (Figure 3, same phases as in Figure 1) are given here. Key phases of early spring (flowering of *Galanthus nivalis* and *Forsythia suspensa*) most frequently advance by +6 days with a very broad distribution. The distribution of shifts in leaf unfolding of different trees is narrower with maximums of +4 and +2 days. Leaf unfolding of *Quercus robur*, but especially *Fagus sylvatica* and flowering of *Malus domestica*

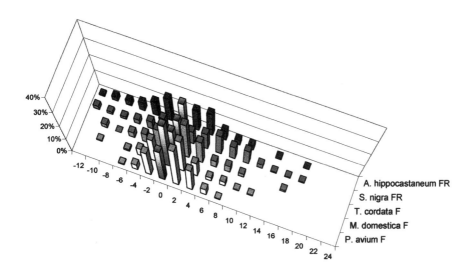

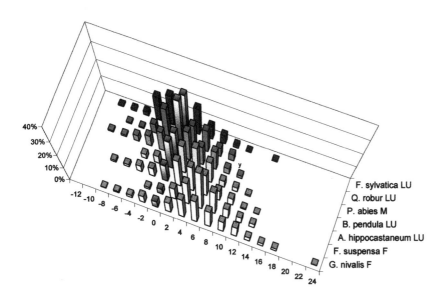

Figure 2. Histograms [%] of differences [days] between 1951-1973 and 1974-1996 subintervals. The columns of 0 which signify differences from −1 to +1 days are marked in floating grey. F Flowering, LU leaf unfolding, FR fruit ripening, LC leaf colouring, GS length of growing season

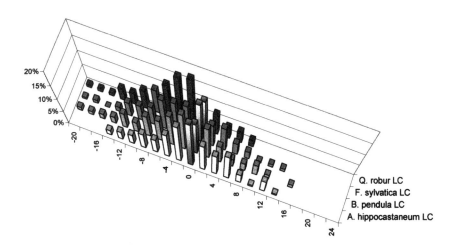

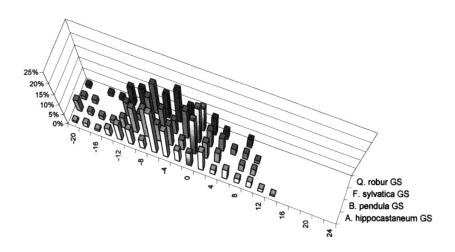

Figure 2 (continued). Histograms [%] of differences [days] between 1951-1973 and 1974-1996 subintervals. The columns of 0 which signify differences from −1 to +1 days are marked in floating grey. F Flowering, LU leaf unfolding, FR fruit ripening, LC leaf colouring, GS length of growing season

Plant phenological changes

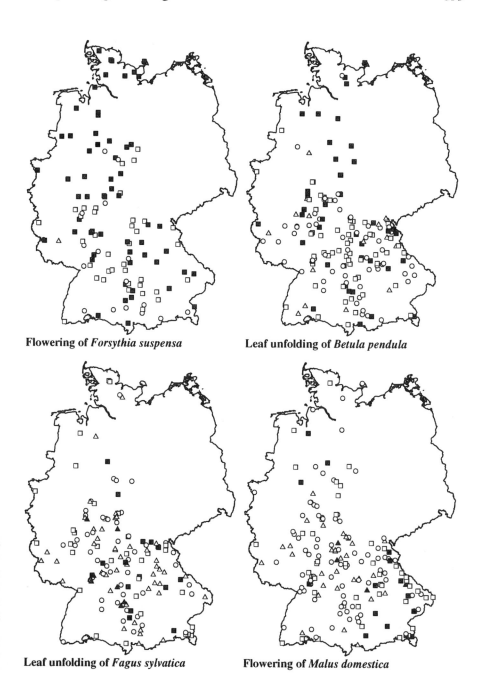

Figure 3. Differences [days] between 1951-1973 and 1974-1996 stations means. Triangle dark grey: −20 to −3 days later in the second subinterval, triangle light grey: -3 to −1 days later, circle: −1 to +1 days of change, rectangle light grey: +1 to +3 days earlier, rectangle dark grey +3 to +20 earlier.

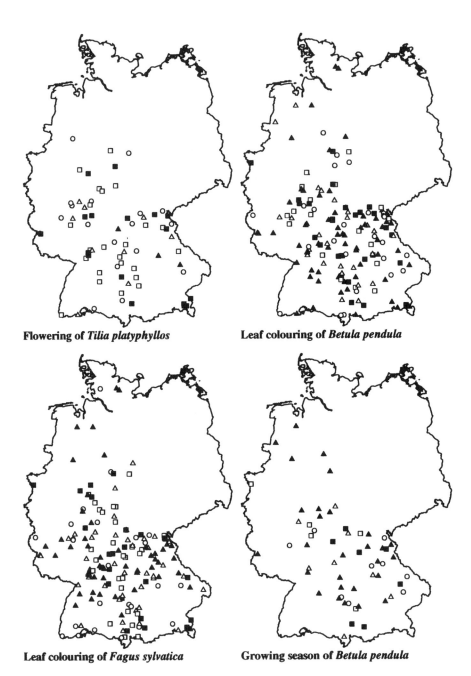

Figure 3 (continued). Differences [days] between 1951-1973 and 1974-1996 stations means. Triangle dark grey: −20 to −3 days later in the second subinterval, triangle light grey: −3 to −1 days later, circle: −1 to +1 days of change, rectangle light grey: +1 to +3 days earlier, rectangle dark grey +3 to +20 earlier.

mostly reveal no or only very small changes. Fruit ripening has a very broad distribution of changes as well as leaf colouring. For the latter the maximum of the distribution lays between 0 and -4 days showing a delay of these phases. The frequency distribution of the length of the growing season is more irregular. The spatial distribution of the differences of subinterval means is shown on the maps of Figure 3: Similar to Figure 1 sign and height of the trends seems not to be dependent on geographical parameters.

SUMMARY

A lot of studies reveal shifts in plant phenological phases in Europe and North America during the last 4 to 5 decades. The immense phenological database of the German Weather Service, collected by more than 2000 volunteers, can be perfectly used as direct biomonitoring data: Phenological phases cover the whole annual cycle from very early spring to late autumn and the corresponding spatial variability of the trends may also be investigated.

The results of the two different methods of trend analyses were similar and reveal a strong seasonal variation (Menzel et al. 2001, see also Table 2):
- Clear advances of key indicators of earliest and early spring (-0.18 to -0.23 days/year)
- Notable advances of the succeeding spring phenophases such as leaf unfolding of deciduous trees (-0.16 to -0.08 days/year)
- Weaker phenological changes during autumn (delayed by +0.03 to +0.10 days/year on average).

In general, the growing season has been lengthened by up to -0.2 days/year (mean linear trends). The mean 1974-1996 growing season was up to 5 days longer than in the 1951-1973 period. This means, the two methods of trend analysis applied lead to the same results (23 years x -0.2 days/year = -4.6 days). There are not any substantial regional differences in the trends as revealed by maps and further statistical analyses. But trends of succeeding phenological phases seem to be stable: At single locations they are mirrored by subsequent phases, but they are not necessarily identical.

Thus, the analysis of phenological observations seems to be a very effective tool in detecting climatic changes in North America and Europe. Compared to NDVI satellite observations, phenological ground observations have a finer temporal resolution that is very useful in detecting these changes which are relatively small compared to the satellites 7 day / 10 day composites.

REFERENCES

Ahas R., 1999, Long-term phyto-, ornitho- and ichthyophenological time-series analyses in Estonia. *Int. J. Biometeorol.* 42: 119-123.

Beaubien E.G. & Freeland H.J., 2000, Spring phenology trends in Alberta, Canada: links to ocean temperature. *Int. J. Biometeorol.* 44 (2): 53-59.

Bradley N.L., Leopold A.C., Ross J. & Huffaker W., 1999, Phenological changes reflect climate change in Wisconsin. *Proc. Natl. Acad. Sci. USA Ecology* Vol. 96: 9701-9704.

Cayan D.R., Kammerdiener S.A., Dettinger M.D., Caprio J.M. & Peterson D.H., 2001, Changes in the onset of spring in the western United States. *Bull. Am. Meteorol. Soc.* 82(3): 399-415.

Chmielewski F.M. & Rötzer T., 2000, Phenological trends in Europe in relation to climatic changes. *Agrarmeteorologische Schrift der Humboldt Universität Berlin* 07.

Defila C., 1991, *Pflanzenphänologie in der Schweiz*. Inaugural-Dissertation. Universität Zürich.

Defila C., 2000, Phytophenological trends in Switzerland. *International Conference Progress in Phenology – Monitoring, Data Analysis, and Global Change Impacts*, Oct. 4-6, 2000, Freising, Germany.

Estrella N., 1999, *Trends der forstlichen Vegetationsperiode von 1951-1996 in Deutschland*. Diploma thesis at the Chair of Bioclimatology and Air Pollution Research, Univ. of Munich.

Gornik W., 1994, Untersuchungen zur Problematik der Mittelwertbildung bei phänologischen Datenreihen. *Arboreta Phaenologica* 39: 5-10.

Jaagus J. & Ahas R., 2000, Space-time variations of climatic seasons and their correlation with the phenological development of nature in Estonia. *Clim. Res.* 15(3): 207-219.

Jones G.V. & Davis R.E., 2000, Climate influences on grapevine phenology, grape composition, and wine production and quality for Bordeaux, France. *Am. J. Enology Viticult.* 51(3): 249-261.

Keeling C.D., Chin F.J.S. & Whorf T.P., 1996, Increased activity of northern vegetation inferred from atmospheric CO_2 measurements. *Nature* 382: 146-149.

Magnuson J.J., Robertson D.M., Benson B.J. et al., 2000, Historical trends in lake and river ice cover in the northern hemisphere. *Science* 289: 1743-1746.

Mahrer T., 1985, *Untersuchung über die herbstliche Laubverfärbung der Buche in der Region Liesthal-Moehlin-Basel auf Grund von langjährigen phänologischen Beobachtungsreihen und ihre Zusammenhänge mit klimatologischen Parametern*. Unveröff. Diplomarbeit der ETH Zürich.

Menzel A., 1997, Phänologie von Waldbäumen unter sich ändernden Klimabedingungen - Auswertung der Beobachtungen in den Internationalen Phänologischen Gärten und Möglichkeiten der Modellierung von Phänodaten. *Forstliche Forschungsberichte* Nr. 164, München.

Menzel A., 1998, *Zeitliche Trends ausgesuchter phänologischer Phasen in Deutschland aus dem Zeitraum 1951-1996*. Unveröff. Bericht an den Deutschen Wetterdienst.

Menzel A., 1999, Phenology as Global Change Bio-Indicator. *Ann. Meteorol.* 39: 41-43.

Menzel A. & Fabian P., 1999, Growing season extended in Europe. *Nature* 397: 659.

Menzel A., 2000, Trends in phenological phases in Europe between 1951 and 1996. *Int. J. Biometeorol.* 44 (2): 76-81.

Menzel A., Estrella N. & Fabian P., 2001, Spatial and temporal variability of the phenological seasons in Germany from 1951-1996. *Global Change Biol.* 6: 1-10.

Myneni R.B., Keeling C.D., Tucker C.J., Asrar G. & Nemani R.R., 1997, Increased plant growth in the northern high latitudes from 1981 to 1991. *Nature* 386: 698-702.

Rapp J. & Schönwiese C.D., 1994, "Thermische Jahreszeiten" als anschauliche Charakteristik klimatischer Trends. *Meteorol. Z.* 3: 91-94.

Rapp J. & Schönwiese C.D., 1995, Atlas der Niederschlags- und Temperaturtrends in Deutschland 1891-1990. Frankfurter Geowissenschaftliche Arbeiten, Serie B *Meteorologie und Geophysik*, Band 5. 255pp.

Rötzer T., Wittenzeller M., Haeckel H. & Nekovar J., 2000, Phenology in central Europe – differences and trends of spring phenophases in urban and rural areas. *Int. J. Biometeorol.* 44: 60-66.

Schnelle F., 1955, *Pflanzen-Phänologie*. Leipzig, Geest & Portig KG.

Schwartz M.D. & Reiter B.E., 2000, Changes in North American Spring. *Int. J. Climatol.* 20 (8): 929-932.

Walkovszky A., 1998, Changes in phenology of the locust tree (*Robinia pseudoacacia* L.) in Hungary. *Int. J. Biometeorol.* 41: 155-160.

High summits of the Alps in a changing climate
The oldest observation series on high mountain plant diversity in Europe

HARALD PAULI, MICHAEL GOTTFRIED & GEORG GRABHERR
University of Vienna, Institute of Ecology and Conservation Biology, Department of Conservation Biology, Vegetation and Landscape Ecology, Althanstrasse 14, A-1090 Wien, Austria

Abstract: High mountain vegetation is considered to respond sensitively to climatic changes. Therefore, historical records on vascular plant species richness from high summits of the Alps were used as reference data to detect climate-induced changes of plant diversity. Reinvestigations on 30 summits provided evidence that species are migrating to higher altitudes.

The longest observation series originates from the nival summit of Piz Linard (3411 m a.s.l., SE-Switzerland). Five investigations on Piz Linard, between 1835 and 1937, showed that the number of species increased successively from one to 10 species. This tenfold rise of the species number was most likely caused by the climate warming since the 19th century. Other causes, like enhanced seed dispersal by animals, fertilising effects by grazing mammals or by air pollution, can largely be excluded.

However, species richness remained unchanged after 1937 according to data from 1947 and 1992, although climate warming continued. The subnival/nival species pool probably reached a saturation in the 1930s. On the other hand, half of the species, which have been already present in 1937, expanded their population size and grew at new sites in 1992. These species were able to fill "empty" micro habitats, where climate warming obviously caused increasingly favourable growing conditions. For potential invaders from the alpine grassland below, conditions were still to harsh, and appropriate ascent corridors appear to be underrepresented at this mountain. On other high summits of the Alps, immigration of alpine grassland species within the 2nd half of the 20th century was observed.

The upwards shift of species on high mountain summits suggests that plants already respond to climate warming. Serious biodiversity losses can be expected in the near future if predicted climate warming holds true.

INTRODUCTION

Complete historical records on vascular plant species richness exist for more than 100 high summits of the Alps. The majority of these peaks exceed 3000 m, reaching the subnival belt above the closed alpine grassland, or the nival belt above the climatic snow line. All these records are older than 40 years and some date back to the 19th century. The earliest summit record originates from Piz Linard, a nival peak of 3411 m a.s.l. north-west of the Lower Engadin valley (SE-Switzerland), which was investigated in 1835 by Oswald Heer, a pioneer of alpine ecology in Switzerland (Heer 1866).

For the following reasons, these old summit records can be considered as unique reference data to determine "fingerprints of climate change" on mountain biota: (1) sampling sites on summit positions are simple to reassign. This facilitates the reinvestigation, provided that the lower limits of the sampling areas were indicated by the historical authors. (2) The sites are located in environments dominated by abiotic, climate-related ecological factors. Therefore, effects of climate change may be more pronounced compared to ecosystems of lower altitudes. (3) Species from lower altitudes may respond to climate warming by an upwards shift of their distribution ranges. High mountain ecosystems are considered to be particularly sensitive for an early recognition of such climate-induced migration processes, due to the steep thermal gradient within short distances.

In fact, early (Braun-Blanquet 1955, 1957) and recent (Hofer 1992, Gottfried et al. 1994, Grabherr et al. 1994, 1995, 2000a, Pauli et al. 1996, 2001) reinvestigations provided evidence for an upwards shift of vascular plants on many of these mountains.

This paper gives a brief overview on results of a comparative study of historical and recent records from 30 selected summits reinvestigated by the authors in the early 1990s. Methods and results of this study were published between 1994 and 2001 in various papers mentioned below. In this paper, particular emphasis is taken on the investigations on Piz Linard, which provided the longest observation series on high mountain plant diversity.

CHANGE OF SPECIES RICHNESS ON 30 SUMMITS

The summits were selected with respect to the accuracy and reliability of the historical reference data. In Figure 1 all 30 summits are shown with their altitude, time of the historical and the recent investigation and the change of species richness. The change of species richness is indicated as the weighted species increase per decade (wSID in %) calculated according to the following formula:

High summits of the Alps in a changing climate

$$wSID = (wnS - wmS)/(eS + wmS) *100/[(pY - hY)/10]$$

where: wSID = weighted species increase per decade, wnS = weighted number of new found species, wmS = weighted number of species not found again (missing species), eS = number of species found in both records, pY = year of the present investigation, hY = year of the historical investigation.

The new found species (wnS) were weighted for abundance: 1 for species with high abundance, 0.5 for species with low abundance, 0.25 for species with very low abundance. This weighting accounts for the fact that some rare species might have been not recognised by the historical author.

Furthermore, values were standardised by a 10-year interval because of the different time spans between old and new records. For details on the method see Gottfried et al. (1994), Grabherr et al. (2000a), Pauli et al. (1996, 1998).

The majority of the summits (21, i.e. 70 %) showed a distinct increase of species richness (Figure 1). On the remaining 9 summits, species richness was stagnating or slightly decreasing.

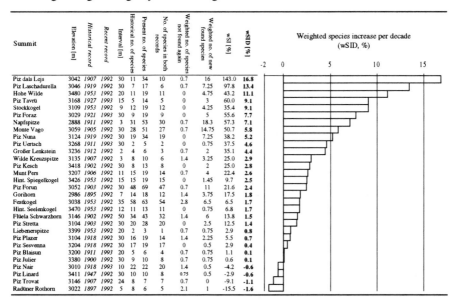

Summit	Elevation [m]	Historical record	Recent record	Interval [m]	Historical no. of species	Present no. of species	No. of species in both records	No. of species not found again	Weighted no. of new found species	wSI [%]	wSID (%)
Piz dals Lejs	3042	1907	1992	30	11	34	10	0.7	16	143.0	16.8
Piz Laschadurella	3046	1919	1992	30	7	17	6	0.7	7.25	97.8	13.4
Hohe Wilde	3480	1953	1992	20	11	19	11	0	4.75	43.2	11.1
Piz Tavrü	3168	1927	1993	15	5	14	5	0	3	60.0	9.1
Stockkogel	3109	1953	1992	9	12	19	12	0	4.25	35.4	9.1
Piz Foraz	3029	1921	1993	30	9	19	9	0	5	55.6	7.7
Napfspitze	2888	1911	1992	3	31	53	30	0.7	18.3	57.3	7.1
Monte Vago	3059	1905	1992	30	28	51	27	0.7	14.75	50.7	5.8
Piz Nuna	3124	1919	1992	30	19	34	19	0	7.25	38.2	5.2
Piz Uertsch	3268	1911	1993	30	2	5	2	0	0.75	37.5	4.6
Großer Lenkstein	3236	1912	1992	2	4	6	3	0.7	2	35.1	4.4
Wilde Kreuzspitze	3135	1907	1992	3	8	10	6	1.4	3.25	25.0	2.9
Piz Kesch	3418	1902	1992	30	8	13	8	0	2	25.0	2.8
Munt Pers	3207	1906	1992	11	15	19	14	0.7	4	22.4	2.6
Hint. Spiegelkogel	3426	1953	1992	15	15	19	15	0	1.45	9.7	2.5
Piz Forun	3052	1903	1992	30	48	69	47	0.7	11	21.6	2.4
Gorihorn	2986	1895	1992	7	14	18	12	1.4	3.75	17.5	1.8
Festkogel	3038	1953	1992	35	58	63	54	2.8	6.5	6.5	1.7
Hint. Seelenkogel	3470	1953	1992	12	11	13	11	0	0.75	6.8	1.7
Flüela Schwarzhorn	3146	1902	1992	50	34	43	32	1.4	6	13.8	1.5
Piz Stretta	3104	1903	1992	30	20	28	20	0	2.5	12.5	1.4
Liebenerspitze	3399	1953	1992	20	2	3	1	0.7	0.75	2.9	0.8
Piz Plazer	3104	1918	1992	30	16	19	14	1.4	2.25	5.5	0.7
Piz Sesvenna	3204	1918	1992	30	17	19	17	0	0.5	2.9	0.4
Piz Blaisun	3200	1911	1993	20	5	6	4	0.7	0.75	1.1	0.1
Piz Julier	3380	1900	1992	30	9	10	8	0.7	0.75	0.6	0.1
Piz Nair	3010	1918	1993	10	22	22	20	1.4	0.5	-4.2	-0.6
Piz Linard	3411	1947	1992	30	10	10	8	0.75	0.5	-2.9	-0.6
Piz Trovat	3146	1907	1992	24	8	7	7	0.7	0	-9.1	-1.1
Radüner Rothorn	3022	1897	1992	5	8	6	5	2.1	1	-15.5	-1.6

Figure 1. Change of species richness on 30 subnival and nival summits of the Eastern Alps; interval [m] = summit area investigated (in elevation metres from the top of the summit), wSID (weighted species increase per decade) = change of species richness determined by species numbers weighted by their abundance and standardised over time (see text for details). For the full data set for reinvestigation purposes see Grabherr et al. (2000a).

The increase of species richness appeared to be dependent on the geomorphological character of the summits, with high rates on summits with little erosion and stable habitats for plants to establish. High rates of increase were neither related with the altitude of the summits nor with a particular bedrock material.

For details on this summit comparison see Gottfried et al. (1994) and Pauli et. al. (1996). For a calculation of minimum plant migration rates on these summits, which were between zero and 4 elevation metres per decade for the most common species, see Grabherr et al. (1995, 2000a). The full data sets for future reinvestigations were published by Grabherr et al. (2000a).

PIZ LINARD – THE OLDEST OBSERVATION SERIES

Method

The area within the uppermost 30 elevation metres on Piz Linard was sampled for species richness for seven times between 1835 and 1992. Data on the distribution of species within the summit area are available by the reinvestigations in 1947 and in 1992 (compare Table 1).

In 1947, a distribution map of all plant species within the 30 m-summit area was drawn. This original distribution map, published by Braun-Blanquet (1957), shows the locations of single individuals or of groups of individuals for each species present in 1947. The distribution map was made by Campell who visited the summit together with Braun-Blanquet. Braun-Blanquet (1957) mentioned that this map was made for future monitoring purposes and good weather conditions during the summit visit in 1947 allowed a detailed species sampling for several hours. Thus, we can consider these records as very reliable data.

In 1992, the species mapping was repeated, again with good weather conditions and 4.5 working hours on the summit. However, it was not possible to determine each individual or groups of individuals within a one-day visit – species were too numerous in 1992. Therefore, we divided the summit area into 14 sectors. The sectors were divided along the geomorphological structure of the summit terrain (see Figure 2). Species abundances were scored in each sector in qualitative terms (see Table 1).

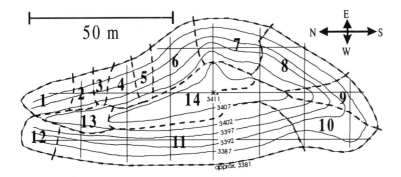

Figure 2. The 30-m summit area of Piz Linard, divided into 14 sectors. For each sector, species presence in 1947 was assigned irrespectively to their abundance (according to the original distribution map in Braun-Blanquet 1957) and species abundance in 1992 was determined (see Table 1). The sectors (broken lines) were divided along the geomorphological structure of the summit terrain.

Results

Changes of species richness

In 1835, only one species (*Androsace alpina*) was noted within the uppermost 30 elevation metres. The number of species increased successively to 10 species within the same summit area until 1937 (Figure 3). After 1937, the number of species obviously remained stable on Piz Linard – no more, but the same 10 species, were found in 1947 (Braun-Blanquet 1957), and the 1992-record also revealed only 10 species (Figure 3).

Changes of species abundance

Table 1 shows the distribution of the species within the 30-m summit area of Piz Linard in 1947 and in 1992 (compare Figure 2). The comparison shows that five species (*Androsace alpina, Ranunculus glacialis, Saxifraga bryoides, Saxifraga oppositifolia,* and *Poa laxa*) have obviously expanded their population size. These 5 species were already well established in some locations on the summit (Braun-Blanquet 1957), however, expanded into formerly empty sectors until 1992. From those species found to be rare or of intermediate abundance in 1947 (Braun-Blanquet 1957) only two have possibly expanded their distribution slightly (i. e. *Draba fladnizensis* and *Gentiana bavarica*), whereas *Cerastium uniflorum* was rare in 1992, like in

1947, but was found at an other location. Two species, which had been rare in 1947, were not found again (*Leucanthemopsis alpina* and *Saxifraga exarata*) and two species (*Cardamine resedifolia* and *Luzula spicata*) were new, but each only at one location. For maps of the historical and present distributions on the Piz Linard summit see Pauli et al. (2001).

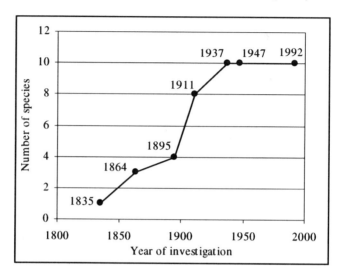

Figure 3. Change of species richness within the 30 uppermost elevation metres of Piz Linard (3411 m a.s.l.). Historical data from Heer (1866), Braun (1913), and Braun-Blanquet (1957), data for 1992 according to Grabherr et al. 2000a.

Discussion

The remarkable increase of species richness on the summit of Piz Linard between 1835 and 1937 provided clear evidence that high mountain plants are going to migrate upwards. With respect to the pronounced glacier retreat in the early 20th century, Braun-Blanquet (1957) suspected that climate warming is the main reason for the raised number of species – glaciers in SE-Switzerland shrinked by about 30 % of their surface area between 1895 to around 1940 (Zingg 1952). In the mid-20th century, Braun-Blanquet (1957) already proposed that mountain summits can be used as indicator habitats for climatic changes. In the early 1990s the findings from Piz Linard were confirmed by the determination of enhanced species numbers on 21 (out of 30) peaks, distributed over a large part of the Eastern Alps.

Table 1. Species distribution over the 14 summit area sections (30-m summit area) on Piz Linard (3411 m a.s.l.). The 1947 data show presence/absence values as assigned to each sector from the original map of Braun-Blanquet (1957), the 1992 data show abundance values as recorded by Pauli et al. (2001, in press). p = species present in the sector, c = species common (common or scattered to common), i = intermediate abundance (scattered), r = rare (rare-scattered, locally scattered, rare or very rare) in the sector.

	Summit area sector	*Ranunculus glacialis* L.	*Poa laxa* Haenke	*Saxifraga bryoides* L.	*Saxifraga oppositifolia* L.	*Androsace alpina* (L.)Lam.	*Draba fladnizensis* Wulf.	*Gentiana bavarica* L.	*Cerastium uniflorum* Clairv.	*Leucanthemopsis alpina* (L.)Heywood	*Saxifraga exarata* Vill.	*Cardamine resedifolia* L.	*Luzula spicata* (L.)DC.
1947	1	p	p	p	p	p				p	p		
	2												
	3												
	4												
	5												
	6												
	7	p	p	p	p	p	p	p					
	8				p	p							
	9												
	10	p	p	p						p			
	11	p	p										
	12												
	13												
	14	p	p	p						p			
Frequency		**5**	**5**	**4**	**3**	**3**	**1**	**1**	**1**	**2**	**1**		
1992	1	c	i	i	i	c	r						r
	2	r											
	3	c	c	c	i	i							
	4	r											
	5	c	c	i	i	i							
	6	r											
	7	c	c	c	i	c	i	i	r				r
	8	r	r	r	r	r		r					
	9	i	i	r	r								
	10												
	11	r	r										
	12	i	i		i	r	r						
	13	i	r	r	i	i							
	14	r	r	r	r	r							
Frequency		**13**	**10**	**8**	**9**	**8**	**3**	**2**	**1**			**1**	**1**

Causes other than climate change, which might potentially influence species richness on high mountain summits, appear to be of minor importance for the following reasons:

(1) Mountain tourism: Mountain climbers may cause impacts by trampling or by supporting the upwards dispersal of diaspores. Most of the summits investigated, however, are not frequently visited by tourists and most hillwakers and climbers keep usually close to the trails or climbing routes. On Piz Linard, visits by tourists were very occasional between 1835 and 1947 – the visit by Heer in 1835 is considered to be the first ascent to the summit at all. On one peak (Piz Julier) which is frequently visited by tourists, species richness was stagnating (Figure 1). Probably, the area of potential habitats for invading species were reduced by trampling impacts. Negative effects on species richness by trampling appear to be much more important than positive effects by an upwards dispersal of seeds by tourists.

(2) Grazing: Domestic livestock such as sheep or goats may cause changes of the species composition due to trampling, grazing, fertilising effects, as well as by epi- and endozoochorous seed dispersal. Yet, most of the investigated summits are subnival peaks which lie above the traditional grazing area. Sheep or goats may reach these peaks only occasionally. Wild-living high mountain ungulates, like *Capra ibex*, might act as seed dispersers even in elevations above 3000 m. Alpine ibex was near to extinction in the 19th century. The re-establishment of *Capra ibex* started in the first half of the 20th century, but substantial increases of ibex colonies were reached in the second half of the century (Giacometti 1991). Therefore, effects caused by *Capra ibex* can be completely neglected for the period 1835 to 1937 in which species richness on Piz Linard changed from one to 10 species.

(3) Birds, such as ptarmigans (*Lagopus mutus*), ring ouzels (*Turdus torquatus*), or nutcrackers (*Nucifraga caryocatactes*), may support seed dispersal in high mountain environments (for the nutcracker compare Grabherr et al. 2000a). In contrast to *Capra ibex*, however, changes of their abundance in the Eastern Alps during the 19th and 20th century are not evident. It is likely that the populations of these bird species remained more or less stable. At least for the last decades, no obvious changes of their distribution ranges were observed in Switzerland by comparing data from the 1970s and 1990s (Schmid et al. 1998). In Tyrol, the distribution of ptarmigan and ring ouzel remained more or less stable according to data from the beginning and the end of the 20th century (Landmann & Lentner 2001). Thus, seed dispersal by birds is not likely to explain the increase of plant species richness on the high summits.

(4) Anthropogenic nutrient deposition: Air pollution in high mountain areas, which might cause a fertilising effect, was of minor importance at least in the 19th and early 20th century. Thus, the increase of species richness on Piz Linard cannot be related to atmospheric nutrient deposits.

On the other hand, it is conspicuous that the number of species on Piz Linard stagnated after 1937 (Figure 3), and has not continued to increase until the 1990s, while climate warming did continue until today. In addition, other high summits of the Eastern Alps have experienced an increase of species richness between the 1950s and the 1990s (e.g. Hohe Wilde, 3480 m; Stockkogel, 3109 m; see Figure 1). However, five of the 10 species which occurred in 1937 and 1947 on Piz Linard have increased in abundance and grew in new summit area sectors in 1992. Four of these 5 species have reached the summit after 1935, but were already well established in 1947. After 1947 they have continued to expand by filling unpopulated micro habitats. The five species *Androsace alpina, Ranunculus glacialis, Saxifraga bryoides, Saxifraga oppositifolia,* and *Poa laxa* are common in the subnival belt of the siliceous central Alps (i.e. in the open plant assemblages between the alpine grassland belt and the mostly bare nival environment), compare Pauli et al. (1999). Obviously, the summit environment of Piz Linard has changed between the 1830s to the late 20th century from harsh nival conditions to subnival conditions, to which the five expanding species are well adapted.

The stagnation of species richness on Piz Linard was probably due to a saturation of the subnival/nival species pool around 1940, followed by an increase of abundance. Potential immigrants from the alpine species pool did not colonise the summit until 1992. Still too harsh habitat conditions on the summit, combined with the absence of appropriate ascent corridors from the alpine grassland belt to the summit area, have probably hindered the appearance of new species. Other peaks of similar altitude, like Mount Hohe Wilde (3480 m, southern Ötztaler Alpen), experienced an increase of species richness from 10 to 19 species between 1953 and 1992 (Figure 1). Most of the invading species on Hohe Wilde can be considered as plants of the alpine grassland belt (see Grabherr et al. 2000a). This mountain has more or less uninterrupted south-facing slopes down to the alpine grassland, which is not the case on Piz Linard.

The obvious increase of species richness (or abundance of species) on Piz Linard as well as on most other peaks investigated, suggests that climate warming is a discernible driving force for the ongoing upwards shift of mountain plants. Recent predictions of atmospheric warming of 1.4 to 5.8 K over the period 1990-2100 (IPCC 2001) may lead to drastic changes of biodiversity patterns in high mountain ecosystems, including the extinction of a variety of species. How severe such "extinction scenarios" are can only be documented by long-term monitoring. The newly emerging research initiative GLORIA (Global Observation Research Initiative in Alpine Environments; Grabherr et al. 2000b) aims to establish an effective observation network towards a large-scale comparison of climate-induced changes of biodiversity patterns and the detection of expected biodiversity losses in high mountain regions (see www.gloria.ac.at).

ACKNOWLEDGEMENTS

This study was supported by the Austrian Academy of Sciences as a contribution to the International Geosphere-Biosphere Programme. Thanks to Martin Pollheimer (Birdlife Austria) for useful remarks on bird distributions and ornithochory.

REFERENCES

Braun J., 1913, Die Vegetationsverhältnisse der Schneestufe in den Rätisch-Lepontischen Alpen. *Neue Denkschr. Schweiz. Naturforsch. Ges.* 48: 348 pp.

Braun-Blanquet J., 1955, Die Vegetation des Piz Languard, ein Maßstab für Klimaänderungen. *Svensk Botanisk Tidskrift* 49(1-2): 1-9.

Braun-Blanquet J., 1957, Ein Jahrhundert Florenwandel am Piz Linard (3414 m). *Bull. Jard. Botan. Bruxelles* Vol Jubil W Robyns (Comm S.I.G.M.A. 137): 221-232.

Giacometti, M., 1991, Beitrag zur Ansiedlungsdynamik und aktuellen Verbreitung des Alpensteinbockes (*Capra i. ibex* L.) im Alpenraum. *Z. Jagdwiss.* 37: 157-173.

Gottfried M., Pauli H. & Grabherr G., 1994, Die Alpen im "Treibhaus": Nachweise für das erwärmungsbedingte Höhersteigen der alpinen und nivalen Vegetation. *Jahrb. Ver. Schutz d. Bergwelt*, München, 59: 13-27.

Grabherr G., Gottfried M. & Pauli H., 1994, Climate effects on mountain plants. *Nature* 369: 448.

Grabherr G., Gottfried M., Gruber A. & Pauli H., 1995, Patterns and current changes in alpine plant diversity. In: F.S. Chapin & C. Körner (eds.) *Arctic and Alpine Biodiversity: Patterns, Causes and Ecosystem Consequences.* Ecological Studies 113, Springer, Berlin.

Grabherr G., Gottfried M. & Pauli H., 2000a, Long-term monitoring of mountain peaks in the Alps. In: C. Burga & A. Kratochwil (eds.) *Biomonitoring: General and Applied Aspects on Regional and Global Scales.* Tasks for Vegetation Science, Vol. 35. Kluwer, Dordrecht.

Grabherr G., Gottfried M. & Pauli H., 2000b, GLORIA: a Global Observation Research Initiative in Alpine Environments. *Mountain Research and Development* 20/2: 190-191.

Heer O., 1866, Der Piz Linard. *Jahrb. Schweiz. Alpin Club* III, Bern: 457-471.

Hofer H.R., 1992, Veränderungen in der Vegetation von 14 Gipfeln des Berninagebietes zwischen 1905 und 1985. *Ber. Geobot. Inst. ETH Zürich* (Stiftung Rübel) 58: 39-54.

IPCC, 2001, *Third assessment report* of Working Group I of the Intergovernmental Panel on Climate Change – Summary for policy makers.

Landmann A. & Lentner R., 2001, Die Brutvögel Tirols: Bestand, Gefährdung, Schutz und Rote Liste. *Ber. nat. med. Ver. Innsbruck*, Suppl. 14: 1-182.

Pauli H., Gottfried M. & Grabherr G., 1996, Effects of climate change on mountain ecosystems - upward shifting of alpine plants. *World Res. Rev.* 8: 382-390.

Pauli H., Gottfried M., Reiter K. & Grabherr G., 1998, Monitoring der floristischen Zusammensetzung hochalpin/nivaler Pflanzengesellschaften. In: A. Traxler (ed.) *Handbuch des vegetationsökologischen Monitorings*, Teil A: Methoden. Umweltbundesamt, Wien.

Pauli H., Gottfried M. & Grabherr G., 1999, Vascular plant distribution patterns at the low-temperature limits of plant life - the alpine-nival ecotone of Mount Schrankogel (Tyrol, Austria). *Phytocoenologia* 29/3: 297-325.

Pauli H., Gottfried M. & Grabherr G., 2001, The Piz Linard (3411 m, the Grisons, Switzerland) – Europe's oldest revisitation site for mountain vegetation studies. In: L. Nagy, G. Grabherr, C. Koerner & D.B.A. Thompson (eds.) *Alpine Biodiversity in Europe – A Europe-wide Assessment of Biological Richness and Change*. Ecological Studies, Springer, in press.

Schmid H., Luder R., Naef-Daenzer B., Graf R. & Zbinden N., 1998, *Schweizer Brutvogelatlas. Verbreitung der Brutvögel in der Schweiz und im Fürstentum Liechtenstein 1993-1996*. Schweizerische Vogelwarte. Sempach.

Zingg T., 1952, Gletscherbewegungen in den letzten 50 Jahren in Graubünden. *Wasser- und Energiewirtschaft* V-VII, Zürich.

Evergreen broad-leaved species as indicators for climate change

GIAN-RETO WALTHER[*], GABRIELE CARRARO[#] & FRANK KLÖTZLI[‡]
[*]*Institute of Geobotany, University of Hannover, Nienburger Str. 17, DE-30167 Hannover, Germany;* [#]*DIONEA SA, Via Lungolago Motta 8, CH-6600 Locarno, Switzerland;* [‡]*Geobotanical Institute, Swiss Federal Institute of Technology (ETH), Zürichbergstr. 38, CH-8044 Zürich, Switzerland.*

Abstract: Conspicuous changes in vegetation composition have been observed in southern Switzerland. More than a dozen exotic species recently started to colonise forest understorey vegetation. All the exotic species share equal characteristics (evergreen broad-leaved or laurophyllous plant functional type) and a synchronous pattern of dispersal in time and space. This stresses the fact that external factors may determine the process. Climatic data from local meteorological stations are evaluated with special regard to limiting climatic parameters for laurophyllous species. With this background, climatic forcing must be considered as one of the major contributing factors for the increasing tendency of exotic species dispersal and their capability to establish in Swiss lowland forests.

INTRODUCTION

In recent decades, an obvious increase in the number and frequency of exotic evergreen broad-leaved species has been observed in parts of Switzerland (Gianoni et al. 1988, Klötzli et al. 1996, Carraro et al. 1999). In the south of the country, especially in areas lower than 600 m a.s.l., the structure and composition of forests has completely changed due to the shift in the shrub layer from indigenous, deciduous to exotic, evergreen broad-leaved species. The particular character of these species, their origin from warm-temperate areas, as well as the temporal synchronisation of their increase in abundance in the last thirty years strongly suggest changes in environmental factors within that period as the driving factor.

The ranges and growth of temperate tree species can often be related to the length of the growing season and especially to absolute minimum temperatures (IPCC 1996). In particular, the limit of evergreen broad-leaved towards deciduous forest on the global scale is to a major degree climatically determined (e.g. Woodward 1987, Klötzli 1988) and since the classical paper of Iversen (1944), the limits of the range of native European, evergreen broad-leaved (= laurophyllous) species such as *Ilex aquifolium*, *Hedera helix* and *Viscum album* are thought to follow winter low temperature isotherms. However, the northern limit of exotic species in Europe appears to be controlled by a combination of growing season length and minimum winter temperature (see Beerling 1993). Thus, in view of the tendency of exotic evergreen broad-leaved species to spread into deciduous forest, we find ourselves confronted with the question as to whether this increase is due to changing climatic conditions.

THE INSUBRIAN REGION AT THE SOUTHERN FOOT OF THE ALPS

The investigation area is situated on the south to south-west exposed slopes of Lago Maggiore, surrounding the peninsula of Locarno and Ascona in the canton Ticino in southern Switzerland. The core investigation area encompasses an altitudinal range from 200 m up to 1030 m a.s.l. on the south to south-west exposed face of Monte Brè near Locarno (8°47' N, 46°11' E). The soils have developed on siliceous rocks and were described by Blaser (1973) as cryptopodzolic rankers. The investigation area is part of the so-called Insubrian region which has particular climatic conditions. The mean annual temperature amounts to almost 12 °C, with mean monthly minimum temperatures in January of 0 °C and mean monthly maxima in July with 25 °C. The annual rainfall of approximately 1800 mm is concentrated on relatively few days but scattered throughout the year, allowing a great number of sunny days but no drought period, in contrast to the typical Mediterranean climate. The most prominent vegetation types in the area belong to deciduous forest associations and are dominated by *Quercus*, *Tilia*, *Fraxinus* and *Castanea* species (see e.g. Oberdorfer 1964, Delarze et al. 1999).

The rather mild but humid climate has allowed the cultivation of a variety of ornamental shrubs and trees in gardens and parks. Since the 17th century, species have been introduced from warm-temperate to subtropical areas of various continents. Although cultivated in gardens and parks for more than 200 years, the spread of evergreen representatives of these introduced species was not reported before the end of the 19th century, although the area was much visited by botanists who described its flora in some detail. The first records date from the turn of the 19th/20th century, but these occurrences were probably not permanent as the age of the oldest individuals in present forest stands is not more than 50 years. Even during the Swiss national forest inventory in the 1960s, there are no records to reveal the occurrence of exotic evergreen broad-leaved species (see e.g. Ellenberg & Rehder 1962, Leibundgut 1962, Ellenberg & Klötzli 1972). Zuber (1979) recorded an early stage of the invasion, with few representatives of exotic evergreen broad-leaved species in the herb layer and also some in the shrub layer. This has since become a common condition. Zubers vegetation relevés were taken at sites arranged on a regular grid with a 100 m spacing, covering the entire south to south-west exposed facing of the Monte Brè. Almost 25 years later these plots were resurveyed and analysed to determine changes in species abundance and frequency.

UNUSUAL ASPECTS OF VEGETATION SHIFTS

Within the time span of two decades, the change of the forest vegetation was most evident in the composition of the shrub and herb layers. Given the duration of the forest cycle, one could not expect great changes in the tree layer. The detailed results have been reported elsewhere (see Walther 2000). Here, we focus on the shift in woody species, with special regard to the group of exotic laurophyllous species and the link to recent climatic change. Amongst the species with the greatest increase in frequency, the group of evergreen broad-leaved species are the most prominent, including both indigenous and exotic species (Figure 1).

An increasing number of exotic evergreen broad-leaved species, including the palm *Trachycarpus fortunei*, have successfully rejuvenated and succeeded in colonising deciduous forests adjacent to gardens and parks (a complete list is given in Walther 1999). Within the past thirty years they have grown up to the shrub layer and, in some parts *Cinnamomum glanduliferum*, even mixed in the canopy. Nowadays, these exotic species must be considered as naturalised and ecologically important components of these ecosystems.

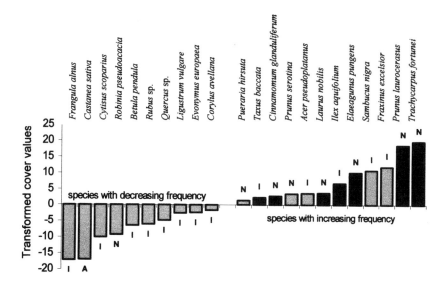

Figure 1. Net changes in frequency between 1975 and 1998 of shrub layer species (differences are given in transformed Braun-Blanquet cover values (see Walther 2000); I = indigenous, A = Archaeophytes, N = Neophytes, ▢ = deciduous, ▬ = evergreen).

A review of the literature (cf. Walther 2001) makes clear that this shift in the structure and composition of the forest has occurred in not more than half of a century, and with increased intensity in the last three decades. In most cases the period, when a species was introduced and when it spread are different. Although the species originate from different continents and were introduced in a time span varying approximately 200 years, the time of spread is concentrated over a few decades and gives the impression of a synchronised and externally triggered process (Figure 2).

In addition, all the invasive species are of the same plant functional type, which makes this invasion process rather unusual and calls for interpretation.

IS CLIMATE CHANGE A TRIGGERING FACTOR FOR THESE VEGETATION SHIFTS?

The inspection of the global temperature time series reveals an accelerated increase in global mean temperatures since the 1970s. Recent decades have been exceptionally warm, with the 1990s being the warmest decade of the millennium (IPCC 2001). A string of 16 consecutive months, where each monthly mean global temperature broke all previous records, occurred from May of 1997 to August of 1998 (Karl et al. 2000). The

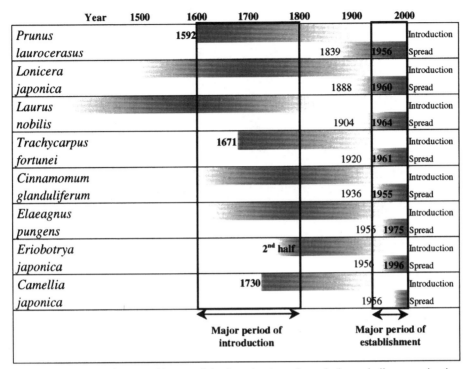

Figure 2. The asynchronous history of the introduction of exotic laurophyllous species in historical time (upper bars) and their synchronous spreading in the last century (lower bars), first on azonal sites (gardens, gorges, etc.), later on zonal sites.

growing season in Europe has extended by 10.8 days since the early 1960s, with consequences for the phenology of European temperate species (Menzel & Fabian 1999). Time series of long-term meteorological measurements in Switzerland have revealed an increase in the annual mean temperature in the period from 1960 to 1990 (Aschwanden et al. 1996). All the meteorological stations included in this study showed rising trends, even though they were scattered throughout the country and located at different altitudes. Also the two meteorological stations close to the investigation area shared this trend.

Since many of the introduced plants originate from warmer regions, these particularly climate-sensitive species depended on the artificial climate of a garden and their populations were restricted to planted individuals in gardens and parks. Unfavourable climate may have prevented the spread of a number of exotic species, but recently the climatic constraints have been reduced for many cultivated plants. This trend is expected to continue, and ornamental gardens and parks are likely to become sources for plant invasions (see Dukes & Mooney 1999). One major question in this context is which climatic parameters are most responsible for determining the range of these species. Although this cannot readily be tested for many taxa because their autecology is not sufficiently known, the group of species sharing the same plant functional type, i.e. evergreen broad-leaved habit, may provide a useful chance to investigate climate change impacts on ecosystems (cf. Diaz & Cabido 1997). Detecting trends in range expansion of this species group requires a thorough analysis of the major controlling factors; these are probably a combination of the absolute minimum temperature, determining whether a species can exist at a particular place or not (see e.g. Sakai & Larcher 1987), and the length of the growing season, which determines the ecophysiological performance, an ecologically important factor for evergreen species in competition with deciduous species.

In recent years, absolute annual minimum temperatures have been well above the level which is considered critical for the establishment of young evergreen broad-leaved species (Figure 3).

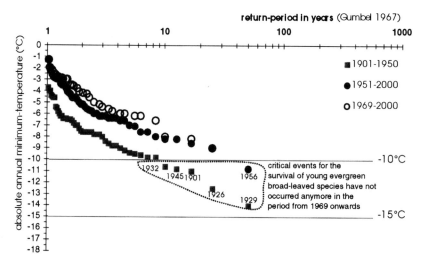

Figure 3. Variation in the return-period of the absolute minimum-temperatures in Lugano, subdivided into three periods (first half of the 20th century, second half of the 20th century and since 1969) (climatic data from SMA Meteo Swiss).

The climatic history of the last century, with recurring critical events for species sensitive to low temperature, may explain the only temporary occurrence of exotic laurophyllous species in the first half of the 20th century. However, since the late 1950s, climatic conditions have allowed the long-term survival of more sensitive exotic evergreen broad-leaved species outside the protected areas of gardens and parks.

An analysis of the length of the growing season, derived from the number of frost-free days, revealed an obvious prolongation of the vegetation period, but only for those species able to photosynthesise during periods of the "winter" season with temperatures above the freezing point (in accord with the findings of Zeller 1951). This excludes the deciduous species and favours evergreen species in particular. In recent years the number of frost days per year has decreased to as little as five days, thus allowing the evergreen broad-leaved species to grow for up to 360 days a year (see Figure 4).

These indications of significantly better climatic conditions for the survival and growth of evergreen broad-leaved species are confirmed by findings derived from the analysis of vegetation change in the area. Tree ring analysis was used to determine the age of camphor trees (*Cinnamomum glanduliferum*) in the investigation area (Sittig 1998). With the absence of the critical minimum temperature values (see Figure 3) and the rising length of the growing period, it is clear that the number of establishment of *Cinnamomum* has increased since the period of milder winter conditions began (Figure 4).

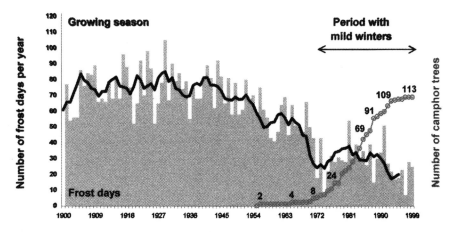

Figure 4. Climatic history vs. recruitment rate of an exotic laurophyllous tree species (*Cinnamomum glanduliferum*).

THE SPATIAL DYNAMICS OF LAUROPHYLLISATION

The relevés which were made at an early stage of colonisation of laurophyllous species and a resurvey of the same sites twenty years later, provided the data base for the spatial analysis of the spread of evergreen broad-leaved species, which has also been called laurophyllisation (cf. Klötzli & Walther 1999). The results are summarised in Figure 5, which

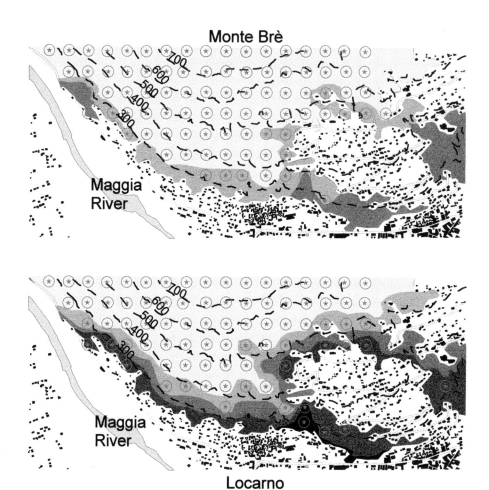

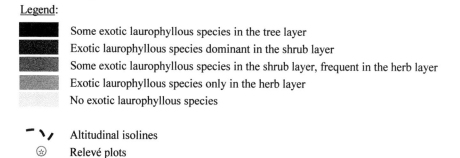

Figure 5. Spatio-temporal range expansion of laurophyllous species at the foot of Mte Brè near Locarno from 1975 (top) to 1998 (bottom).

shows the extent of laurophyllisation using a five point scale to describe the abundance of evergreen broad-leaved species. The area covered with exotic laurophyllous species, divided into the different abundance classes, can be visualised using GIS (Figure 5).

Within a few decades, the exotic laurophyllous species have shifted more and more into the forest as well as grown up into the canopy layer. By comparing the stage of colonisation reflected by the relevés of the two different surveys, the migration rate of the moving front of laurophyllous species into the forest could be estimated (Figure 6).

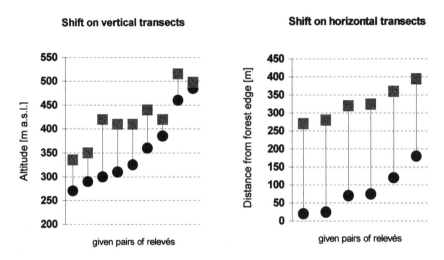

Figure 6. Migration rates of non-indigenous laurophyllous species (● 1975; ■ 1998).

The gradient in decreasing species abundance and frequency, obvious with increasing altitude, suggests that the present altitudinal range limit is approximately 600 m a.s.l.. However, there is a parallel horizontal range shift which leads to the assumption, that the exotic laurophyllous species have not yet attained equilibrium with the environmental conditions given in the area. Hence, further range expansion to more remote areas of the forest is expected in the coming years.

CONCLUSIONS

In contrast to other cases of invasions, with exotic species originating from the same climatic zone as the one of the habitat being invaded, the extraordinary aspect of this example is the fact that the invasive species do not originate from the same climatic zone but from an adjacent warmer biome, therefore crossing the boundary from deciduous to evergreen broad-leaved vegetation.

It is often difficult to assign only one cause to observed changes in natural ecosystems. However, neither management methods nor any other environmental factor is known to have changed over a relevant time period in a way that would particularly favour evergreen broad-leaved species. The obvious change in the local climate history is the only factor known so far that coincides in time with the trigger of the invasion process and the pronounced amplification in abundance and frequency of exotic laurophyllous species. The twofold set of climatic parameters, warmer absolute minimum temperatures and the lower frequency of frost-days, strongly suggest an important influence of climate change on the increased dispersal rate and establishment of exotic evergreen broad-leaved species.

This example from Switzerland provides a glimpse of the capacity of climate sensitive transitions zones to respond to changes in environmental conditions, e.g. climate change. Similar trends have been observed in other regions e.g. in Japan and Georgia, with both, indigenous and exotic evergreen broad-leaved species expanding their ranges into deciduous forest habitats (see Klötzli & Walther 1999). Hence, this functional group of species appears to respond sensitively to changing climatic conditions and to serve as bioindicator of climate change in ecotonal areas in temperate regions. Three decades with changed climatic conditions have profoundly influenced the composition and structure of the vegetation. Thus, with regard to climate change, not only shifts in indigenous species composition and species ranges must be expected, but also new combinations of species contributing to the structure and function of some ecosystems. The new species – such as palms and other exotic evergreen broad-leaved species in the present case – may play key structural and functional roles in "post-climate-change communities" (Williams 1997).

REFERENCES

Beerling D.J., 1993, The impact of temperature on the northern distribution limits of the introduced species *Fallopia japonica* and *Impatiens glandulifera* in north-west Europe. *J. Biogeogr.* 20: 45-53.

Blaser P., 1973, Die Bodenbildung auf Silikatgestein im südlichen Tessin. *Mitteil. Schweiz. Anst. Forstl. Versuchsw.* 49(3): 251-340.

Carraro G., Klötzli F., Walther G.-R., Gianoni P. & Mossi, R., 1999, *Observed changes in vegetation in relation to climate warming*. Final report NRP 31, vdf Hochschulverlag, Zürich, 87pp.

Delarze R., Gonseth Y. & Galland P., 1999, *Lebensräume der Schweiz*. Ott, Thun, 413pp.

Diaz S. & Cabido M., 1997, Plant functional types and ecosystem function in relation to global change. *J. Veg. Sci.* 8(4): 463-474.

Dukes J.S. & Mooney H.A., 1999, Does global change increase the success of biological invaders? *TREE* 14(4): 135-139.

Ellenberg H. & Rehder H., 1962, Natürliche Waldgesellschaften der aufzuforstenden Kastanienflächen im Tessin. *Schweiz. Zeitschr. Forstw.* 113: 128-142.

Ellenberg H. & Klötzli F., 1972, Waldgesellschaften und Waldstandorte der Schweiz. *Schweiz. Anst. Forstl. Versuchsw.* 48(4).

Gianoni G., Carraro G. & Klötzli F., 1988, Thermophile, an laurophyllen Pflanzenarten reiche Waldgesellschaften im hyperinsubrischen Seenbereich des Tessins. *Ber. Geobot. Inst. ETH*, Stiftung Rübel, Zürich 54: 164-180.

Gumbel E.J., 1967, *Statistics of extremes.* Columbia University Press.

IPCC, 1996, *Climate change 1995.* Impacts, adaptations and mitigation of climatic change: Scientific-technical analyses. Contribution of working group II to the second assessment report of the intergovernmental panel on climate change.

IPCC, 2001, *Climate change 2001: Impacts, adaptations and vulnerability.* Contribution of working group II to the third assessment report of the intergovernmental panel on climate change. Cambridge University Press, Cambridge.

Iversen J., 1944, *Viscum, Hedera* and *Ilex* as climate indicators. *Geol. Fören. Förhandl.* 66(3): 463-483.

Karl T. R., Knight R. W. & Baker B., 2000, The record breaking global temperatures of 1997 and 1998: Evidence for an increase in the rate of global warming? *Geophys. Res. Lett.* 27(5): 719-722.

Klötzli F., 1988, On the global position of the evergreen broad-leaved (non-ombrophilous) forest in the subtropical and temperate zones. *Veröff. Geobot. Inst. ETH*, Stiftung Rübel, Zürich 98: 169-196.

Klötzli F., Walther G.-R., Carraro G. & Grundmann A., 1996, Anlaufender Biomwandel in Insubrien. *Verh. Ges. Ökol.* 26: 537-550.

Klötzli F. & Walther G.-R. (eds.), 1999, *Conference on recent shifts in vegetation boundaries of deciduous forests, especially due to general global warming.* Proceedings of the Centro Stefano Franscini, Mte Verita, Ascona. Birkhäuser, Basel, 342pp.

Leibundgut H., 1962, Waldbauprobleme in der Kastanienstufe Insubriens. *Schweiz. Zeitschr. Forstw.* 113: 164-188.

Menzel A. & Fabian P., 1999, Growing season extended in Europe. *Nature* 397: 659.

Oberdorfer E., 1964, Der insubrische Vegetationskomplex, seine Struktur und Abgrenzung gegen die submediterrane Vegetation in Oberitalien und in der Südschweiz. *Beitr. naturk. Forsch. SW-Deutschl.* 23(2): 41-187.

Sakai A. & Larcher W., 1987, *Frost survival of plants.* Ecological Studies 62, Springer, Berlin.

Sittig E., 1998, *Dendroökologische Rekonstruktion der Einwanderungsdynamik laurophyller Neophyten in Rebbrachen des südlichen Tessins.* M.Sc.-Thesis, Univ. Marburg, 138pp.

Walther G.-R., 1999, Distribution and limits of evergreen broad-leaved (laurophyllous) species in Switzerland. *Bot. Helv.* 109(2): 153-167.

Walther G.-R., 2000, Climatic forcing on the dispersal of exotic species. *Phytocoenologia* 30(3-4): 409-430.

Walther G.-R., 2001, Laurophyllisation – a sign of a changing climate? In: C.A. Burga & A. Kratochwil (eds.) *Biomonitoring: General and applied aspects on regional and global scales.* Tasks for vegetation science 35: 207-223.

Williams C.E., 1997, Potential valuable ecological functions of non-indigenous plants. In: J.O. Luken & J.W. Thieret (eds.) *Assessment and Management of Plant Invasions*, 26-34.

Woodward F.I., 1987, *Climate and plant distribution.* Cambridge University Press, Cambridge.

Zeller O., 1951, Über die Assimilation und Atmung der Pflanze im Winter bei tiefen Temperaturen. *Planta* 39: 500–526.

Zuber R.K., 1979, Untersuchungen über die Vegetation und die Wiederbewaldung einer Brandfläche bei Locarno (Kanton Tessin). *Beiheft zu Zeitschr. Schweiz. Forstverein* 65.

The expansion of thermophilic plants in the Iberian Peninsula as a sign of climatic change

EDUARDO SOBRINO VESPERINAS*, ALBERTO GONZÁLEZ MORENO[#], MARIO SANZ ELORZA*, ELIAS DANA SÁNCHEZ[°], DANIEL SÁNCHEZ MATA[‡] & ROSARIO GAVILÁN[‡]

*Departamento Producción Vegetal: Botánica y Protección Vegetal, Escuela Técnica Superior Ingenieros Agrónomos, 28040 Madrid, Spain. [#]Instituto Nacional de Investigación y Tecnología Agraria y Alimentaria (INIA), C.ª La Coruña, km 7.5, 28040 Madrid, Spain. [°]Departamento de Biología Vegetal y Ecología, Universidad de Almería, La Cañada, 04120 Almería, Spain.
[‡]Departamento de Biología Vegetal II, Facultad de Farmacia, Ciudad Universitaria, 28040 Madrid, Spain

Abstract: Investigations conducted in the Iberian Peninsula over the last three decades show how some thermophilic plant species have extended their range. The native thermophilic species, *Dittrichia viscosa* (L.) W. Greuter, and *Sonchus tenerrimus* L., have expanded, apparently without any direct intervention by man. In addition, alien species have crossed biogeographical barriers to expand their range into new areas. These include taxa from the neotropics and the Cape Province of South Africa which have colonised areas close to the Mediterranean coast (*Ageratina adenophora* (Spreng.) King. & H. Rob., *Araujia sericifera* Brot., *Tropaeolum majus* L. and *Arctotheca calendula* (L.) Levyns). By analysing thermometric data series from 10 meteorological stations, we were able to correlate the expansion of the ranges of these species with mean temperature increases, particularly those of mean minimum temperatures. If other thermophilic species (native and introduced) also increase their range, this could have serious consequences for plant biodiversity.

INTRODUCTION

Today, the general consensus is that present climatic change is mainly the result of the accumulation in the atmosphere of various greenhouse gases (carbon dioxide, methane, carbon tetrachloride, nitrous oxide, methyl bromide, methane and halocarbons) all derived from human activity. In particular, the concentration of CO_2 in the atmosphere has increased from 290 ppm$_v$ at the start of the industrial era to 315 ppm$_v$ in 1958, and is currently estimated at some 360 ppm$_v$. It has been estimated that during the 20th century most of Europe has experienced a mean annual temperature increase of about 0,6 °C. However, warming does not occur equally all over the globe due to the many factors that affect the distribution of energy. Thus, the effects of global warming on ecosystems can vary greatly depending on the geographical area. According to Lines (1998), warming is expected to be greater at the poles than in equatorial areas.

The main direct effects on vegetation of an increased CO_2 concentration in the atmosphere are likely to be enhanced photosynthetic activity and a diminished respiration rate. These effects would be more marked in plants with C_3 metabolism than in C_4 plants. In general, we can expect the size and number of most plant organs to increase unless limited by other environmental factors such as water, nutrients, temperature and solar radiation. However, these factors and the CO_2 concentration interact in a complex way, and in order to predict growth and productivity the interactions between these factors need to be understood in depth (Acok 1990).

Living communities are sensitive to the phenomena defining climate, and particularly to temperature (Anguita 1988). The most obvious aspect of climatic change is the increase in temperature; it is less easy to determine effects due to changes in the amount and distribution of clouds and rainfall, though these may also be important. Changes in ecological communities are likely to be especially spectacular in certain biocenoses such as coral reefs (Buddmeier & Gatuso 2000). In general, animals, being capable of movement, can respond quickly to environmental changes and, in a somewhat limited manner, can avoid or take advantage of the new conditions. Several cases of animal species rapidly colonising new habitats have been described, among which those of birds and fish are most outstanding, due to their ability to move and also because of the lack of clear migratory barriers. In Spain, various phenomena have been attributed to climatic change. These include the failure of many storks (*Ciconia ciconia*) to migrate in winter (7.594 non-migrant storks were recorded in the national winter census of 1995; Gomez-Manzaneque 1997), the nesting of traditionally African species such as the *Elanus caeruleus* (Ferrero 1997), the sporadic presence of African birds such as the maribu (*Leptostilos crumeniferus*) in the wetlands of the Guadalquivir river (Martín-Rodríguez 2001), and the presence of the African *Mycteria ibis* in Alicante (Soler & Traver 2001). Similarly, tropical fish species such as *Capros aper* have been caught in the Mediterranean (Lloris 1999), and changes have also been detected in the distribution of marine fish species in the northeast Atlantic (Cushing 1982) and off the African coast (Lloris 1986). Global warming may even have its repercussions on human health due, for instance through its effects on mosquitoes that transmit diseases such as malaria, dengue, yellow fever or different types of encephalitis (Epstein 2000)

Despite the wealth of animal examples, it is the higher plants (phanerogams) which suffer the full consequences of climate, given that they are normally anchored to the ground. Indeed, this makes them ideal indicators of climate, since at times their distribution may be related to a set of complex physical factors yet on other occasions may be explained by a single limiting factor such as temperature. The latter appears to be the case for *Ilex aquifolium* L., for which the northern limit in Europe corresponds to the 0 °C isotherm (Strasburger et al. 1986). It is probable that global warming will influence the vegetation of several areas of the Earth, but that changes will usually occur in a gradual manner. However, the increase or reduction of certain temperature-sensitive species as a consequence of the expansion-retraction of their distribution areas could occur much more rapidly. These sensitive species provide a useful means of systematically monitoring climatic changes over large areas in contrast to the more scattered information provided by meteorological stations.

Over the last thirty years, it has been possible to detect the advance of the ranges of several thermophilic species in the Iberian Peninsula, including native and tropical and subtropical species. The first results of this ongoing work have led us to propose the following hypotheses for the present study:

1) Global warming has led to the expansion of the distribution areas of certain native thermophilic species (*Dittrichia viscosa* and *Sonchus tenerrimus*) during the last 30 years.

2) The increased distribution areas of *Dittrichia viscosa* (short-day species) and *Sonchus tenerrimus* (species indifferent to the photoperiod) are mainly related to increased temperatures in the winter months which allow them to extend their flowering period in areas that were previously too cold.

3) The introduction and subsequent naturalisation of several alien species of neotropical origin (*Ageratina adenophora, Araujia sericifera, Tropaeolum majus, Arctotheca calendula*) in coastal Mediterranean areas has been favoured by the rise in temperatures.

METHODS AND MATERIALS

During the past 30 years, we have systematically investigated the distribution of certain plant species in the northwest quadrant of the Iberian Peninsula and the Mediterranean coast. Special efforts have been made to identify invasive alien species, which are increasingly abundant in both natural and anthropogenic ecosystems. Based on these observations, we selected two groups of species to illustrate this expansion phenomenon: (i) native thermophilic species (*Dittrichia viscosa* (L.) W. Greuter, *Sonchus tenerrimus* L.) which have expanded from the Mediterranean coast to include colder areas with no apparent direct intervention by man, and (ii) tropical and subtropical species that have been able to cross biogeographical barriers and colonise areas close to the Mediterranean coast (*Ageratina adenophora* (Spreng.) King. & H. Rob., *Araujia sericifera* Brot., *Tropaeolum majus* L. and A*rctotheca calendula* (L.) Levyns). The advance of native species was investigated by surveying two transects: Tarragona (0 m above sea level) – Lérida – Zaragoza – Logroño (Ebro valley), and Zaragoza – Calatayud – Guadalajara – Alcalá de Henares – Madrid – Majadahonda (730 m above sea level). The latter transect intersects with the former running from the Mediterranean coast to the central Iberian Peninsula. For alien species, direct observations were made along the Mediterranean coast from the warm southern areas of Málaga and Almería, to the more northerly locations of Tarragona and Barcelona. Plant material collected in field trips was placed in the herbarium of the Real Jardín Botánico de Madrid (MA). Data on locality and phenological state were recorded for each collection.

In addition to direct field observations, data were also derived from the herbaria of MA and the Faculty of Pharmacy of the Universidad Complutense, Madrid (MAF), as well as from the literature. The taxonomical terminology used was derived from Flora europaea (Tutin et al. 1964-1980), Flora dels Països Catalans (Bolòs & Vigo 1995) and various floras for the countries of origin of alien species.

The study was designed to establish correlations that might serve to confirm the proposed hypotheses. This involved contrasting areas of expansion with lines referring to the number of days below freezing point (Font-Tullot 1983) for the two time periods, prior to and after 1970. The presence of native, thermophilic species in traditionally colder areas was considered a sign of expansion. In the case of *Dittrichia viscosa,* a figure is provided indicating the colonisation of localities of high altitude also according to the two time-frames selected, since this effect was particularly clear in this species.

The expansion of thermophilic plants in the Iberian Peninsula 167

Thermometric information was obtained from data series (means of minimum, maximum and mean temperatures) provided by the Instituto Nacional de Meteorología and derived from 9 meteorological stations. These measurement points were carefully selected to represent the characteristic climatic features of the study locations (Figure 1). Only stations with sufficiently long temperature records were considered. Two of the stations are located on the coast [Reus (R) and Cambrils (Cam)] and a further two relatively close to the coast [Sevilla (S) and Nerja (N)] but within the isoline of under 10 days of freezing per year (Font-Tullot 1983). The six remaining stations are located on the two transects used to estimate the expansion of selected alien species. Temperature data for each locality were subjected to linear regression and in the case of Seville, a best-fit polynomial regression was also applied.

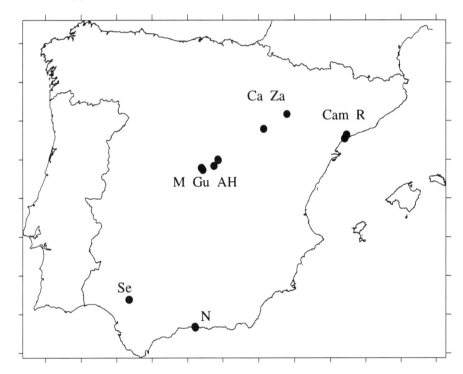

Figure 1. Meteorological stations used to supply temperature data: Sevilla (Se); Cambrils (Cam); Reus (R); Guadalajara (Gu); Madrid. CU (M); Zaragoza (Za); Nerja (N); Alcala de H. (AH); Calatayud (Ca).

RESULTS AND DISCUSSION

Thermometric datasets

Figures 2 to 4 show the linear regression analysis of maximum, minimum and mean temperature datasets for the thermometric data series obtained from the meteorological station of Seville (used as the generic model). In all three cases, there is a tendency towards increasing temperatures in recent years, although mean minimum temperatures show the steepest slope and contribute most to the increase. When the linear regression analysis of mean minimum temperatures was compared to best-fit polynomial regression, a clear increase from 1970 was observed (Figure 5).

A similar rising tendency was also shown by data from the 9 remaining meteorological stations, with greatest increases recorded for mean minimum temperatures. We therefore predicted that this factor would exert the greatest influence on the increased distribution areas of the species under study. Nevertheless, temperature increases were not identical for each location (Table 1). The greatest increases in temperature have occurred in Seville and Nerja (Málaga). Both these localities are in the southern part of the peninsula and show characteristically high temperatures; Seville presents a historical mean annual temperature (tm) of 18,0 °C and Málaga one of 18,2 °C (Elias & Ruiz-Beltran 1973). The differences are consistent with the trends reported by IPCC (1992), which indicate that the warming of the last decades is primarily attributable to an increase in minimum night temperatures, with only a small contribution made by maximum or daily temperatures.

Table 1. Temperature changes (Δ °C) from 1970, and corresponding regression equations.

Meteor. station	Fig.	Variable	Regression Equation	Data series	Δ°C from 1970
Sevilla (Se)	2	Ø Max.	Y = 0,0777x - 3,0285	1923-2000	0,303
Sevilla	3	Ø Min.	Y = 0,2664x - 10,391	1923-2000	1,041
Sevilla	4	ØMean	Y = 0,1727x - 6,7334	1923-2000	0,673
Sevilla	5	ØMean	Y = 0.0033x2 - 0.0854x - 3.3352	1923-2000	0,341
Cambrils (Cam)	7	Ø Min.	Y = 0,0076x - 0,2621	1932-2000	0,262
Reus (R)	8	Ø Min.	Y = 0,0295x - 0.8996	1940-2000	0,899
Guadalajara(Gu)	11	Ø Min.	Y = 0.0137x - 0.6147	1911-2000	0,615
Madrid CU (M)	13	Ø Min.	Y = 0,0302x - 0,9214	1940-2000	0,921
Zaragoza (Za)	9	Ø Min..	Y = 0,0195x - 0.5938	1940-2000	0,594
Nerja C (N)	16	Ø Min.	Y = 0.0172x - 0.5259	1940-2000	0,525
Canal. (AH)	12	Ø Min.	Y = 0,0224x - 0,6846	1940-2000	0,685
Calatayud (Ca)	10	Ø Min.	Y = 0.0138x - 0.42	1940-2000	0,420

The expansion of thermophilic plants in the Iberian Peninsula

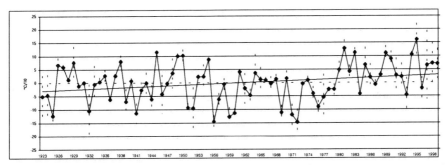

Figure 2. Climatic data series with mean annual maximum temperature anomalies, standard deviations, and regression line, Sevilla BAT (Se).

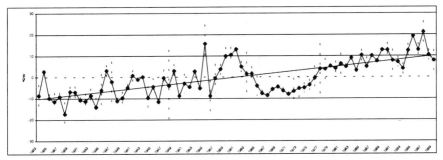

Figure 3. Climatic data series with mean annual minimum temperature anomalies, standard deviations and regression line, Sevilla (Se).

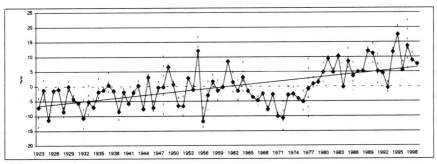

Figure 4. Climatic data series with mean annual mean temperature anomalies, standard deviations and regression line, Sevilla (Se).

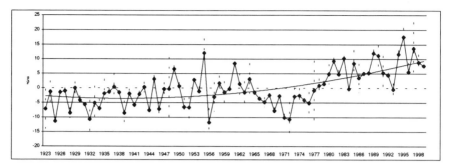

Figure 5. Climatic data series with mean annual mean temperature anomalies, standard deviations and polinomial regression, Sevilla (Se).

Native thermophilic species of expanding distribution area

Sonchus tenerrimus L. (Compositae)

This native, thermophilic herbaceous species occurs as a therophyte or chamaephyte, and produces brown achenes, 2,5-3,5 mm in length (Bolòs & Vigo 1995). It flowers in January to December and is considered to be indifferent to the length of the photoperiod. However, Gallego (1987) reports it to flower from April to May and occasionally in November in western Andalucía. Our observations over the last 2 years indicate flowering in the months of November and December in central and northeast areas of Spain, where this species was previously absent. New localities includes Majadahonda (Madrid), Ciudad Universitaria (Madrid), Alcalá de Henares (Madrid), La Almunia de Doña Godina (Zaragoza), Zaragoza, Tamarite de Litera (Huesca) and Lérida.

According to Bolòs & Vigo (1995), Gallego (1987), Samo (1994) and Stübing & Peris (1998) it inhabits wastelands, rocky areas, stone walls, nitrogenated, stony soils, rock and wall fissures and disturbed ground along roadsides. According to the different herbarium data and our own collections, until around 1970 this plant followed a thermophilic distribution pattern along the Mediterranean coast as shown in Figure 6, approximately corresponding to an imaginary line marking the zone of less than 10 days a year under 0 °C (Font-Tullot 1983). The presence of these species in the interior Iberian Peninsula indicated by Bolòs & Vigo (1995) would appear to refer to data later than 1970, with the exception of the reference (C. Vicioso 1906, MA-139881) for Calatayud. In our survey, we found that the species had spread mainly along roadside habitats. It now has an almost continuous distribution along the N-II motorway from Zaragoza to Calatayud, but it also colonises ruderal environments and appears as a weed in gardens and crops. The presence of pappus efficiently aids its dispersal by wind. Gallego (1987) describes its introduction into California, southern Australia, New Zealand, Mexico and South Africa. Figure 6 shows the expansion of its distribution from 1970 to the present day. We attribute its increased range to the rise in minimum temperatures, which permitted its expansion during winter, autumn and early spring months.

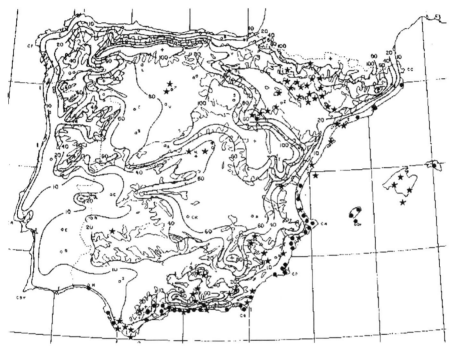

Figure 6. Distribution of *Sonchus tenerrimus* in the Iberian Peninsula (● before 1970 / ★ after 1970), on a map with lines referring to the number of days below freezing point (Font-Tullot 1983).

Worthy of note are various new occurrences of *Sonchus tenerrimus* in recent years. These include its appearance as a rare ruderal in the Castillo Templario of Lérida in 1997, where it flowered in the month of December, in Tamarite de Litera (Huesca) in 1999, in Haro (La Rioja) in 1998, as a garden weed in Zaragoza in 1990, along the edges of the N-II motorway in Calatayud (Zaragoza) in 1999, as an arable weed in Alcalá de Henares (Madrid) in 2000, as a ruderal in the Ciudad Universitaria (Madrid) in 1999, and at a height of close to 800 m in Majadahonda (Madrid) in 1999. Most recent references report its presence at high altitudes in the locality of Torla, in the Central Pyrenees in the vicinity of the Parque Nacional de Ordesa, and on Monte Perdido at a 1.000 m. In the autumn and winter 2000, it was frequently found in the cities of Zaragoza and Lérida, and as a roadside ruderal between Zaragoza and Lérida and between Zaragoza and Calatayud. The mountains of the Iberian Range to the SE of Calatayud appear to limit its expansion, but it makes a re-appearance on the plain S of Guadalajara in Alcalá de Henares (Madrid). Given the diverse habitats it colonises, it is clearly a highly adaptable species and has even been found as an epiphyte growing on several garden palms of the coastal town Salou (Tarragona). It is also commonly found growing on *Phoenix canariensis* in Almería.

The lines of expansion of this species are thought to have followed main roads, with three principal axes: 1) Tarragona – Lérida – Zaragoza, 2) Zaragoza – Tudela – Haro, 3) Zaragoza – Calatayud – Alcalá de Henares – Madrid – Majadahonda. It may be that Calatayud acted as a secondary dispersion centre to judge from the report by Vicioso in (C. Vicioso 1906, MA-139881).

The dispersal strategy of *Sonchus tenerrimus* is based on its growth, flowering and fruiting during the autumn and winter months (September-December), for which present conditions are ideal. Fruiting may also occur at the end of winter and in early spring, when there is no water deficit, e.g. under the conditions of the Mediterranean climate and the milder minimum temperatures recorded along the transect: Cambrils (Tarragona) (Figure 7), Reus (Tarragona) (Figure 8), Zaragoza (Figure 9), Calatayud (Zaragoza) (Figure 10), Guadalajara (Figure 11), Alcalá de Henares (Figure 12), Madrid-Ciudad Universitaria (Figure 13).

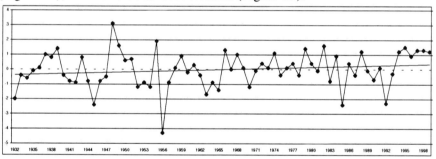

Figure 7. Climatic data series with mean annual minimum temperatures, Cambrils (Cam), and regression line.

The expansion of thermophilic plants in the Iberian Peninsula 173

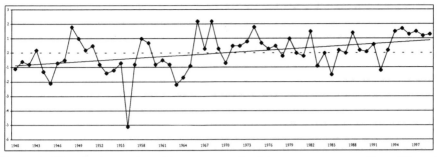

Figure 8. Climatic data series with mean annual minimum temperatures, Reus (R), and regression line.

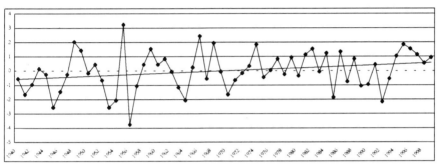

Figure 9. Climatic data series with mean annual minimum temperatures, Zaragoza (Za), and regression line.

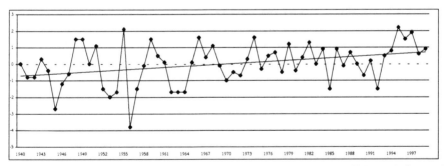

Figure 10. Climatic data series with mean annual minimum temperatures, Calatayud (Ca), and regression line.

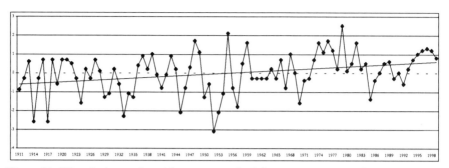

Figure 11. Climatic data series with mean annual minimum temperatures, Guadalajara (Gu), and regression line.

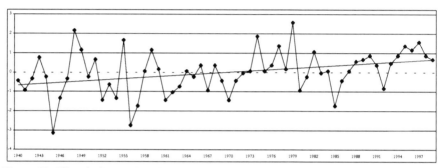

Figure 12. Climatic data series with mean annual minimum temperatures, Canaleja (AH), and regression line.

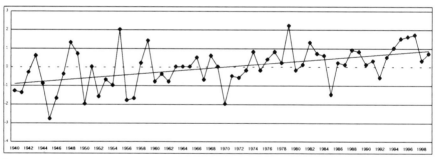

Figure 13. Climatic data series with mean annual minimum temperatures, Madrid CU (M), and regression line.

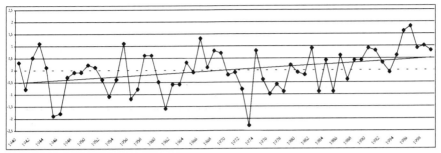

Figure 16. Climatic data series with mean annual minimum temperatures, Nerja C. (N), and regression line.

Dittrichia viscosa (L.) Greuter (Compositae)

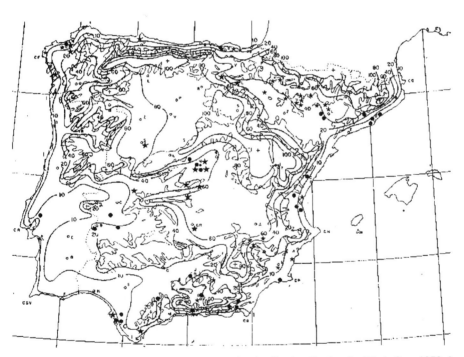

Figure 14. Distribution of *Dittrichia viscosa* in the Iberian Peninsula (● before 1970 / ★ after 1970) on a map with lines referring to the number of days below freezing point (Font-Tullot 1983).

This is a bushy, erect plant with hairy achenes of 2 mm length. *Dittrichia viscosa* flowers exclusively in autumn and winter and is probably a short-day plant (Bolòs & Vigo 1995).

Sanz Elorza (2001) describes its preference for roadside ditches, embankments and disturbed ground in general, being a pioneer in the colonisation of dry, altered terrain. It is mainly found in the regions occupied by Quercetalia ilicis Br.-Bl. ex Molioner 1934 em. Rivas-Martinez 1975 and occasionally reaches those of Quercion pubescenti-sessiliflorae Br.-Bl. 1932 in sheltered enclaves. Its altitude limit is currently 0-1.000 m, but in the past three decades its area has extended towards higher altitudes (Figure 15), perhaps aided by the fact that it flowers in autumn and winter, and that its natural habitat is one of disturbed soil, which to some extent facilitates dispersal. Figure14 shows the advance of this species from 1970 until the present, once again indicating a correlation between range and increasing minimum temperatures.

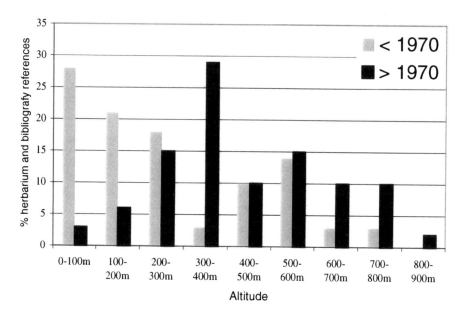

Figure 15. Percentage distribution of *Dittrichia viscosa* citations bibliography by altitude intervals (grey: before 1970; black: after 1970).

Alien tropical and Capensis species undergoing expansion

Ageratina adenophora (Spreng.) King. & H. Rob. (Compositae)

This composite is a native of Mexico and is invasive along the Spanish Mediterranean coast and was recently detected in the eastern Andalusian coast in the regions Nerja (Figure 16) and Frigiliana of the Málaga province (Sanz-Elorza & Sobrino 1999a). More specifically, it has colonised the valleys of the rivers Chillar and Higuerón, where it inhabits cold or shady environments in regions subjected to strong or moderate anthropogenic influence. It also occurs in the Canary Islands (Ceballos & Ortuño 1976), the Azores and Madeira (Erickson et al. 1979), where it is invasive in wet areas, pine woods and degraded sub-tropical forrested regions of medium to low altitude. Because it is so aggressive, it has been considered a threat to the rich endemic flora of the Canary Islands (Gómez Campo 1996) and in particular to the unique plant communities of the Parque Nacional de la Caldera de Taburiente on the island of La Palma (Palomares 1998). In its area of origin, it inhabits pine and oak woods and tropical forests 1000 to 2000 m high (Arber 1984). Its presence in the Iberian Peninsula is very recent and it is currently in phase of expansion, as shown in Figure 17.

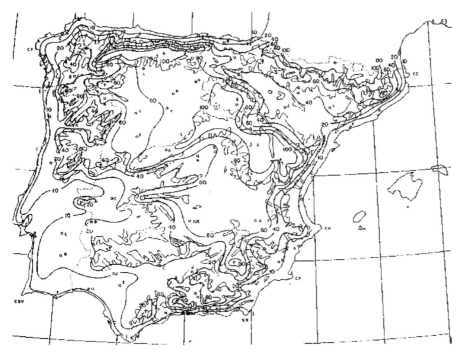

Figure 17. Distribution of *Ageratina adenophora* in the Iberian Peninsula, on a map with lines referring to the number of days below freezing point (Font-Tullot 1983).

Araujia sericifera Brot. (Asclepideaceae)

This is a green-stemmed liana which flowers in summer. It produces a large fruit, 8-12 cm in length, within a green follicle, and with seeds bearing silky appendages.

This naturalised, neotropical plant comes from warm areas of eastern South America (Bailey 1976). According to Casasayas (1984), it was introduced into Europe as an ornamental plant and for the production of fibre from its fruit. It has naturalised in many areas and grows over diverse materials such as metal-netting fences, tree trunks, reeds and in abandoned gardens. We were even able to find it in Cambrils (Tarragona) on *Quercus ilex* subsp. *ballota* in the few remaining fragments of climax vegetation in the Baix Camp (Tarragona). Bolòs & Vigo (1995) describes its advance over suburban lands and in vegetable gardens of coastal Catalan regions. However, it should be noted that prior to 1980 only two localities (Port Bou and the Ebro Delta) had been reported (Malagarriga 1976, Balada & Folch 1977). In the orange plantations of Valencia it is now referred to as a weed that competes with the orange trees. It has also been described in the País Vasco, Castellón, Alicante, Murcia and Granada, but except for the latter, in locations close to the coast. Figure18 provides these locations on the map of Font-Tullot (1983) and indicates that they are all found in coastal areas and are more recent than 1970.

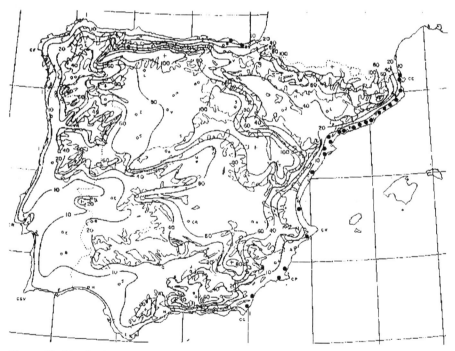

Figure 18. Distribution of *Araujia sericifera* in the Iberian Peninsula, on a map with lines referring to the number of days below freezing point (Font-Tullot 1983).

In view of the ancient use of this species in gardening, it seems probable that its recent spread in both in natural and cultivated areas (as a weed in citrus plantations) is a product of increased minimum temperatures. The increased survival of this neotropical species does not appear to be related to flowering, since *Araujia sericifera* flowers during the summer. The spread of this species seems to have commenced on the Catalan coast and extended towards the southern coast, since it is currently hardly used as a garden species. These geographical jumps may be explained by human transport of diaspores aided by the silks in the seed that allow efficient wind dispersal.

Araujia sericifera has even been reported to be cultivated in a few gardens of the central Iberian Peninsula at 700 m above sea level in Aravaca (Madrid) in 1998 (MAF-154931). This may also be the consequence of the increase in minimum temperatures in winter months involving a virtual lack of temperatures below zero and low intensity frost during the last three years (1998, 1999, 2000) (Meteorological Station of the Unidad de Botánica Agrícola de la Escuela Técnica Superior Ingenieros Agrónomos, Madrid) (data not shown).

Tropaeolum majus L. (Tropaeolaceae)

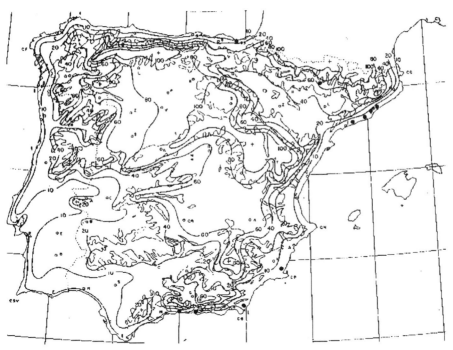

Figure 19. Distribution of *Tropaeolum majus* in the Iberian Peninsula, on a map with lines referring to the number of days below freezing point (Font-Tullot 1983).

This is an annual or evergreen herbaceous plant of South American origin found from Peru to Colombia. It is thought that Spanish colonisers introduced it into Europe (Chittenden 1986). Since then, it has been cultivated as an ornamental plant for its beautiful flowers. It often escapes from culture and appears close to populated areas in ruderal settings. It has been cited as naturalised in the Catalan regions of Baix Llobregat,

Barcelonès, Maresme, Tarragonès and Valles Occidental (Casasayas 1989), in the Huesca province (Sanz-Elorza 2001), on the east Málaga coast (direct observation), on the Granada coast (MAF-129559), in Almería (Sagredo 1987), in Murcia (Sanchez-Gomez et al. 1998), in Castellón (Samo 1995), and in the Pais Vasco (Aizpuru et al. 2000). On the island of Tenerife, it is highly invasive on roadsides (direct observation). Figure 19 shows the distribution of *Tropaeolum majus* before and after 1970, reflecting its spread over the last three decades. Its naturalisation was apparently related to the increase in minimum temperatures which permits its survival. This figure even includes an inland sighting in Huesca (Sanz-Elorza 2001).

Arctotheca calendula (L.) Levyns (Compositae)

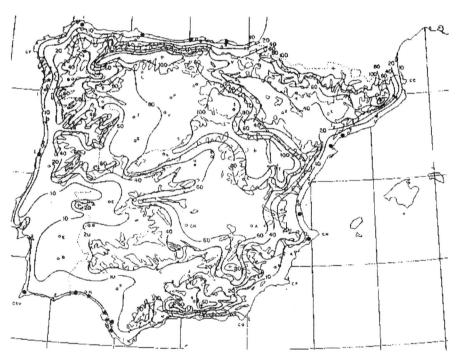

Figure 20. Distribution of *Arctotheca calendula* in the Iberian Peninsula (★ before 1970 / ● after 1970), on a map with lines referring to the number of days below freezing point (Font-Tullot 1983).

This alien South African species from the Cape province of South Africa flowers in spring (Bond & Goldebatt 1984). It is an annual, herbaceous creeping therophyte that grows as rosettes, and is acaulous or has a short decumbent stem. It produces achenes covered with a woolly substance for dispersion (Bolòs & Vigo 1995).

The species was introduced into the Iberian Peninsula as early as quoted in Coutinho (1913) in the south of Portugal. Subsequently, it was found by Gros (1925 in Casasayas 1989) on beaches in Cádiz. It was later identified in the NE Iberian Peninsula, but only as an ephemeral according to the definition of Kornas (1978) and in the vicinity of l´Ametlla de Mar. After this, it was detected in abundance in Cambrils by Sanz-Elorza & Sobrino (1999b), and further south in Valencia by Carretero (1985) and in Alicante (Soler et al. 1995). It has also spread to coastal areas in the north of Spain, in Asturias (Mayor et al. 1975), Lugo and Cantabría (Lainz & Loriente 1981), País Vasco (Lizaur et al. 1983), La Coruña (MAF 150386) and Pontevedra (MAF- 93750). In all these cases, it became naturalised in coastal areas with strong marine influence. This set of localities is shown in Figure 20, where it may be seen that its distribution area was initially limited to the warmest areas of the SE Iberian Peninsula. Since 1970, however, it has spread not only to the Mediterranean and southern Atlantic coasts, but also in coastal zones further north. The species is widely planted in gardens such and has probably escaped on several occasions, an idea supported by the distance between the different localities.

CONCLUSIONS

The data presented here allow us to confirm the hypotheses proposed in the introduction, and even to detail the significant effects of rising mean temperatures in terms of increased distribution areas of the plants examined. The following conclusions highlight the serious nature of this tendency:

1) A correlation was shown between temperature increases detected in thermometric data series corresponding to the last 30 years and the expansion of native, thermophilic species (*Sonchus tenerrimus*, *Dittrichia viscosa*) towards colder inland areas where they were previously absent.

2) The increase of the mean minimum temperatures is closely related with the decrease of the continentality following the Rivas-Martínez´s Continentality Index: $Ic = T_{max} - T_{min}$ (Rivas-Martínez et al. 1999).

3) Mean minimum temperatures have shown the greatest increase and are consequently thought to be most important for the expansion of the two native species. The fact that both are able to flower in autumn and winter (*Sonchus tenerrimus* is indifferent to the photoperiod and *Dittrichia viscosa* is a short-day plant) has facilitated the expansion.

4) The alien species considered have extended their range in coastal regions considerably over the last three decades. This probably reflects the

increase in mean minimum temperatures which allows these species to survive under conditions that were previously limiting.

5) The present study describes the increased range of only two thermophilic species with special characteristics within the Mediterranean area and does not, therefore, point to a general trend of thermophilic coastal species advancing towards inland areas. However, if invasibility depends on being able to surpass threshold temperatures characteristic for each species, and if the Earth's temperature continues to rise, the Mediterranean coasts may act as a source for plant species which invade inland areas. This could have major consequences for plant biodiversity, since new competitive interactions would arise which could certainly negatively affect both the number of endemic Iberian species and the scarce natural ecosystems that still exist.

6) Tropical alien species introduced into coastal areas might also surpass limiting thermal thresholds if temperatures continue to rise. This is also likely to affect plant species composition, both in coastal and inland areas of the Iberian Peninsula, and lead to increased prominence of invasive species.

REFERENCES

Acok B., 1992, Effects of carbon dioxide on photosynthesis, plant growth, and other processes. In: B.A. Kimball, N. Rosemberg & L. Hartwell (eds.) *Impact of carbon dioxide, trace gases, and climate change on global agriculture*. ASA, CSSA, SSA. Madison, Wi.

Aizpuru I., Aseguinolaza C., Uribe Echevarría P.M., Urrutia P. & Zorraquín I., 2000, *Claves ilustradas de la flora del País vasco y territorios limítrofes*. Gobierno Vasco, Vitoria.

Anguita F., 1988, *Origen e historia de la Tierra*. Ed. Rueda, Alcorcón (Madrid).

Arber A., 1984, *Flora Novo Galiciana*. A descriptive account of the vascular plants of Western Mexico. Vol. 12 Compositae. The University of Michigan Press. 1157pp.

Aseguinolaza C., Aizpuru I., Catalan P., Gómez D. & Uribe-Echevarría P.M., 1984, *Catalogo florístico de Alava, Vizcaya, y Guipúzcoa*. Gobierno Vasco. Vitoria.

Bailey L.H., 1976, *Hortus Third: A concise dictionary of plants cultivated in the United States and Canada*. Macmillan, New York.

Balada R & Folch R., 1977, Cataleg floristic del delta de l'Ebre. *Treb. Inst. Catalana Hist. Nat.* 8: 69-101.

Bolòs de O. & Vigo J., 1995, *Flora dels Països Catalans*. Vol. 3. Ed. Barcino, Barcelona.

Braun-Blanquet J., 1932, Zur Kenntnis nordschweizerischer Waldgesellschaften. *Beith. Bot. Centralb. (Dresden)* 49B:7-42.

Buddemeier R.W & Gatuso J.P., 2000, Degradación de los arrecifes coralinos. *Mundo Científico* 217: 44-48.

Carretero J.L., 1985, Aportaciones a la flora exótica valenciana. *Collect. Bot.* 16(1): 133-136.

Casasayas T., 1989, *La flora alóctona de Catalunya*. Tesis Doctoral. Universidad de Barcelona, Barcelona.

Ceballos L. & Ortuño F., 1976, *Vegetación y flora forestal de las Canarias occidentales*. Instituto Nacional para la Conservación de la Naturaleza. Madrid.

Chittenden F.J., 1986, *Dictionary of Gardening* (2nd Synge, P.M. ed.). the Royal Horticultural Society. Clarendon Press. Oxford.

Coutihno, 1913, *A Flora de Portugal*. París, Lisboa, Río de Janeiro.

Cushing D.H., 1982, *Climate and Fisheries*. Academic Press, London.

Epstein P.R., 2000, Salud y calentamiento global de la atmósfera y océano. *Investigación y Ciencia* 289: 16-24.

Erickson O., Hansen A. & Sunding P., 1979, *Flora of Macaronesia*. Checklist of vascular plants. Tenerife.

Elias F. & Ruiz-Beltran L., 1973, *Clasificación agroclimática de España*. Servicio Metereológico Nacional, Madrid.

Ferrero J.J., 1997, El elanio azul, la rapaz que llego de Africa. *Biológica (Madrid)* 15: 36-42.

Font Tullot I., 1983, *Climatología de España y Portugal*. Instituto Nacional de Meteorología, Madrid.

Gallego M.J., 1987, *Sonchus* L. In: B. Valdes, S. Talavera & E. Fernandez-Galiano (eds.) *Flora Vascular de Andalucia Occidental* 3: 85-88. Ketres Ed., Barcelona.

Gómez Campo C., 1996, *Libro rojo de especies vegetales amenazadas de las Islas Canarias*. Gobierno de Canarias, Tenerife.

Gómez de Barreda D., 1979, *Araujia sericifera* Brot., mala hierba en los agrios españoles. *Levante Agrícola* 205: 13-15.

Gómez Manzaneque A., 1997, Cigüeña blanca, cada vez más cerca del hombre. *Biológica (Madrid)* 8: 28-39.

IPCC, 1992, *Climate Change*. WMO/UNEP, Geneve. Cambridge University Press, Cambridge.

Kornas J., 1978, Remarks on the analysis of a synanthropic plant flora. *Acta Bot. Slov. Acad. Sci. Slov. Ser. A* 3: 385-393.

Laínz M. & Loriente E., 1981, Contribuciones al conocimiento de la flora montañesa. *Anales Jard. Bot. Madrid* 38(2): 469-475.

Lines A., 1998, Síntesis acerca del efecto invernadero y sus efectos. In: Instituto de Ingeniería de España (ed.) *Energía y cambio climático*. Ministerio de Medio Ambiente, Madrid.

Lizaur X., Escobar L. & Izaguirre B., 1983, Contribución al conocimiento de la flora vascular guipuzcuana. *Munibe-Ciencias* 35(1-2): 35-44.

Lloris D., 1986, Ictiofauna demersal y aspectos biogeograficos de la costa sudoccidental de Africa. *Monograf. Biol. Mar.* 1: 9-432.

Lloris D., 1999, Cambio climático. ¿actividad humana o natural? *Mundo Científico* 197: 61-65.

Malagarriga H.T., 1976, Catalogo de las plantas superiores del l´Alt Empordà. *Acta Phytotax. Barcinon.* 18.

Martín Rodríguez J., 2001, Aves africanas en la España meridional. *Quercus* 180: 40.

Mayor M., Díaz T.E. & Navarro F., 1975, Adiciones al catálogo florístico del cabo de Peñas (Asturias). *Rev. Fac. Cienc. Univ. Oviedo* 15-16: 137-138.

Molinier R., 1934, Études phytosociologiques et écologiques en provence occidentale. *Ann. Mus. Hist. Nat. Marseille* 27(1): 1-273.

Palomares S., 1988, *Führer für den Besuch des Nationalparks der Caldera de Taburiente*. Ministerio de Medio Ambiente. Madrid.

Rivas-Martínez S., 1975, Observaciones sobre la sintaxonomía de los bosques acidófilos europeos. Datos sobre la Quercetalia robori-petraeae en la Península Ibérica, *Coll. Phytosociol.* 3: 255-260.

Rivas-Martínez S., Sánchez-Mata D. & Costa M., 1999, North American Boreal and Western Temperate Forest Vegetation. *Itinera Geobot.* 12: 5-136.

Samo A.J., 1994, *Catalogo florístico de la Provincia de Castellón*. Servei de Publicacions, Diputacio de Castelló.

Sanz Elorza M., 2001, *Flora y vegetación arvense y ruderal de la provincia de Huesca*. Tesis doctoral. Escola Tecnica Superior d'Enginyeria Agraria. Universitat de Lleida.

Sanz-Elorza M. & Sobrino E., 1999a, *Ageratina adenophora* (Spreng.) King. & H. Rob. (*Compositae*), alóctona nueva para la flora ibérica y europea. *Anales Jard. Bot. Madrid* 57(2): 424-425.

Sanz-Elorza M. & Sobrino E., 1999b, *Diferencias en la capacidad de acogida de elementos florísticos alóctonos entre las zonas costeras y del interior en el Mediterráneo occidental*. Actas Congreso Sociedad Española de Malherbología, Logroño, 83-88.

Soler R. & Traver M.C., 2001, Un tantalo africano en Alicante: *Quercus* 180: 40.

Strasburger E., Noll F., Schenck H., Shimper A.F.W., von Denffer D., Bresinky A., Ehrendorfer F & Ziegler H., 1986, *Tratado de Botánica*. Ed. Marín, Barcelona.

Stübing G. & Peris J.B., 1998, *Plantas silvestres de la Comunidad valenciana*. Ed. Jaguar, Madrid.

Tutin T.G., Heywood V.H., Burges N.A., Valentine D.H., Walters S.M., Webb D.A.& Moore D.M., 1964-1980, *Flora europaea*. 5 Vols, Cambridge.

Climate change and coastal flora
Does climatic change affect the floristic composition of salt marshes and coastal dunes at the German coasts?

DETLEV METZING & ALBRECHT GERLACH
AG Pflanzenökologie, FB 7, Universität Oldenburg, D-26111 Oldenburg, Germany

Abstract: The azonal vegetation of coastal dunes and salt marshes is composed of taxa adapted to the special abiotic and edaphic conditions of the coastal habitats. The stability of dune and marsh ecosystems depends on the presence and abundance of particular plant species. Changing environmental conditions as the predicted climate change following the greenhouse effect may dislocate the area boundaries of several plant taxa. To evaluate the climatic sensibility of particular plant taxa, climatic envelopes were estimated for vascular plant species occurring at the German coast of the Northern and Baltic Seas. Correlation of distribution patterns and climatic data allows a prediction of area dislocations caused by climatic change for particular taxa. First signs of climate change impact are discussed.

INTRODUCTION

Climate is a main factor limiting the distribution areas of plant species at a broad continental scale (Woodward 1987, Jäger 1992, Dahl 1998). Consequently climate change following the greenhouse effect will dislocate area boundaries of plant taxa (Savidge 1970, Jäger 1995). Shifts of area ranges will not only affect the biodiversity of particular areas, but may also affect the function of ecosystems and reduced availability of natural resources (Peters 1992, Kapelle et al. 1999).

Coastal regions, which are characterized by high physical and biological diversity, are particularly vulnerable to global change effects (Holligan & Reiners 1992, Sterr 1998). This paper aims at predicting in selected examples such effects on the distribution areas of plant species for the coastal region of Germany.

Importance of coastal flora

The main biotopes of the German coast areas are salt marshes, dune complexes, shingle, strandlines, and sea-cliffs. The azonal vegetation of dunes and salt marshes is composed of plant taxa adapted to the special abiotic and edaphic conditions of the coastal habitats. The presence of these plants in this area is of high importance for several reasons:

Coastal development

The geomorphology of the coast is influenced by plants to a large extent. The development of barrier islands along the North Sea coast would not have been possible without plant species fixing the wind-blown sand. *Elymus farctus* affects the development of small foredunes, whereas mainly *Ammophila arenaria* initiates the formation of higher dunes ("biogenic dune formation", Gerlach 1993). Further species take part in stabilising the dune system with the sequence of white, grey, and brown dunes (Ellenberg 1996). In salt marshes the plants can support the sedimentation of fine material during tidal inundation. The resulting accretion may compensate even future rises of the sea level – at least partly (Dijkema 1992). Dense vegetation cover may prevent soils from erosion caused by tidal or stormy events. Thus plants play an important role in coastal defence.

Biodiversity

Plants are a significant element in the biodiversity reservoir of the coastal region. Several plant taxa, mainly highly specialised in the unique conditions of the salt marshes or dunes (e. g. salinity, moving sand, wind stress) are restricted to the coastal area in Germany. Some of them are threatened in the hinterland due to intensive agriculture, but are still quite common in coastal habitats. The function and high productivity of the salt marshes is essential for coastal ecosystems, for example as feeding ground or for breeding purposes of many bird species. Today the unique diversity and the ecological value of coastal areas are protected by law in nature conservation reserves or national parks to a large extent.

Climate change and coastal flora

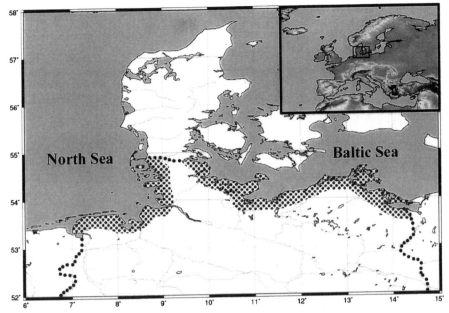

Figure 1. Study area: German coastal regions of the North Sea and the Baltic Sea.

Use

The establishment of nature conservation reserves and the appearance of landscape formed by plants are of high value for tourism. Due to the high productivity of the marshes these were used as pastures, mainly by cattle and sheep, or for hay-making. In the last decades, however, on many sites grazing has been reduced for economic reasons or increasing nature conservation areas (Dijkema 1984).

The importance of plants for biodiversity, development and usage of the coastal region leads to the question of whether possible future changes may have an effect on the presence of plant taxa in these areas. Undoubtedly such a change is the global warming which will not only provoke increasing temperature and changed precipitation in coastal areas, but also the dynamics of inundation, sedimentation and erosion will change (Sterr 1993, 1998). However, surface temperature and precipitation are the main factors influencing the distribution on a larger scale and therefore climate change will change the distribution range of plant species. This paper deals with the question of whether climate change will effect shifts of species ranges of plant taxa occurring at the German Coast.

METHODOLOGY

225 vascular plant taxa occurring in coastal habitats of salt marshes and coastal dunes in Germany were chosen to estimate their climatic sensitivity (Metzing, in prep.). In a next step the taxa with a range boundary in the German and adjacent coastal areas were selected for further study (Table 1), because it can be assumed, that a dislocation of their distribution limit may affect the occurrence in the study area. Data about the distribution patterns for the European area were taken from the literature (Meusel et al. 1965-1995, Jalas et al. 1972-1994, Hulten & Fries 1986).

Table 1. Plant species occurring at the German coast area with distribution limits in the study area. Taxa of the upper group have northern/eastern distribution limits at the German coast and adjacent regions, whereas taxa of the lower group have southern/western distribution boundaries there. Nomenclature follows Wisskirchen & Haeupler (1998).

Alopecurus bulbosus	*Ilex aquifolium*
Anthoxanthum aristatum	*Juncus anceps*
Atriplex portulacoides	*Juncus maritimus*
Beta vulgaris (subsp. *maritima*)	*Juncus pygmaeus*
Brassica oleracea	*Leontodon saxatilis*
Bromus thominii	*Limonium vulgare*
Calystegia soldanella	*Oenanthe lachenalii*
Carex trinervis	*Oenothera ammophila*
Cerastium diffusum	*Parapholis strigosa*
Cochlearia anglica	*Phleum arenarium*
Cochlearia danica	*Polypodium interjectum*
Crambe maritima	*Sagina subulata*
Cynodon dactylon	*Silene otites*
Dianthus carthusianorum	*Spartina anglica*
Elymus athericus	*Stellaria crassifolia*
Festuca polesica	*Trifolium ornithopodioides*
Genista anglica	*Tuberaria guttata*
Genista germanica	*Ulex europaeus*
Glaucium flavum	
Alopecurus arundinaceus	*Linnaea borealis*
Arnica montana	*Melilotus dentatus*
Atriplex calotheca	*Odontites litoralis*
Atriplex longipes	*Petasites spurius*
Bassia hirsuta	*Potamogeton filiformis*
Blysmus rufus	*Puccinellia capillaris*
Empetrum nigrum	*Rumex longifolius*
Galium sterneri	*Salix hastata*
Gentianella campestris	*Vaccinium uliginosum*
Gentianella uliginosa	*Vaccinium vitis-idea*
Juncus balticus	

The distribution areas of the selected taxa were digitised and transferred to a geographic information system (software: Idrisi for Windows 2.0) to analyse the relation to climatic data. Climate data (temperature, precipitation) were obtained from the IPCC Data Distribution Centre (1999; for dataset construction see New et al. 1999) for the period 1961-1990 with a spatial resolution of 0.5° (an area of about 32 x 55 km in the study area). With the overlay operation it was possible to plot the climatic amplitude covered by species areas. This method allows the climatic envelope of each species to be determined, which is defined by the climatic space that corresponds to the geographic boundaries within a plant taxon is considered to grow and reproduce under natural conditions (Box et al. 1993, 1999).

To predict dislocation of distribution areas, two scenarios are used, which were provided by the IPCC (Houghton et al. 1996) and regionalised for the German coasts by von Storch et al. (1998). According to these scenarios an increase of annual temperature of 2.5 K and a 15 % increase of winter precipitation up to the year 2050 are assumed in the worst case (+ 1.5 K and 7.5 %, resp. best case). Future distribution areas are modelled in GIS based on present area ranges and predicted climate change. We use climatic envelopes for the variables annual, January and July mean temperature. Due to possible inaccuracy inherent in outline distribution (Stott 1981) and eventual errors in transferring distribution areas to GIS with different map projection systems only values inside the 95 % confidence interval are used for each variable.

RESULTS

Climatic envelopes and species' range shifts

Plotting the climate space covered by a species against the climate space of definite geographical regions allows for a first estimation of whether future climatic conditions will fit the climatic envelope of the species. In Figure 3 the climatic space covered by *Blysmus rufus* (Saltmarsh Flat Sedge, Cyperaceae) is shown in regard to annual mean temperature and rainfall. The species is distributed more to the north, so the southern limit is located in the study area just like the margin of the species' climatic envelope. It is obvious, that future climate at the German coast (as predicted in the climate scenario) will result in no overlap with the climatic envelope of *Blysmus rufus* anymore. The increase of precipitation is insignificant in regard to the existing variability in the German coastal area. Moreover, in a canonical ordination it can be shown that temperatures are the best fitting climate

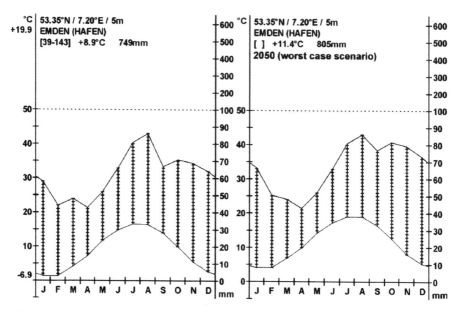

Figure 2. Predicted climatic change for the German coast up to 2050. The model, based on an IPCC szenario, assumes an increase in temperature of 2.5 K in the worst case and 15 % more precipitation between September and February, here visualised by climatic diagrams for Emden, NW Germany (source: Lieth et al. 1999, modified). Left: Present climate, right: predicted climate for 2050. The maximum of rain falls in summer, an arid period is lacking.

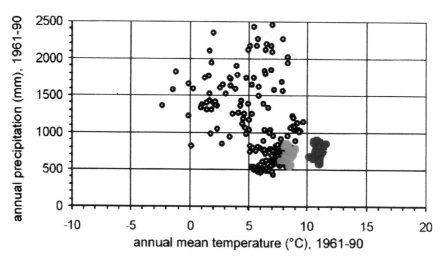

Figure 3. Scatter diagram of the climatic envelope for *Blysmus rufus*. Each point represents a grid rectangle with a geographical extension of 0.5° x 0.5° where the species is present. The present climatic space (time period 1961-1990) of the German coastal area is indicated by light grey spots. The dark grey spots reflect the climate which will prevail at the German coast in 2050 according to the worst case scenario: 2.5 degree increase in annual temperature and 15 % more precipitation in winter.

Climate change and coastal flora

variables for distribution of the considered species in Europe (Metzing, in prep.).

In the study area a temperature gradient from east to west exists in the winter, with lower values in the east, whereas the temperature decreases from south to north in summer (Figure 4). Mean temperatures of the coldest and warmest months provide a good indication of the temperature curve shape over the year. The significance of summer and/or winter temperatures for distribution limits has been outlined by Jeffree & Jeffree (1994). The diagram for *Blysmus rufus* (Figure 5) shows, that a future lacking overlap of climatic envelope of the species and study area is caused by summer temperatures rather than by winter temperatures. It can be assumed, that increasing summer temperatures will provoke a shift of the southern distribution limit to the north, resulting in a loss of this species for the German coastal flora. By comparison, the occurrence of another species, the Sea Rush *Juncus maritimus* (Juncaceae), may benefit from the predicted climate change: the climatic space of the study area will shift towards the climate envelope centre of *Juncus maritimus* (Figure 6). This species is present at the German coast and it can be assumed, that it will not be lost here by changing climate.

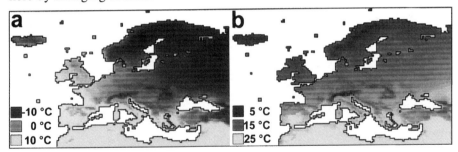

Figure 4. Mean temperature in Europe for a) January and B) July: a temperature gradient from west to east exists in the winter, whereas the gradient is oriented from south to north in the summer.

Using the diagrams illustrating the climatic envelopes it can be estimated whether the climatic conditions in a given area may be sufficient for a given species even when the climate changes. But they tell us only roughly, in which direction dislocations of potential distribution boundaries will happen. A more detailed idea is given by distribution areas modelled with the GIS, based on the predicted climate change. To test the used model, such maps were also prepared for present climatic conditions. "Present" means the time period 1961-1990 of the underlying climatic data. Moreover, it is the time span, in which the used distribution maps were published. The example of *Blysmus rufus* is shown again in Figure 7. Figure 7a shows the present distribution area of the species in Europe. Figure 7b-d displays the modelled

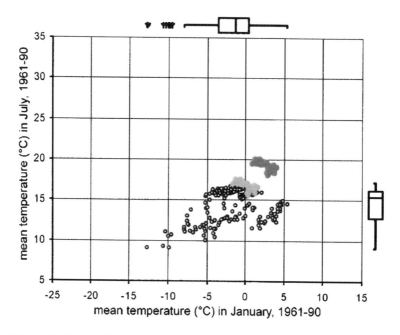

Figure 5. Scatter diagram of the climatic envelope for *Blysmus rufus* (explanation see Figure 3).

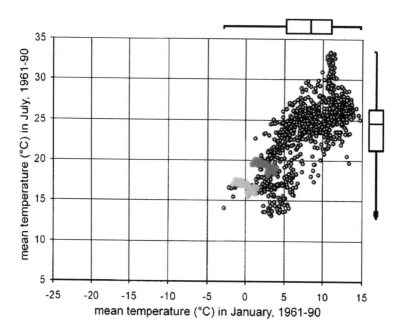

Figure 6. Scatter diagram of the climatic envelope for *Juncus maritimus* (explanation see Figure 3).

species range (climatic envelope shape) for different scenarios (Figure 7b present climate, Figure 7c best case scenario, Figure 7d worst case scenario). The modelled area reflects the present distribution area quite well, although the isolated presence in the western White Sea (NW Russia) is not reproduced by the model. The discrepancy for inland areas in Central Europe, where the species should be present according to the model, can be explained ecologically. *Blysmus rufus* is a species restricted to saline areas, which are present only at the coast and in a few inland salt marshes, however, as the present area is well predicted concerning the coastline, the model is appropriate for the purpose of predicting suitable distribution areas. Even for the best case scenario there will be no suitable climatic conditions at the German coast for this species. The worst case scenario predicts a retreat of *Blysmus rufus* to the coastline of Norway and perhaps N Scotland.

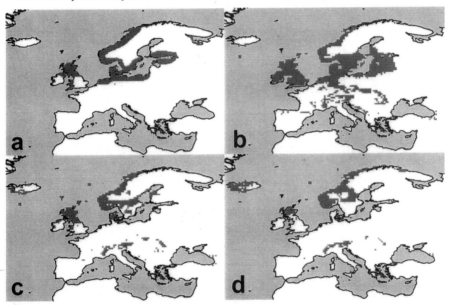

Figure 7. a) Distribution area of *Blysmus rufus* in Europe; modelled species range (climatic envelope shape) for different scenarios: b) present climate, c) best case scenario, d) worst case scenario).

Another example of a species which may get lost for the German flora according to the model is *Odontites litoralis* (Scrophulariaceae). The southern distribution limit of this species, which is restricted to coastal areas, may shift northwards resulting in a disappearance of the species in Germany as well as in the Baltic Sea area and the British Isles (Figure 8). *Odontites litoralis* is ecologically restricted to coastal salt marshes. Therefore the inland areas indicated as suitable areas by the climatic envelope model are no potential areas and can be neglected.

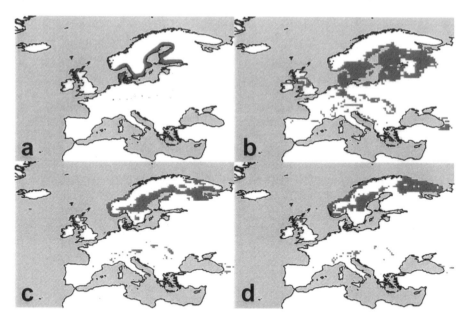

Figure 8. a) Distribution area of *Odontites litoralis* in Europe (after Bolliger 1996); modelled species range (climatic envelope shape) for different scenarios: b) present climate, c) best case scenario, d) worst case scenario).

Climate change will not only effect a loss of biodiversity. Species with a more boreal range may disappear from the study area, whereas species with a more southerly range may spread out into this area. To answer the question of where a potential species pool for such immigration is located, Figure 9 displays these areas where the present climatic conditions are similar to those of the German coast in the future (as predicted by the climate scenarios). Assuming the best case scenario, suitable conditions for species which currently have their northeastern distribution boundary at the coast of Belgium or NE France, will exist at the German coast in 2050. This equals a distance of about 400 km (cf. Figure 9a). It is assumed that dispersal along the coastline is more probable than long-distance dispersal on a continental scale (e. g. from the Black Sea). For the worst case scenario there are no suitable climatic conditions near the coastline in Central Europe, but most suitable areas are found in N France. For southerly distributed species the northern distribution limit can be explained by the minimum heat sum (Woodward 1988). The heat sum is closely correlated to the annual mean temperature as well as to the January and July temperatures. Considering this fact, potential candidates for immigration to the study area of the German coast may originate from the northern coast of France, close to the area already indicated by the model. This requires a range shift of about 600 km along the coast.

Climate change and coastal flora

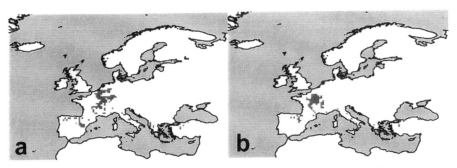

Figure 9. Congruence of present climate conditions with the climate predicted for the German coastal area for two climate change scenarios; a) best case, b) worst case. The darker areas show the area with highest congruence (for three variables, annual mean temperature, mean temperatures in January and July), where the present climate corresponds with that predicted for the German coast in 2050.

A probable expansion of distribution limits to the German coast will be exemplified by two species here. The Rock Samphire, *Crithmum maritimum* (Apiaceae), also a species restricted to coastal areas, is no member of the German flora up to date. The species has a south-western distribution limit at the coast of the Netherlands (Figure 10). The model predicts a north-western expansion of the suitable distribution area, so that an immigration of the species to the German coast may be assumed.

Although the Gorse, *Ulex europaeus* (Fabaceae) is already sub-spontaneously present in dune areas at the German coast, the natural area has no overlap with the German coastal area (Figure 11). The potential shift of the species range in eastern direction may provoke an increase in species abundance at the German coast.

Fingerprints?

It is evident that climate change will affect the location of climatic envelopes. Area dislocations are highly probable – some species will disappear, others may become extinct at the German coast. For about 25 % of the considered taxa a shift of suitable distribution area can be expected up to 2050 for the study area (Metzing, in prep.). Although the rate of surface temperature increase enhanced in the most recent decades, the temperature has ascended appreciably since the end of the 19th century (Houghton et al. 1996, Parry 2000). Consequently it may be assumed, that the process of area dislocation has already started. Are there already "fingerprints" of climate change in the coastal flora?

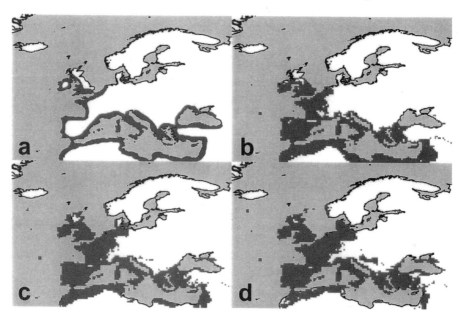

Figure 10. a) Distribution area of *Crithmum maritimum* in Europe; modelled species range (climatic envelope shape) for different scenarios: b) present climate, c) best case scenario, d) worst case scenario).

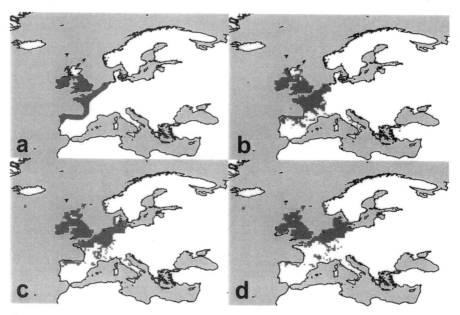

Figure 11. a) Distribution area of *Ulex europaeus* in Europe; modelled species range (climatic envelope shape) for different scenarios: b) present climate, c) best case scenario, d) worst case scenario).

The Mediterranean-Atlantic Yellow Horned Poppy, *Glaucium flavum* (Papaveraceae) has a northeastern distribution limit at the western coast of the Skagerrak (Denmark, S Norway). However, it occurs only sporadically there and is less common than in N France or S England. In the Netherlands the species was reported more frequently in the past decades (Weeda et al. 1985). In North Frisia a population became established recently (at least since 1985), which is the first stable locality in the German Wadden Sea (Borcherding 1999).

The above mentioned *Crithmum maritimum* has a Mediterranean-Atlantic distribution, too. It is common at the shores of N France and S England, most northern findings are known from the Netherlands, where it was reported around the last turn of the century for the first time (Weeda et al. 1987). The species did not belong to the German flora so far (Wisskirchen & Haeupler 1998). However, a first finding of this species has been reported on the island Helgoland in the North Sea for 2000 (Kremer & Wagner 2001). It is early for a definite statement concerning a permanent establishment of *Crithmum* at the German coast. The change of climatic conditions, however, may favour the occurrence of this species in the German coastal area.

Juncus maritimus, which has already been presented above as a potential winner of climate change has been documented in Germany at the Baltic Sea coast as well as on the southern margin of the North Sea since the 19th century, but was lacking at the northwestern coast of Schleswig-Holstein (N Germany) (Figure 12). It was found there near Eiderstedt in 1935 for the first time and also on the North Frisian Islands in the fifties. Raabe (1970) assumed, that the documented expansion was an effect of the habitat dynamics rather than of climate change and that the species may have been overlooked due to the nonpermanent existence of suitable habitats (wet and sandy saline swamps). In the light of our current knowledge, however, its expansion at the northern distribution limit as a possible impact of climatic change cannot be excluded.

Possible area dislocations of ecological key species may be of special importance. Such a species is the heath forming Crowberry, *Empetrum nigrum* (Empetraceae), which occurs on the German islands as a dominant plant at the northern slopes of the older dunes or in moist dune valleys. The most southern populations at the coast are found on the Dutch Wadden Sea islands. The species is Arctic-Alpine distributed and has a preference for cooler regions and a high relative air humidity in the summer (Bell & Tallis 1973). In hot and dry summer periods these heath mats already now decease from the margin, especially at the sites exposed to the sun, indicating that *Empetrum* is near the physiological limit there (Gerlach 1993, Mühl 1993). An increase in temperature and relative dryness in the summer, as predicted for the future, will enhance this process certainly. A regress of the species in the coastal dunes and shift of the southern distribution limit to northern direction may change the today's appearance and ecology of the older dunes considerably.

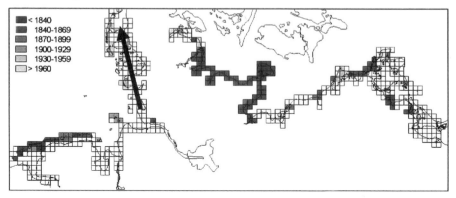

Figure 12. First documented findings of *Juncus maritimus* at the German coast in six time periods. The arrow indicates the northward expansion after 1935.

DISCUSSION

How will plants respond to climate change?

These "fingerprints" of climate change impact on coastal vascular plants may appear less evident in comparison with highly significant responses of other organisms. The indication of species range shifts due to climate change in animals can be explained by their ability to migrate (e. g. Parmesan 1996, Bairlein & Winkel 1998, Parmesan et al. 1999, Ott 2000). The spores of bryophytes with short life cycles can be dispersed by wind over large distances within short time periods. Therefore, they may respond to climatic fluctuations much faster than most vascular plants (Frahm 1997). At the coast of the study area the temperature increase of 2.5 K equals a coastline distance of about 600 km. By way of comparison in mountain regions with a steep temperature gradient the distances which plants have to overpass following suitable climate conditions are much shorter (Grabherr et al. 1994). Hence the plants there will show a faster reaction to climate change than under the conditions of long-distance dispersal. With respect to the time span of 50 years a distance of 600 km seems to be very long regarding the potential migration rates (Huntley 1991). It can be assumed, however, that seed dispersal by sea current, tidal movement, and wind drift along the coastline (at least for the species of the salt marshes, fore- and primary dunes) will happen faster than for species of mainland biotopes, as potential dispersal routes at the coast are not fragmented as they are in cultivated landscapes. Birds travelling along the coast may act as dispersers for certain species, too. At the German coast the landmass between the Northern and the Baltic Sea is a geographical barrier, which may prevent or delay species area expansion eastwards for those species, which are not able to disperse diaspores over long distances (e. g. species with nautohydrochory).

The speed, at which coastal species will react to climate change depends on characteristics of their life history, seed dispersal, population sizes, distribution patterns, genetic variation, availability and fragmentation of suitable habitats as well as on competition, too (Sætersdal & Birks 1997, Woodward 1988). Therefore, even species with similar distribution areas may respond to increasing temperature or changed precipitation individualistic. The models simulate the direction in which species ranges will shift, but do not predict, how fast this will happen. Moreover, climate is not the only distribution-limiting factor, although it is the most important one. Other factors such as aerial nitrogen input, habitat destruction, and competition by neophytes may overlie climatic effects. Future will show, whether the presented "fingerprints" are really indicators of climate change or just caused by natural fluctuations. To detect and monitor species shifts caused by climate change, national or regional floristic mapping programs should be continued and additional monitoring programs are required (Stohlgren et al. 2000). For the special case of *Empetrum nigrum* in the brown dunes monitoring by remote sensing methods should be appropriate to detect a probable regress of Crowberry areas.

ACKNOWLEDGEMENTS

This study was financially supported by the BMBF (German Ministry of Education and Research), 01 LK 9602/2.

REFERENCES

Bairlein F. & Winkel W., 1998, Vögel und Klimaveränderungen. In: J.L. Lozán, H. Graßl & P. Hupfer (eds.) *Warnsignal Klima*. Wissenschaftliche Auswertungen, Hamburg.

Bell J.N. & Tallis J.H., 1973, Biological flora of the British Isles. *Empetrum nigrum. J. Ecol.* 61: 289-305.

Bolliger M., 1996, Monographie der Gattung *Odontites* (Scrophulariaceae) sowie der verwandten Gattungen *Macrosyringion, Odontitella, Bornmuellerantha* und *Bartsiella*. *Willdenowia* 26: 37-168.

Borcherding R., 1999, Der Gelbe Hornmohn (*Glaucium flavum*). *Wattenmeer Int.* 17: 27.

Box E.O., Crumpacker D.W. & Hardin E.D., 1993, A climatic model for location of plant species in Florida, U.S.A. *J. Biogeogr.* 20: 629-644.

Box E.O., Crumpacker D.W. & Hardin E.D., 1999, Predicted effects of climatic change on distribution of ecologically important native tree and shrub species in Florida. *Clim. Change* 41: 213-248.

Dahl E., 1998, *The Phytogeography of Northern Europe*. University Press, Cambridge.

Dijkema K.S., 1984, Salt Marshes in Europe. *Nature and Environment Series* No. 30. Council of Europe, Strasbourg.

Dijkema K.S., 1992, Sea level rise and management of salt marshes. *Wadden Sea Newsl.* 1992-2: 7-10.

Ellenberg H., 1996, *Vegetation Mitteleuropas mit den Alpen*. Ed. 5. E. Ulmer, Stuttgart.

Frahm J.-P. & Klaus D., 1997, Moose als Indikatoren von Klimafluktuationen in Mitteleuropa. *Erdkunde* 51: 181-190.

Grabherr G., Gottfried M. & Pauli H., 1994, Climate effects on mountain plants. *Nature* 369: 448.

Gerlach A., 1993, Biogeochemistry of nitrogen in a coastal dune succession on Spiekeroog (Germany) and the impact of climate. *Phytocoenologica* 23: 115-127.

Holligan P.M. & Reiners W.A., 1992, Predicting the responses of the coastal zone to global change. *Adv. Ecol. Res.* 22: 211-255.

Houghton J.J., Meiro Filho L.G., Callander B.A., Harris N., Kattenberg A. & Maskell K. (eds.) 1996, *Climate Change 1995 - The Science of Climate Change*. University Press, Cambridge.

Hulten E. & Fries M., 1986, *Atlas of North European Vascular Plants*. 3 Vols. Koeltz, Königstein.

Huntley B., 1991, How plants respond to climate change: migration rates, individualism and the consequences for plant communities. *Ann. Bot.* 67(suppl.): 15-22.

IPCC Data Distribution Centre, 1999, *Providing Climate change and related scenarios for impacts assessments*. CD-ROM, Version 1.0, April 1999. IPCC-DDC, Norwich.

Jäger E.J., 1992, *Kausale Phytochorologie und Arealdynamik*. Habil. thesis, Martin-Luther-Universität, Halle-Wittenberg.

Jäger E.J., 1995, Klimabedingte Arealveränderungen von anthropochoren Pflanzen und Elementen der natürlichen Vegetation. *Angew. Landschaftsökol.* 4: 51-57.

Jalas J. & Suominen J., 1972-1994, *Atlas Florae Europeae*. Vols. 1-10. Committee for Mapping the Flora of Europe & Societas Biologica Fennica Vanamo, Helsinki

Jeffree E.P. & Jeffree C.E., 1994, Temperature and the biogeographical distributions of species. *Funct. Ecol.* 8: 640-650.

Kappelle M., Van Huuren M.M.I. & Bass P., 1999, Effects of climate change on biodiversity: a review and identification of key research issues. *Biodivers. Conserv.* 8: 1383-1397.

Kremer B.P. & Wagner A., 2001, *Crithmum maritimum* L. - Neu für Deutschland. *Florist. Rundbr.* 34: 1-8.

Lieth H., Berlekamp J., Fuest S. & Riediger S., 1999, *Climate Diagram World Atlas*. CD-ROM. Backhuys, Leiden.

Meusel H. et al., 1965-1995, *Vergleichende Chorologie der Zentraleuropäischen Flora*. Vols. 1-3 (Text- und Kartenteil). G. Fischer, Jena.

Mühl M., 1993, Zur Synsystematik der Krähenbeerheiden auf den Ostfriesischen Inseln. *Drosera* '93: 11-32.

New M., Hulme M. & Jones P., 1999: Representing twentieth-century space-time climate variability. Part I: Development of a 1961-90 mean monthly terrestrial climatology. *J. Clim.* 12: 829-856.

Ott J., 2000, Die Ausbreitung mediterraner Libellenarten in Deutschland und Europa - die Folge einer Klimaveränderung? *NNA-Ber.* 2: 13-35.
Parmesan C., 1996, Climate and species' range. *Nature* 382: 765-766.
Parmesan C., Ryrholm N., Stefanescu C., Hill J.K., Thomas C.D., Descimon H., Huntley B., Kaila L., Kullberg J., Tammaru T., Tennent W.J., Thomas J.A. & Warren M., 1999, Poleward shifts in geographical ranges of butterfly species associated with regional warming. *Nature* 389: 579-583.
Parry M. (ed.), 2000, *Assessment of Potential Effects and Adaptations for Climate Change in Europe: Summary and Conclusions.* Jackson Environment Institute, Norwich.
Peters R.L., 1992, Conservation of Biological Diversity in the Face of Climate Change. In: R.L. Peters & T.E. Lovejoy (eds.) *Global Warming and Biological Diversity.* Yale University Press, New Haven & London.
Raabe E.-W., 1970, Die Wanderung von *Juncus maritimus* an der jütischen Westküste. *Kieler Notizen* s. vol.: 9-11.
Sætersdal M. & Birks H.J.B., 1997, A comparative ecological study of Norwegian mountain plants in relation to possible future climatic change. *J. Biogeography* 24: 127-152.
Savidge J.P., 1970, Changes in plant distribution following changes in local climate. In: F. Perring (ed.) *The Flora of a Changing Britain.* E.W. Classey, Hampton.
Sterr H., 1993, Der Einfluß von Klimavarianz auf die rezente Morphodynamik der deutschen Ostseeküste. In: H.-J. Schellnhuber & H. Sterr (eds.) *Klimaänderung und Küste.* Springer, Berlin & Heidelberg.
Sterr H., 1998, Gefährdung in den Küstenregionen. In: J.L. Lozán, H. Graßl & P. Hupfer, (eds.) *Warnsignal Klima.* Wissenschaftliche Auswertungen, Hamburg.
Stohlgren T.J., Owen A.J., Lee M., 2000, Monitoring shifts in plant diversity in response to climate change: a method for landscapes. *Biodivers. & Conserv.* 9: 65-86.
Stott P., 1981, *Historical Plant Geography.* G. Allen & Unwin, London.
Storch H. von, Schnur R. & Zorita E., 1998, *Szenarien & Beratung. Anwenderorientierte Szenarien für den norddeutschen Küstenbereich.* Abschlußbericht FKZ 01 LK 9510/0.
Weeda E.J., Westra R., Westra C. & Westra T., 1985, *Nederlandsche Oecologische Flora. Wilde Planten en hun relaties 1.* IVN, Amsterdam.
Weeda E.J., Westra R., Westra C. & Westra T., 1987, *Nederlandsche Oecologische Flora. Wilde Planten en hun relaties 2.* IVN, Amsterdam.
Wisskirchen R. & Haeupler H., 1998, *Standardliste der Farn- und Blütenpflanzen Deutschlands.* E. Ulmer, Stuttgart.
Woodward F.I., 1987, *Climate & Plant Distribution.* University Press, Cambridge.
Woodward F.I., 1988, Temperature and the distribution of plant species. In: S.P. Long & F.I. Woodward (eds.) *Plants and Temperature.* Society of Experimental Biology, Cambridge.

Sizing the impact: Coral reef ecosystems as early casualties of climate change

OVE HOEGH-GULDBERG
Centre for Marine Studies, University of Queensland, Brisbane, 4072 QLD Australia

Abstract: The environment surrounding the world's coral reefs is rapidly changing. Already tropical oceans are almost one degree warmer than a century ago and current rates of warming are among the highest recorded (e.g. possibly as high as 5 °C per century, Northern tropical Pacific Ocean, NOAA 2000). Mass bleaching events (known only since 1979) are caused by small excursions in sea temperature above average summer maxima. These events are increasingly followed by the widespread mortality of reef-building corals and other symbiotic invertebrates. Mass coral bleaching and mortality begin with the destruction of the dark reactions of the photosynthetic reactions of the symbiotic algae. In some cases (e.g. recent events in NW Australia, Okinawa, Seychelles), communities of corals have been reduced from 50-80 % to less than 5 % living cover. Thermal mortality events have also increased in scale and intensity over the past two decades. There is little evidence that corals or their symbionts are acclimating or adapting fast enough to these changes. Projections from General Circulation Models of tropical sea temperatures reveal that the thermal tolerance of these organisms will be exceeded on an annual basis by 2030-2050. This leads to the inescapable conclusion that reef-building corals will become rare on coral reefs within the next 50 years. Evidence of the scale of impact expected within the next few decades is already available. In the warm conditions of 1998 alone, 10-16 % of the world's reef-building corals died. The impacts of thermal stress, coupled with carbon dioxide induced reductions in seawater alkalinity (causing reduced calcification by corals) and the impact of non-climate related stresses, suggest that even mild global climate change will have major impacts on the health of tropical marine ecosystems and coastlines. These changes in turn are expected to have major influences on the fisheries, coastal protection and tourism associated with the nations that depend on these ecosystems.

INTRODUCTION

The rate of climate change represents the ultimate challenge for organisms that generally occupy ecological spaces that are defined by the physical and biological variables of the environment (Andrewartha & Birch 1954). While these variables may fluctuate within a set range, natural selection largely operates on populations of organisms such that their gene pools dictate physiologies that are largely optimal for a given environment. If conditions change, organisms may either tolerate or acclimate to the new conditions, or they may experience mortality as lethal limits and their capacity for acclimation are exceeded. In the first case, ecosystems may change little as organisms adjust their physiologies to the new conditions. In the second case, substantial shifts in the distribution and abundance of organisms are likely to occur. Between these two extremes, organisms may shift their distributional ranges as the new conditions vary (e.g. Parmesan et al. 1999). The last response will depend on the rate at which the environmental envelope moves away from the current range of a species. If the environmental envelope migrates too quickly, organisms may decline in numbers or become locally extinct.

There is currently great interest in projecting how ecosystems will change under the climate change scenarios. Most atmospheric models suggest that even mild changes in gas concentration will lead to substantial changes in climate (IPCC 2001). Gaining a perspective on how climate change is likely to impact populations and ecosystems is complicated, however, by higher order interactions that go beyond that of the direct effects of environmental change. For example, if an organism is a rare member of an assemblage, impacts might be expected to be less than that if the organisms provides habitat for a multitude of other species. Equally, rare organisms may play pivotal roles that may have unexpected influences on the outcome of a change in their abundances (Glynn 1985a, b).

Coral reefs are the most diverse marine ecosystem on the planet (Connell 1978). Tens of thousands of plants, animals and protists live in an estimated 600,000 square kilometres of reef. The biodiversity of this region is extraordinary and rivals that associated with highly diverse terrestrial assemblages like rainforests. The environmental conditions surrounding coral reefs vary little over both short and long time-scales when compared to terrestrial habitats. Sea temperatures, for example, usually vary less than 0.5 °C over the course of a day and experience long lag times in terms of seasonal changes (e.g. McClanahan 1988, Tanner 1996). There is also considerable evidence that these environments have responded less to past climate change, due to the large thermal capacity and hence thermal inertia of oceanic waters. The reconstruction of paleoceanographic temperatures (based on faunal and $\delta^{18}O$ ratios) suggests that tropical sea surface temperatures (SST) have remained more or less constant (within 2 °C of current tropical sea temperatures) over the past 18,000 years (Thunell et al. 1994). By contrast, sea temperatures of more temperate areas of the globe have experienced far greater changes (up to 10 °C, Thunell et al. 1994) during the same period. In keeping with this, current climate change is occurring faster at high latitude and more poleward regions of the earth.

Climate change represents a major challenge to the earth's marine ecosystems. The heat content of the world's oceans has increased by approximately 2×10^{23} joules between the mid-1950s and mid-1990s, with the global volume mean temperature increase for the period since the 1950s being 0.31 °C for the upper 300-meters (Levitus et al. 2000). Most of this warming trend is attributable to human activities, particularly the release of greenhouse gases into the atmosphere (IPCC 2001). Analysis of temperature data over the past century reveals that the world's tropical oceans are now approximately 1 °C warmer than they were at the beginning of the 1900s (Levitus et al. 2000). Recent estimates of the rate of rise put this at an even higher rate (0.5 °C per decade in the northern hemisphere tropical oceans, Strong et al. 2000) although some of these rates require further confirmation from non-satellite sources. These trends are projected to continue (IPCC 2001), with substantial changes being projected for the earth's oceans over the next several decades (Hoegh-Guldberg 1999).

The urgency of understanding the impacts of climate change has been highlighted by recent mass bleaching events (Glynn 1993, Brown 1997, Hoegh-Guldberg 1999). During mass bleaching events, reef-building corals loose their brown symbionts (unicellular dinoflagellate algae) and turn stark white in colour (bleaching). These events, which have been followed by large-scale mortalities among coral communities, are increasing in intensity and frequency. There is now a strong case for these events being associated with the warming trends in tropical oceans, and to be one of the strongest "fingerprints" of global climate change (Goreau & Hayes 1994, Brown 1997, Hoegh-Guldberg 1999). This article explores mass coral bleaching and its connection to the changing conditions of the world's oceans. It then describes recent work that has projected how the frequency and intensity of these events is likely to change over the next 100 years. The last part of this article will explore the crucial question as to how projected rates of oceanic warming are likely to effect the distribution and function of nearshore tropical communities and their human dependents. As will be argued, if climate continues to affect changes in sea temperature, the change within tropical nearshore communities are likely to be enormous.

PLANT-ANIMAL SYMBIOSIS AND CORAL REEFS

Clear and nutrient-poor waters are typical of tropical oceans (Odum & Odum 1955). Despite this, coral reefs are characterised by high primary productivity, which contrast the low values of surrounding waters. This contrast was recognised by early coral reef biologists like Charles Darwin who puzzled over the "conditions favourable to their vigorous growth" (Darwin 1842). Coral reefs form rich and complex food chains that support large populations of fish and other megafauna. By contrast, the open oceanic waters even a small distance away from coral reefs are usually devoid of significant populations of animal and plant life. Almost a century of research has sort to understand the mystery underlying why coral reefs are such hot spots of primary productivity. It appears that the many symbioses that exist between reef species play a key role in maintaining these high rates of productivity.

The recycling of nutrients is a major theme within the symbioses that abound coral reefs (Muscatine & Porter 1977). Under normal circumstances the primary productivity of tropical marine plants is dependent directly on the concentration of nutrients in the water column (Odum & Odum 1955). Because of the low ambient concentrations of these nutrients (due to thermal stratification and nutrient loss, Atkinson 1992, Furnas & Mitchell 1999), the water column above coral reefs is very often devoid of the essential inorganic nutrients required for the production of organic compounds by plants and other primary producers. The water column of tropical oceans is consequently largely devoid of plant-like organisms that form the food of animals and bacteria. Symbioses, especially those involving endosymbiosis (cells within cells) between plants and animals circumvent the problem of low external nutrients by involving the direct passage of nutrients from primary producer to consumer without involving the problem presented by the dilute concentrations of the water column. The efficient internal recycling of nutrients within symbioses as well as detrital pathways (e.g. Johnstone 1989, Johnstone et al. 1990) is a critical feature of coral reef ecosystems.

One of the most prominent symbioses on coral reefs is that that exists between corals (Phylum Cnidaria) and dinoflagellate algae ("zooxanthellae", Division Pyrrophyta, Trench 1979). While many other coral reef organisms form symbioses with dinoflagellates (e.g. clams, sponges, flatworms), corals are the most abundant form of this symbiosis. Corals and their symbiotic dinoflagellates are intertwined to the point where the dinoflagellates are located within the cells of the animal host (except in Tridacnid clams). Many millions of zooxanthellae may occupy a single polyp (Figure 1). Zooxanthellae sit within the tissues of corals and photosynthesise, passing as much as 95 % of what they make in the sunlight to the animal host (Muscatine 1990). Zooxanthellae translocate amino acids, sugars, carbohydrates and small peptides to the host invertebrate across a host-symbiont barrier. These translocated materials are a rich source of nutrients and energy for the host coral (Muscatine 1973, Trench 1979, Swanson & Hoegh-Guldberg 1998). The photosynthetic energy passed to the coral by the zooxanthellae is crucial to the physiology and ecology of reef-building corals. Corals with the highest growth rates have symbiotic dinoflagellates and only symbiotic lay down significant amounts of calcium carbonate or limestone. Reef-building or scleractinian corals are often referred to as the "frame-builders" of coral reef systems with other organisms serving to weld the structure together (e.g. calcareous red algae). Reef-building corals (and their symbionts) have been the principal organisms behind coral reef ecosystems for several hundred million years (Veron 1986, 2000).

Figure 1. Structure of reef-building corals.
A. Close-up of a polyp of the reef-building coral *Pocillopora damicornis*.
B. Symbiotic dinoflagellates (Z = "zooxanthellae", N = nematocysts of coral) isolated from a reef-building coral.   Photographer: O. Hoegh-Guldberg

The benefits of the primary production of corals and the associated primary producers flow down a complex food chain (Odum & Odum 1955). Coral reefs tend to be associated with rich populations of fish, birds, turtles and marine mammals (e.g. Maragos 1994, Kepler et al. 1994). Traditionally, these organisms have supported small human populations. Recent massive increases in tropical human populations, however, put coral reef ecosystems under threat of over-exploitation and degradation, which according to some, may have started as much as 300 years ago (e.g. Caribbean reefs, Jackson 1997). A wide range of anthropogenic stresses are now affecting coral reef ecosystems, including over-fishing, increased sedimentation and eutrophication. While outside the terms of this review, these stresses are widely recognised as being responsible for much of the decline in the health of coral reef ecosystems over the past century.

CLIMATE CHANGE AND TROPICAL MARINE ECOSYSTEMS

Changes in the carbon dioxide concentration of the atmosphere are forcing three major changes in the environment surrounding tropical marine ecosystems. The first is the change in the carbonate alkalinity, which is due to the change in the concentration of carbon dioxide in the atmosphere (Gattuso et al. 1998, Kleypas et al. 1999a). The second is the change in sea temperature, a result of changes to the heat balance of the earth due to increased concentrations of greenhouse gases like carbon dioxide (IPCC 2001). The third involves changes in sea level as a result of thermal expansion of the world's oceans and melting polar ice caps. While the latter appears to be debatable as to its impact on reef systems (i.e. healthy reef systems may well keep up with projected sea level change, Done 1999), there is a strong case for a reduced carbonate alkalinity and increased sea temperature having important negative effects on the health of coral reef ecosystems.

The distribution of reef-building corals varies as a function of latitude, with the highest abundance of calcifying corals located closest to the equator (Veron 2000). In concert with these changes, light, temperature and the carbonate alkalinity of seawater decreases in a poleward direction, making it increasingly difficult for calcifying organisms to grow and precipitate calcium carbonate for their skeletons. These three factors are thought to exert a controlling influence on the distribution of coral reefs (Kleypas et al. 1999a). While light is expected to show minimal changes under climate change, both oceanic alkalinity and sea temperature are expected to change quite substantially as carbon dioxide and other greenhouse gases accumulate in the earth's atmosphere.

Reduced ocean alkalinity and calcification

The total carbonate alkalinity of seawater is expected to decrease in the ocean as carbon dioxide increases (Gattuso et al. 1998, Kleypas et al. 1999b). Increasing carbon dioxide above a solution will lead to changes in the acidity and available carbonate pool (aragonite saturation state). This pool of ions is the resource in seawater from which corals and other calcifying organisms draw the ions to precipitate calcium carbonate. The aragonite saturation state directly determines calcification rates of organisms like corals (Langdon 2000). Gattuso et al. (1998) and Kleypas et al (1999b) calculated that projected carbon dioxide concentrations would potentially decrease the aragonite saturation state in the tropics by 30 percent by 2050 (under an IS92a scenario). Langdon (Langdon 2000, Langdon et al. 2000) has recently shown that corals growing in large tank systems within the Biosphere II habitat calcify at 40 % lower rates when carbon dioxide levels are doubled.

How these changes in total alkalinity will affect the distribution of coral reef ecosystems is complex. Natural reef systems represent a balance between calcification, and the physical and biological eroding influences that remove deposited calcium carbonate. If one compares reef deposition rates (up to 20 cm yr^{-1}) with actual rates of growth (1 cm yr^{-1}, Done 1999), it is clear that more than 90 % of the calcium carbonate that is laid down in a coral reef structure is removed by physical and biological erosion. Reef calcification rates don't have to decrease much for net calcification rates to tip the balance in favour of the disappearance of coral reef calcium carbonate. There is now considerable speculation about weakened reef infrastructure as a result of a decreased carbonate alkalinity. Coastal erosion and the loss of reef integrity are two further impacts expected if corals do not produce net carbonate under a reduced aragonite saturation state. Reduced coastal protection due to weakened or rapidly eroding coral reefs could add substantially to the potential costs associated with warming tropical seas.

Sea temperature

Coral reefs prosper in the shallow, sunlit waters of the tropics in which water temperatures range between 18 °C and 30 °C. In colder waters, hermatypic (or symbiotic) corals may survive but consolidated reef systems fail to form. In addition to seasonal changes in sea temperature, interannular variability is contributed to by changes in oceanic circulation, which can vary both stochastically and cyclically. The best known examples of how large scale changes to water circulation can dramatically alter sea temperatures are the effect El Niño-Southern Oscillation (ENSO) events on sea temperature. ENSO events have occurred approximately every 5-14 years (20 have occurred since 1900) over the last couple of thousand years and involve a weakening or reversal of the easterly trade winds that blow across the Pacific Ocean. This in turn. leads to changes in the intensity of oceanic current systems in both hemispheres. In addition to substantial influences on climate (e.g. drought in Africa and Australasia, increased rainfall in the south-eastern USA), equatorial waters intrude more southerly and bring unusually warm waters (1-5 °C) in contact with ecosystems formerly unexposed to these sea temperatures. Correlated with the arrival of this warm water are usually striking changes such as reduced productivity and loss of kelp forests from some temperate coastlines (Glynn 1988), and the reduction of reef-building coral communities along some tropical coastlines (e.g. loss of reef systems in Galapagos and other localities in the eastern Pacific and Caribbean following the 1982-83 ENSO event Glynn & Colgan 1992, Goreau 1992).

There is now abundant evidence that average sea temperature is changing rapidly across the planet. Average temperature in the upper 300 m of the ocean has increased by 0.31 °C from the mid 1950s to the mid 1990s (Levitus et al. 2000). Rapid changes are occurring in many parts of the ocean – some rates being unprecedented even for interglacial transitions (e.g. 0.5 °C per decade, Strong et al. 2000, Lough 1999). A recent survey of rates of change in six sites within the world's tropical oceans has revealed that oceans are now almost a degree warmer than they were a century ago and are currently warming at rates that range between 0.5 and 2.5 °C per century (Table 1, Hoegh-Guldberg 1999). Warming trends over the past 20 years are complicated by the occurrence of temporary cooling trends due to volcanic activity and other influences like cloud formation (Street et al. 2000, Myhre et al. 2001). However, as indicated by Levitus et al. (2000), the overall trend since the 1950s has been one of a warming of oceanic waters through out the world's ocean.

Projections for general circulation models from the Max-Planck-Institut für Meteorologie (Germany), Commonwealth Scientific and Industrial Research Organisation (CSIRO, Australia) and the Hadley Centre (United Kingdom) all show the same upward trend in sea temperature for a scenario (IS92a, IPCC 1992) developed around the doubling of carbon dioxide and equivalents by the end of the present century. These models all reveal that the rate of change in tropical oceans is likely to increase rather than decrease (Hoegh-Guldberg 1999).

Table 1. Rates of warming detected by regression analysis within Trimmed Monthly Summaries from the Comprehensive Ocean-Atmosphere Data Set (COADS, up to Dec 1992) and IGOSS-nmc blended data (January 1993-April 1999). Data obtained from the Lamont Doherty Earth Observatory server (http://rainbow.ldgo.columbia.edu/). Data were included only if all months were recorded (hence shorter periods for some parts of the world). All trends were highly significant with the possible exception of Rarotonga. GBR, Great Barrier Reef (Table adapted from Hoegh-Guldberg 1999).

Locations	Period examined	Rate of change per 100 years	Significant of trend
Jamaica	1903-99	1.25	< 0.001
Phuket	1904-99	1.54	< 0.001
Tahiti	1926-99	0.69	0.003
Rarotonga	1926-99	0.84	0.05
Southern GBR	1902-99	1.68	< 0.001
Central GBR	1902-99	1.55	< 0.001
Northern GBR	1902-99	1.25	< 0.001

Figure 2 summaries the general trends for 3 key tropical regions. In all cases, sea temperatures increase by 2-3 °C by the end of the century. Projections of future sea temperature for these studies were derived from the ECHAM4/OPYC3 General Circulation Model (Roeckner et al. 1996) for the period 1850-2100. Two runs were done using IS92a scenarios with and without the cooling effect of aerosols taken into account. The global coupled atmosphere-ocean-ice model (Roeckner et al. 1996) was developed by the Max-Planck-Institut für Meteorologie (Germany) and is used by the United Nations for climatology simulations. Horizontal resolution in these models is roughly equivalent to 2.8° x 2.8° latitude-longitude. This model has been used in climate variability (Roeckner et al. 1996, Bacher et al 1997, Christoph et al. 1998), climate prediction (Oberhuber et al. 1998) and climate change studies with a high degree of accuracy (Timmermann et al. 1999, Roeckner et al., 2000). As noted, the model was run with and without the cooling effect of aerosols. Changes in aerosols were determined as follows: Observed concentrations of sulfate aerosols were used up to 1990 and thereafter changes according to the IPCC scenario IS92a. The tropospheric sulfur cycle was also incorporated but with only the influence of anthropogenic sources considered. Natural biogenic and volcanic sulfur emissions are neglected, and the aerosol radiative forcing generated through the anthropogenic part of the sulfur cycle only. The space/time evolution in the sulfur emissions has been derived from Örn et al. (1996) and from Spiro et al. (1992).

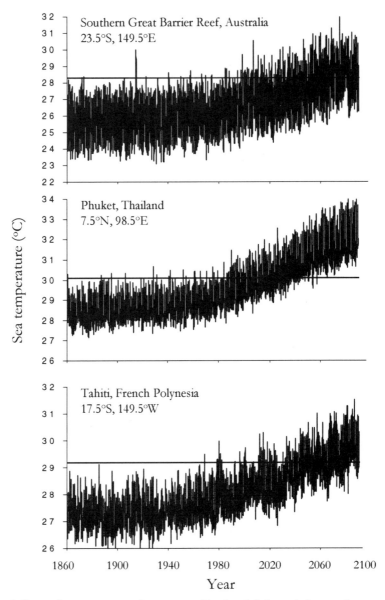

Figure 2. Sea surface temperature data generated by the global coupled atmosphere-ocean-ice model (ECHAM4/OPYC3, Roeckner et al. 1996, communicated by A. Timmermann of KNMI, Netherlands). Temperatures were generated for each month from 1860 to 2100, and were from a model forced by greenhouse gas concentrations that conform to the IPCC scenario IS92a (IPCC 1992). Effects of El Niño-Southern Oscillation (ENSO) events are included. Horizontal lines indicate the thermal thresholds of corals at each site. Data were generated for three regions: Southern Great Barrier Reef, Australia (23.5° S, 149.5° E) Tahiti, French Polynesia (17.5° S, 149.5° W) and Phuket, Thailand (7.5° N, 98.5° E).

The estimates of the rate of climate change explored by Hoegh-Guldberg (1999) are conservative relative to current scenarios released by the Intergovernmental Panel on Climate Change during its third assessment report (IPCC 2001). Changes in greenhouse gases that were used in these models were derived from the IPCC mid-range IS92a scenario (IPCC 1992). In this scenario, observed concentrations of greenhouse gases were used up to 1990, and changes outlined in the IPCC scenario IS92a (IPCC 1992) were implemented thereafter. The midrange emission scenario, IS92a, is the central estimate of climate forcing by greenhouse gases and assumes a doubling of 1975 CO_2 levels by the year 2100, with sulphate aerosol emissions, which have a cooling effect, remaining at 1990 levels. Greenhouse gases are prescribed as a function of time: CO_2, CH_4, N_2O and also a series of industrial gases including CFCs and HCFCs.

RECENT IMPACTS OF WARMER SEAS

Tropical oceans are now almost a degree warmer than they were 100 years ago (Hoegh-Guldberg 1999). Mass coral bleaching events are a graphic example of what a small upward excursion in sea temperature can have on the health of key organisms like reef-building corals in tropical near shore environments. Coral bleaching occurs when corals rapidly loose the cells and/or the pigments of the zooxanthellae that populate their tissues. Mass coral bleaching events are events in which corals across vast areas of coral reef undergo bleaching. These events have occurred six times since 1979 but no events have been reported in the literature prior to this time. Although confounded to some extent by the fact that the last 3 decades have been one of intense activity in the marine sciences and hence observation frenquency, the relative absence of reports during the 1950s, 60s and 70s is curious given that biologists were very active in both the tropical intertidal and subtidal during this period (Glynn 1993). It is also peculiar that reports of mass bleaching from indigenous fishermen in areas like the Pacific are only from the last 15 years (Hoegh-Guldberg 1994, Y. Loya, pers. comm.). Most authors now agree that the incidence of mass-bleaching events is on the increase (Glynn 1993, Goreau & Hayes 1994, Hoegh-Guldberg & Salvat 1994, Brown 1997, Hoegh-Guldberg 1999, Wilkerson 1999).

Mass-bleaching events occur suddenly across vast areas of the world's tropical oceans and are correlated with small positive sea temperature anomalies (1-3 °C) during the summer months (Hoegh-Guldberg & Smith 1989, Gates 1990, Glynn & D'Croz 1990, Glynn 1991, 1993, Brown 1997, Hoegh-Guldberg 1994, 1999). There is now little doubt that the primary cause of mass bleaching events is associated with these small positive temperature anomalies that cause the light reaction of dinoflagellate photosystems to experience an increased sensitivity to photoinhibition (Iglesias-Prieto et al. 1992, Jones et al. 1998, Hoegh-Guldberg 1999). This phenomenon is similar to the mechanism that underlies heat stress in higher plants and is probably a feature of thermal stress in phototrophic organisms (Jones et al. 1998).

There are two factors that explain why the thermal anomalies have been increasing in frequency in tropical coastal habitats. The first factor is the increase in the mean temperature of tropical oceans over the past 100 years. The second factor is associated with the timing and intensity of El Niño Southern Oscillation (ENSO) events (Glynn 1988, 1991, 1993, Hoegh-Guldberg 1999). These two components combine to produce short periods during the summer months in which sea temperatures rise above the thermal tolerance of the symbiotic dinoflagellates of reef-building corals (Figure 3).

Several factors other than sea temperature influence the outcome once thermal anomalies have been reached. Strong et al. (2000) identify the important contribution that both the magnitude of a thermal anomaly and the exposure time have on the intensity of a bleaching event. The NOAA satellite-derived Degree Heating Week (DHW) calculates the accumulated thermal stress that coral reefs experience. One DHW is equivalent to 1 week of sea surface temperature that is 1 °C above the expected summertime maximum sea (surface) temperature. Degree Heating Months (DHM, using monthly data as opposed to weekly date) show similar results. Treating sea temperature data from November 1981 to August 2000 (Figure 3C) successfully detects (post priori) key bleaching events in three sites in the Pacific. Also evident from these graphs is the extent to which the intensity of events appears roughly encoded. Severe events such as the 1998 event in Palau (mass mortalities) are recorded with much higher values than bleaching events reported to be less severe such as 1998 events in French Polynesia in the same year (which was very mild, personal observation). In French Polynesia, the 1994 as opposed to 1998 was widely reported to be the most intense year for mass coral bleaching (Salvat 1991, Hoegh-Guldberg & Salvat 1995, Drollet et al. 1994, Hoegh-Guldberg 1999).

While the size and length of a thermal anomaly is highly predictive of mass coral bleaching, other factors are likely to play a role. Light is a crucial secondary factor that may be important ecologically. The "Photoinhibitory Model" of coral bleaching proposed by Jones et al. (1998) and expanded by Hoegh-Guldberg (1999) reveals that the extent of damage to photosystems within the symbiotic dinoflagellate is highly light dependent, such that little damage will occur at high temperature if the dinoflagellates are shaded. This leads to variability in susceptibility between and within coral colonies (Jones et al. 1998, Knowlton et al. 1992) and coral species (e.g. host behaviour,

Coral reef ecosystems as early casualties of climate change 215

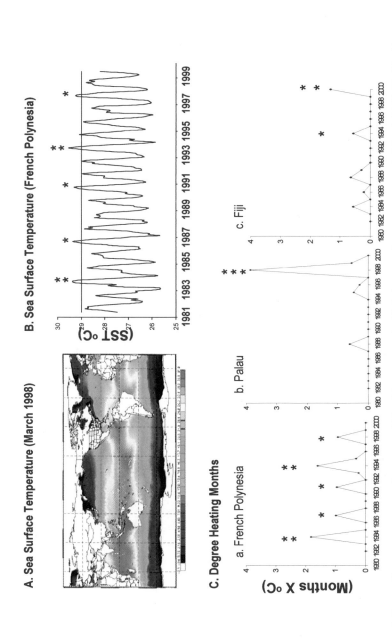

Figure 3. Sea surface temperature and mass coral bleaching events. **A**. Global sea surface temperatures in March 1998. **B**. Weekly sea surface temperature data for Tahiti, French Polynesia (17.5° S, 149.5° W). Horizontal line indicates the minimum temperature above which bleaching events occur (threshold temperature). **C**. Degree heating months for three Pacific sites adapted from Hoegh-Guldberg et al. (2000). Data are IGOSS-nmc blended data courtesy of the Lamont-Doherty Climate Center at Columbia University. Asterisks indicate bleaching events reported in the literature (* = mild, ** = moderate, *** = severe).

Hoegh-Guldberg 1999; host shade compounds like GFP-like pocilloporins, Dove et al. 2001) and the important influence of whether reefs are shaded by cloud or not (Hoegh-Guldberg 1999, 2001). Despite these important influences, these factors are secondary to the primary effect of increased sea temperature.

ECOLOGICAL OUTCOMES OF MASS BLEACHING

The ecological outcomes of changes to the mortality schedules of key groups of organisms like reef-building corals are of enormous importance to perspectives on how climate change is likely to impact the earth's ecosystems. Outcomes following mass coral bleaching, a direct result of warmer than normal conditions, vary depending on the size and length of the thermal anomaly, and other factors such as cloud cover and the species composition of the reefs being affected (as discussed above). Mortality estimates vary from close to zero in cases of mild bleaching (Harriot 1985) to close to 100 % (Brown & Suharsono 1990, Hoegh-Guldberg 1999, Glynn 1990). Strong et al. (2000) and Toscano et al. (1999) have observed that DHWs of greater than 10 in the past have resulted in severe bleaching and mortality among reef-building corals in the Indian Ocean. Similar results hold for Pacific reefs. The 1998 event in Palau resulted in the wide spread loss of corals (J. Bruno, pers. comm.), as did a similar event in Okinawa (Y. Loya, W. Loh & K. Sakai, pers. comm.), Seychelles (Spencer et al. 2000) and Scott Reef (Smith & Heyward, in press). In order to verify the fact that DHMs greater than 2.3 resulted in severe bleaching and significant mass mortality of corals, DHMs were calculated for the 1998 event in Palau, Okinawa, Seychelles (using data of Spencer et al. 2000) and Scott Reef (off Western Australia). IGOSS data were downloaded for these locations and the DHM associated with these mortality events calculated. The DHMs associated with these severe events were 3.9, 3.0, 3.1 and 2.6 for Palau, Okinawa, Seychelles and Scott Reef respectively. By contrast, the DHMs for the southern, mid and northern sectors of the Great Barrier Reef (calculated for SST data for sectors of the Great Barrier Reef that were away from the more affected inshore sections) were 1.7, 1.4 and 0.5 respectively. Bleaching on the Great Barrier Reef was far less severe than in the four other locations (Berkelmans & Oliver 1999).

Projections of sea temperature, if combined with known tolerances of reef-building corals and their zooxanthellae, provide a way to project how the intensity and frequency of mass coral bleaching are likely to change under climate change. Hoegh-Guldberg (1999) projected how the incidence of mass coral bleaching will change as seas warm for six tropical sites across the planet. The results of this study project that the conditions that cause mass bleaching and mortality events will become annual by the year 2030-70 at all tropical sites globally (Figure 4). These projections differ little among general circulation models (Hoegh-Guldberg 1999) and hold for models derived by the Max-Planck-Institut für Meteorologie (Germany), Commonwealth Scientific and Industrial Research Organisation (CSIRO, Australia) and the Hadley Centre (United Kingdom).

The intensity of bleaching events can be hindcast using anomaly and duration data with considerable success (Figure 3C, Hoegh-Guldberg et al. 2000). General circulation model data when combined with the Degree Heating Month methodology of Strong et al. (1996) provide some powerful perspectives on the levels of thermal stress that are likely in coral reef areas within the next century. The current projections of how sea temperatures are likely to change under IS92a scenarios suggest that sea temperatures will be warm enough such that years in which degree heating months exceed 2.3 will be commonplace by the middle of this century (Figure 5). As these events are associated with mass mortality events that are on the scale of events that occurred in Palau, Okinawa and the Indian Ocean, this can only raise the prospect that coral reefs are likely to loose the majority of their living coral cover by this time. Recovery rates of coral reefs are likely to range between 15-50 years, and little recovery is likely if the conditions that cause major mortality events are projected to become yearly events.

A range of responses is expected within coral communities that experience increasing levels of thermal stress. Over short time scales, individual corals and their symbionts would be expected to physiologically acclimate to changing sea temperatures. Beyond this, differential mortality of corals facing thermal stress will lead to selection for coral genotypes that are more thermally tolerant. This response will lead eventually to the evolution of coral populations that are more tolerant of the new thermal regimes. Over short distances, new genetic varieties of corals and symbionts might be expected to flow from warmer areas. The degree to which genotypes will flow between regions depends on two factors (Hoegh-Guldberg 1999). The first is the degree to which coral populations are interconnected such that genes can flow from one reef to another. The second is whether populations of corals will exist to act as source reefs. Little is known about the former issue for coral reef organisms in general. Recent studies suggest that panmixis (genetic connectivity) exists over reasonably large scales for reef-building corals (Ayre & Hughes 2000, Ridgway et al. 2001, Rodriguez-Lanetty et al. 2001). This, unfortunately, does not answer the key question, which is the extent to which larvae flow from one area to another over short-term timescales. Panmixis may exist between populations with relatively small numbers of migrants moving

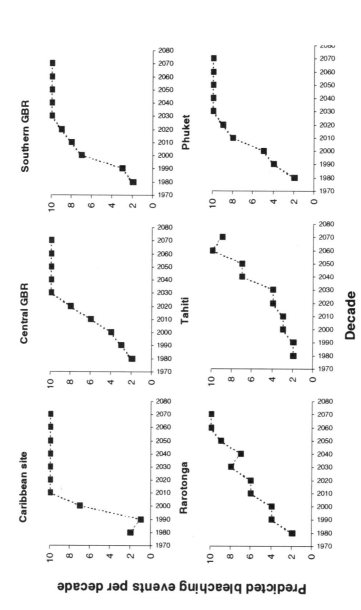

Figure 4. Projected rates of mass bleaching events (events per decade) at six sites. Data were obtained by counting the number of times per decade that predicted temperatures (see Figure 3) exceeded thermal threshold levels (bleaching events) of corals growing off Jamaica in the Caribbean (17.5° N, 76.5° W), Central Great barrier Reef (Central GBR), Southern Great Barrier Reef (southern GBR), Rarotonga (21.5° S, 159.5° W), Tahiti (17.5° S, 149.5° W) and Phuket, Thailand (7.5° N, 98.5° E) Data used are from the ECHAM4/OPYC3 model based on an IS92a scenario that includes an aerosol effect. Model run by the Max Planck Institute and provided by Axel Timmermann. Figure adapted from Hoegh-Guldberg (1999).

Coral reef ecosystems as early casualties of climate change

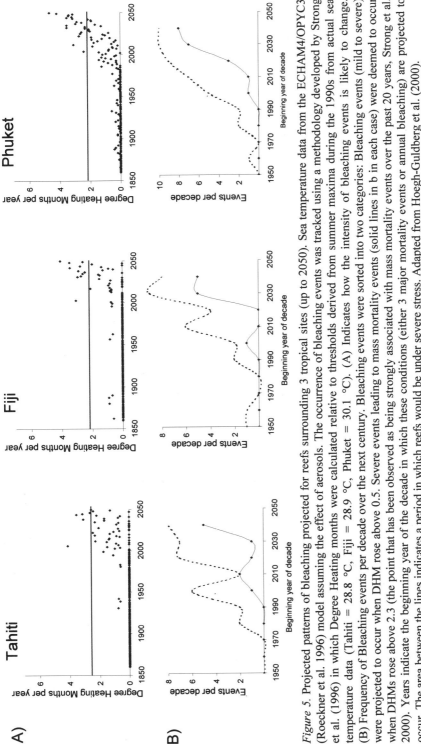

Figure 5. Projected patterns of bleaching projected for reefs surrounding 3 tropical sites (up to 2050). Sea temperature data from the ECHAM4/OPYC3 (Roeckner et al. 1996) model assuming the effect of aerosols. The occurrence of bleaching events was tracked using a methodology developed by Strong et al. (1996) in which Degree Heating months were calculated relative to thresholds derived from summer maxima during the 1990s from actual sea temperature data (Tahiti = 28.8 °C, Fiji = 28.9 °C, Phuket = 30.1 °C). (A) Indicates how the intensity of bleaching events is likely to change. (B) Frequency of Bleaching events per decade over the next century. Bleaching events were sorted into two categories: Bleaching events (mild to severe) were projected to occur when DHM rose above 0.5. Severe events leading to mass mortality events (solid lines in b in each case) were deemed to occur when DHMs rose above 2.3 (the point that has been observed as being strongly associated with mass mortality events over the past 20 years, Strong et al. 2000). Years indicate the beginning year of the decade in which these conditions (either 3 major mortality events or annual bleaching) are projected to occur. The area between the lines indicates a period in which reefs would be under severe stress. Adapted from Hoegh-Guldberg et al. (2000).

between reefs (Ridgway et al. 2001). This is a critical factor (as opposed to the demonstration of panmixis) in determining how rapidly reefs will be repopulated in the wake of a mortality event.

The second factor influencing how rapidly reefs will be repopulated is whether populations of healthy corals will be protected in "refugia" as the oceans become considerably warmer. This issue must be considered in the light of the projections presented in Figure 5 and other studies (Hoegh-Guldberg 1999, Hoegh-Guldberg et al. 2000). Over the shorter term, some reefs may be left to act as sources of genetic material for regeneration as probably happened in the 1997-98 event. As reef waters warm to well over the known thermal tolerances of reef-building corals today (i.e. by the middle of this century), the idea that populations of reef-building corals will be left unaffected is largely untenable. An additional concern is the fact that thermal events may lead to large decreases in the reproductive potential of reef-building corals, as seen on the Great Barrier Reef in 1998. Sub-lethal temperatures have been shown to slow the development of gonads within corals and interrupt a number of other key processes (e.g. fertilization, Hoegh-Guldberg, Harrison & Ward, in review). The significance of these results is considerable and suggests that although corals may recover from some bleaching events, they may have severely decreased reproductive outputs. There are also a number of field studies that indicate very low recruitment during the warm periods associated with the 1998 global cycle of mass bleaching (e.g. Western Australia, Heyward, A. & L. Smith, pers. comm.).

Done (1999) discusses how coral reef ecosystems might respond to climate change outlines four possible scenarios. The scenarios are: (i) Tolerance, in which corals and other symbiotic organisms acclimatize to the changes in aragonite saturation state and changes in sea temperature. Little changes within coral reef ecosystems under this model as organisms respond to changes in the environment by adjusting their physiologies to cope with the new conditions. (ii) Faster turn-over, during which coral populations experience a change in their mortality schedule with reduced life expectancy. Species composition within the communities remains constant but populations shift to an age structure skewed toward younger individuals. (iii) Strategy shift, in which changes in the species composition of coral dominated communities are foreseen in which hardier species (e.g. *Porites* spp.) replace less hardy species (e.g. *Acropora* spp., *Pocillopora* spp.). In the last scenario, (iv) Phase shift, coral communities reduced to remnant, non-dominant communities. In this case, reef substrates are replaced by other communities (e.g. macroalgae, algal turf communities).

A little discussed area of investigation is that of the range expansion of coral reefs to higher latitudes. As waters warm at higher latitudes, the potential for the spread of corals, and hence the communities that depend on them, is considerable. Unknown is the exact dependence of reef-building corals on water temperature (Kleypas et al. 1999b). While their distribution is correlated with the earth's warmer waters, reef-building corals are also strongly correlated with light conditions as a function of latitude. If corals are truly limited by water temperature, then increases should be seen in the range and vigor of corals at higher latitudes with time.

From the large-scale mortality events of 1997-98, it is clear that small upward deviations in mean summer temperatures will push communities beyond their physiological tolerances. According to the Global Coral Reef Monitoring Network survey in 1997-98, 50 % of living corals were removed from reefs in the Indian Ocean alone (Wilkerson 2000). In some examples, such as reefs in the Seychelles and Maldives (Spencer et al. 2000), as much as 99 % of the coral communities were removed. While the Indian Ocean reefs were exceptionally hard hit by the thermal anomalies of 1997-98, reefs in all coral reef realms experienced moderate to severe mortalities (Wilkerson 2000). Between 10 and 16 % (global average) of living corals was also removed from reefs worldwide during this period. These mortality events appear to be unprecedented in the last 6,000 years (Aronson et al. 2000). The many examples of almost complete mortality events from 1997-98 strongly suggests that scenarios (i) and (ii) are not occurring. Shifts in species composition within coral communities may be occurring as suggested by several authors (Gleason 1993, Hoegh-Guldberg & Salvat 1995). Different mortalities between coral taxa following bleaching (e.g. *Acropora* spp. > *Pocillopora* spp. > *Porites* spp., Gleason 1993, Salvat 1991, Hoegh-Guldberg & Salvat 1995, Marshall & Baird 2000) are now a common observation within mass bleaching events. The major changes in species composition of reefs as reported for some areas of the Caribbean by Hughes (1994; note other factors probably also contributed here), Shulman & Robertson (1996), Aronson et al. (2000) and Precht & Aronson (1999) further suggest that scenario (iii) and probably (iv) are already occurring. Given that thermal stress will rise well above that needed to cause a large scale mortality event on the scale of those seen in Okinawa, Palau and the Indian Ocean by the middle of the century (Figure 5, Hoegh-Guldberg et al. 2000), the likelihood of major changes in the abundance of coral dominated communities under climate change is high.

Ultimately, the final form of ecosystems following sustained climate change is hard to predict. Reasoning that corals will "evolve" more thermally tolerant varieties of corals are largely unfounded at this point in time. Evidence of symbiont switching as described by the "Adaptive Bleaching Hypothesis" (Buddemeier & Fautin 1993) is also largely lacking.

The key observation that shows symbionts are switched in for more thermally tolerant varieties has never been demonstrated convincingly, despite claims to the converse (e.g. Baker 2001). Baker (2001), for example, drew his conclusions drawn from an experiment in which corals are transplanted between two depths on a Panamanian coral reef. His corals bleached less and died more often when they were transported from the shallows to the deep site, as compared to when they were transported from the deep site to the shallows. An important observation according to Baker, however, was that the corals transplanted to the shallow site changed their mix of symbiont genotypes while those transplanted to the deeper site did not. From these facts, Baker linked the higher mortality of the corals transported to deeper waters to the fact that they did not bleach and hence (his interpretation) did not vary their symbiont genotypes. In his words "without bleaching, suboptimal host-symbiont combinations persist, leading eventually to significant host mortality."

Baker's experiment, however, is seriously confounded. Ignoring the separate concern that Baker used light rather than temperature stress and that his methods were unable to detect new as opposed to previously rare genotypes, the corals within Baker's chronic and acute treatments were allowed to recover under very different conditions. The corals classified as chronically stressed recovered in deep water under light levels that are near the lower limit for coral growth (20-23 m). Baker's acutely stressed corals, however, recovered under more illuminated conditions at the shallow water site (2-4 m). It is impossible, therefore, to separate changes due to the mix of symbiont genotypes from those associated with the environment under which corals recovered. As much of the metabolism of corals is driven by light (Muscatine 1973), it is perhaps not surprising that stressed corals died more after transplantation from a site where energy is abundant to one in which light and hence energy is scarce. This is by far the more parsimonious explanation for why colony mortality was slightly increased in corals transferred from shallow to deep-water locations. It cannot be concluded, therefore, that bleaching and consequent switching of algal partners confers an advantage to the coral-algal symbiosis.

It is important in the final part of this section, to reiterate that corals are unlikely to become extinct due to current anthropogenic pressures on them. Genetic selection for more tolerant varieties is likely to happen, especially under such direct selection regimes as seen in recent bleaching events. The issue, however, comes down to one of the rate of change. While corals may evolve over time, the rate at which this is likely to happen is likely to fall behind current rates of oceanic warming. As the selection regime continues to ramp further upward, coral communities are likely to become increasingly remnant. The effect of this in turn on the thousands of animals, plants and protests that depend on them for the framework of coral reefs is likely to be large. A massive loss of biodiversity within these ecosystems is almost certain to result as climate change progresses.

RAMIFICATIONS OF CLIMATE CHANGE FOR HUMAN USERS OF REEFS

More frequent episodes of coral bleaching and coral mortality are likely to result in serious socio-economic impacts in countries where dependency on the health of coral reefs. Exact understandings of how human users are likely to be affected descriptions are lacking although some inferences can be draw from the current role that reefs play in the lives of tropical human populations. Coral reefs represent centrally important resources through their role in tourism, fishing, building materials and coastal protection (Carte 1996). Globally, 100 million people depend in part or wholly on coral reefs for their livelihood, and ~15 % (0.5 billion people) of the world's population live within 100 km of coral reef ecosystems (Pomerance 1999). Though differences between tropical regions are enormous, the majority of the tropical coastal societies have low incomes and a large dependence on coral reef fisheries (Hoegh-Guldberg et al. 2000).

Estimating socio-economic losses due to coral bleaching is complex and understandings of the impacts are in their infancy. Cesar et al. (2000), for example, estimated losses to the Maldives' economy from severe coral bleaching and mortality as approximately 1 %. This was surprising in that live coral cover had decreased from over 50 % to less than 5 %. Investigation of this situation revealed that the main reason for the tiny impact was a successful shift of tour operators from dive tourism to the Indian 'honeymooner' market. The results of Cesar et al. (2000) for the Maldives are also low when compared to the East African study of Westmacott et al. (2000). These authors reported a decrease of up to 19 % among dive tourists arriving in Zanzibar after severe bleaching that affected these areas in 1998. The difference between the Maldives and Zanzibar appears to be that the marketing possibilities in the Maldives allow substitution of different sub-groups of the tourism market and the net outlay/quality ratio for travel to Zanzibar (Zanzibar being more questionable to the European tourist; see comments by Hoegh-Guldberg et al. 2000). While it is not the intention of this review to address the socio-economic aspects of climate change on reefs, it is relevant to indicate the complex implications of the loss of coral dominated ecosystems for human users.

CONCLUSION

Mass coral bleaching is triggered by exposure to elevated sea temperatures and is due to thermal stress that causes the photosynthesis of the dinoflagellates of reef-building corals to fail. High light levels exacerbate the effect of increased water temperature. Tropical oceans are warming rapidly and are bringing coral reefs closer to their thermal threshold. Exceeding the thermal threshold of reef-building corals results in the almost complete expulsion of symbiotic dinoflagellates from reef-building corals. Reef-building corals die if the intensity and duration of the thermal stress is large enough. Most evidence points to a dramatic increase in mass coral bleaching events since 1979, when they were first reported in the primary literature. Mass coral bleaching events affect thousands of square kilometres of ocean and can result in the total loss of corals from reefs. Between 10 and 16 % of living reef-building corals were lost in a single event in 1997-98. Projections of sea temperatures from several major general circulation models (CSIRO, Hadley, Max Planck) indicate that the frequency and intensity of mass coral bleaching events will increase steadily over the next 20-70 years, with thermal stresses (measured via Degree Heating Months, DHMs) rising well above that has caused major mortality events like that of Palau, Okinawa and the Seychelles (DHM > 2.3) in 1998. Inspection of the past behaviour of reefs under thermal stress and the likely trends in sea temperature indicate that coral reef communities are already changing and that a major loss of reef integrity and biodiversity in tropical near shore communities is almost certain. The study of the implications for human users of coral reef ecosystems is in its infancy and should be a priority of future studies.

REFERENCES

Andrewartha H.G. & Birch, L. C., 1954, *The distribution and abundance of animals*. University of Chicago Press, Chicago, USA

Aronson R.B., Precht W.F., MacIntyre I. G. & Murdoch J.T., 2000, Coral bleach-out in Belize. *Nature* 405: 36.

Aronson R.B. & Precht W.F., 1999, White band disease and the changing face of Caribbean Coral Reefs. In: J.W. Porter (ed.) Special Issue on Aquatic Diseases, *Hydrobiolgia*, in press.

Atkinson M.J., 1992, Productivity of Enewetak reef flats predicted from mass transfer relationships. *Cont. Shelf Res.* 12: 799-807.

Ayre D.J. & Hughes T.P., 2000, Genotypic diversity and gene flow in brooding and spawning corals along the Great Barrier Reef, Australia. *Evolution* 54: 1590-1605.

Bacher A., Oberhuber J.M. & Roeckner E., 1997, ENSO dynamics and seasonal cycle in the tropical Pacific as simulated by the ECHAM4/OPYC3 coupled general circulation model. *Climate Dynamics* 14: 431-450.

Baker A.C., 2001, Ecosystems: Reef corals bleach to survive change. *Nature* 411: 765.

Berkelmans R. & Oliver J.K., 1999, Large Scale Bleaching of Corals on the Great Barrier Reef. *Coral Reefs*, in press.

Brown B. E. & Suharsono, 1990, Damage and recovery of corals reefs affected by El Nino related seawater warming in the Thousand Islands, Indonesia. *Coral Reefs* 8: 163-170.

Brown B.E., 1997, Coral bleaching: causes and consequences. *Coral Reefs* 16: 129-138.

Buddemeier R.W. & Fautin D.G., 1993, Coral Bleaching as an Adaptive Mechanism. A testable hypothesis. *BioScience* 43, No.5: 320-326.

Carte B.K., 1996, Biomedical potential of marine natural products. *BioScience* 46: 271-86.

Cesar H., Waheed A., Saleem M. & Wilhelmsson D., 2000, Assessing the impacts of the 1998 coral reef bleaching on tourism in Sri Lanka and Maldives. In: S. Westmacott, H. Cesar & L. Pet-Soede (eds.) *Assessment of the socio-economic impacts of the 1998 coral reef bleaching in the Indian Ocean*. CORDIO programme, Report prepared for the World Bank, Washington DC, USA.

Christoph M., Barnett T.P. & Roeckner E., 1999, The Antarctic Circumpolar Wave in a Coupled Ocean-Atmosphere GCM, *Journal of Climate*, in press.

Connell J.H., 1978, Diversity in Tropical Rainforests and Coral Reefs. *Science* 199: 1302-1310.

Darwin C.R., 1842, *The structure and Distribution of Coral Reefs*. Smith Elder and Company, London.

Done T.J., 1999, Coral community adaptability to environmental change at the scales of regions, reefs, and reef zones. *American Zoologist* 39: 66-79.

Dove S.G., Hoegh-Guldberg O. & Ranganathan S., 2001, Major colour patterns of reef-building corals are due to a family of GFP-like proteins. *Coral Reefs* 19: 197-204.

Drollet J.H., Faucon M. & Martin P.M.V., 1994, A survey of environmental physico-chemical parameters during a minor coral mass bleaching event in Tahiti in 1993. *Marine and Freshwater Research* 45: 1149-1156.

Furnas M.J. & Mitchell A.W., 1999, Wintertime carbon and nitrogen fluxes on Australia's Northwest Shelf. *Estuarine Coastal and Shelf Science* 49(2): 165-175.

Gates R.D., 1990, Seawater temperature and sublethal bleaching in Jamaica. *Coral Reefs* 8: 193-197.

Gates R.D., Baghdasarian G. & Muscatine L., 1992, Temperature stress causes host cell detachment in symbiotic cnidarians: Implications for coral bleaching. *Biol. Bull.* 182: 324-332.

Gattuso J.-P., Frankignoulle M., Bourge I., Romaine S. & Buddemeier R.W., 1998, Effect of calcium carbonate saturation of seawater on coral calcification. *Global Planetary Change* 18: 37-47

Gleason D.F. & Wellington G.M., 1993, Ultraviolet radiation and coral bleaching. *Nature* 365: 836-838.

Gleason M.G., 1993, Effects of disturbance on coral communities: bleaching in Moorea, French Polynesia. *Coral Reefs* 12: 193-201.

Glynn P.W., 1993, Coral reef bleaching ecological perspectives. *Coral Reefs* 12: 1-17.

Glynn P.W., 1984, Widespread coral mortality and the 1982-83 El Nino warming event. *Environ. Conserv.* 11: 133-146.

Glynn P.W., 1985a, Corallivore population sizes and feeding effects following El Nino 1982-83 associated mortality in Panama. *Proc. 5th Int. Coral Reef Symp.* 4: 183-188.

Glynn P. W., 1985b, El Nino-associated disturbance to coral reefs and post disturbance mortality by *Acanthaster planci*. *Mar. Ecol. Prog. Ser.* 26: 295-300.

Glynn P.W., 1988, El Nino-Southern Oscillation 1982-1983: nearshore population, community, and ecosystems responses. *Ann. Rev. Ecol. Syst.* 19: 309-345.

Glynn P.W., 1990, Coral mortality and disturbances to coral reefs in the tropical eastern Pacific. In: P.W. Glynn (ed.) *Global ecological consequences of the 1982-83 El Nino-Southern Oscillation*. Elsevier, Amsterdam, pp. 55-126.

Glynn P.W., 1991, Coral reef bleaching in the 1980s and possible connections with global warming. *Trends Ecol. Evol.* 6: 175-179.

Glynn P.W. & D'Croz L., 1990, Experimental evidence for high temperature stress as the cause of El Niño coincident coral mortality. *Coral Reefs* 8: 181-191.

Glynn P.W. & Colgan M.W., 1992, Sporadic disturbances in fluctuating coral reef environments: El Nino and coral reef development in the Eastern Pacific. *Amer. Zool.* 32: 707-718.

Goreau T.J., 1992, Bleaching and Reef Commumity Change in Jamaica: 1951-1991. *Amer. Zool.* 32: 683-695.

Goreau T.J. & Hayes R.L., 1994, Coral Bleaching and Ocean "Hot Spots". *AMBIO* 23: 176-180.

Harriot V.J., 1985, Mortality rates of scleractinian corals before and during a mass-bleaching event. *Mar. Ecol. Prog. Ser.* 21: 81-88.

Hoegh-Guldberg O. & Salvat B., 1995, Periodic mass bleaching of reef corals along the outer reef slope in Moorea, French Polynesia. *Mar. Ecol. Prog. Ser.* 121: 181-190.

Hoegh-Guldberg O. & Smith G.J., 1989, The effect of sudden changes in temperature, irradiance and salinity on the population density and export of symbiotic dinoflagellates from the reef corals *Stylophora pistillata* Esper 1797 and *Seriatopora hystrix* Dana 1846. *J. Exp. Marine Biol. & Ecol.* 129: 279-303.

Hoegh-Guldberg O., 1994, *Mass-bleaching of Coral Reefs in French Polynesia*. Greenpeace International. Report, 36 pp.

Hoegh-Guldberg O., 1999, Climate Change, coral bleaching and the future of the world's coral reefs. *Mar. Freshwater Res.* 50: 839-66.

Hoegh-Guldberg O., Hoegh-Guldberg H., Stout D.K., Cesar H. & Timmermann A., 2000, *Pacific in Peril*. Biological, economic and social impacts of climate change on Pacific coral reefs. Special Report for Greenpeace International ISBN 1 876 221 10 0, 103pp.

Hughes T.P., 1994, Catastrophes, phase shifts, and large-scale degradation of a Caribbean coral reef. *Science* 265: 1547-1551.

Iglesias-Prieto R., Matta J., Robins W.A. & Trench R.K., 1992, Photosynthetic response to elevated temperature in the symbiotic dinoflagellate *Symbiodinium microadriaticum* in culture. *Proc. Natl. Acad. Sci. USA* 89: 10302-10305.

IPCC, 2001, Intergovernmental Panel on Climate Change, *Third Assessment Report*, in press.

IPCC, 1992, *Climate Change 1992*. The Supplementary Report to the IPCC Scientific Assessment. Edited by J.T. Houghton, B.A. Callander & S.K.V. Varney. Cambridge University Press, 200pp.

Jackson J.B.C., 1997, Reefs since Columbus, *Coral Reefs* 16 (Suppl.): 23-32.

Johnstone R.W., 1989, *The significance of coral lagoon sediments in mineralisation of organic matter and inorganic nitrogen fluxes*. PhD Thesis, University of Sydney, Sydney, Australia.

Johnstone R.W., Koop K. & Larkum A.W.D., 1990, Community metabolism and nitrogen recycling in sediments of a coral reef lagoon; One Tree Is., Great Barrier Reef. *Mar. Ecol. Prog. Ser.* 66: 273-283.

Jones R., Hoegh-Guldberg O., Larkum A.W.L. & Schreiber U., 1998, Temperature induced bleaching of corals begins with impairment of dark metabolism in symbiotic dinoflagellates. *Plant Cell Environ.* 21: 1219-1230.

Kepler C.B., Kepler A.K. & Ellis D.H., 1994, The natural history of Caroline Atoll, Southern Line Islands: Part II. Seabirds, other terrestrial animals, and conservation. *Atoll-Research-Bulletin* 0 398 I-III: 1-61.

Kleypas J.A., McManus J.W. & Menez L.A.B., 1999a, Environmental limits to reef development: Where do we draw the line? *Amer. Zool.* 39: 146-159.

Kleypas J.A, Buddemeier R.W., Archer D., Gattuso J.-P., Langdon C. & Opdyke B. N., 1999b, Geochemical Consequences of Increased Atmospheric Carbon Dioxide on Coral Reefs. *Science* 284: 118-120.

Knowlton N., Weil E., et al., 1992, Sibling Species in *Montastraea annularis*, Coral Bleaching, and the Coral Climate Record. *Science* 255: 330-333.

Langdon C., 2000, Direct effect of elevated atmospheric CO_2 on the health of coral reefs. *Proceedings of the JAMSTEC International Coral Reef Symposium.* February 23-24, 2000, 176pp.

Langdon C., Takahashi T., Sweeney C., Chipman D., Goddard J., Marubini F., Aceves H., Barnett H. & Atkinson M. J., 2000, Effect of calcium carbonate saturation state on the calcification rate of an experimental coral reef. *Global Biogeochemical Cycles* 14: 639-654.

Levitus S., Antonov J.I., Boyer T.P., Stephens C., 2000, Warming of the World Ocean. *Science* 287: 2225-2229.

Lough J.M., 1999, *Sea surface temperatures on the Great Barrier Reef: a contribution to the study of coral bleaching.* Final Report, Great Barrier Reef Marine Park Authority.Maragos, 1994.

McClanahan T.R., 1988, Seasonality in East Africa's coastal waters. *Mar. Ecol. Prog. Ser.* 44: 191-199.

Meesters E.H. & Bak R.P.M., 1993, Effects of coral bleaching on tissue regeneration potential and colony survival. *Mar. Ecol. Prog. Ser.* 96: 189-198.

Muscatine L., 1990, The role of symbiotic algae in carbon and energy flux in reef corals. *Coral Reefs* 25: 1-29.

Muscatine L., 1973, Nutrition of corals. In: O.A. Jones & R. Endean (eds.) Biology and Geology of Coral Reefs, Vol. 2. Œ115. Academic Press, New York, 77pp.

Muscatine L. & Porter J.W., 1977, Reef corals: mutualistic symbioses adapted to nutrient-poor environments. *Bioscience* 27: 454-459.

Myhre G., Myhre A. & Stordal F., 2001, Historical evolution of radiative forcing of climate. *Atmos. Environ.* 35: 2361-2373.

Odum H.T. & Odum E.P., 1955, Trophic Structure and Productivity of Windward Coral Reef community on Eniwetok Atoll. *Ecol. Monogr.* 253: 291-320.

Parmesan C., Ryrholm N., Stefanescu C., Hill J.K, Thomas C.D., Descimon H., Huntley B., Kaila L., Kullberg J., Tammaru T., Tennet W.J., Thomas J.A. & Warren M., 1999, Poleward shifts in geographical ranges of butterfly species associated with regional warming. *Nature* 399: 579-583.

Pomerance R., 1999, *Coral Bleaching, Coral Mortality, And Global Climate Change.* Report presented by Deputy Assistant Secretary of State for the Environment and Development to the U.S. Coral Reef Task Force, 5 March 1999, Maui, Hawaii.

Precht W.F. & Aronson R.B., 1999, *Catastrophic mortality of corals in the Belizean shelf lagoon associated with the 1998 bleaching event.* 28th Benthic Ecology Meeting, Baton Rouge, LA, 75pp.

Ridgway T., Hoegh-Guldberg O. & Ayre D., 2001, Panmixia in *Pocillopora verrucosa* from South Africa. *Marine Biology*, in press.

Rodriguez-Lanetty M, Loh W., Carter D. & Hoegh-Guldberg O., 2001, Latitudinal variability in symbiont specificity within the widespread scleractinian coral *Plesiastrea versipora. Marine Biology*, in press.

Roeckner E., Bengtsson L., Feichter J., Lelieveld J. & Rodhe H., 1999, Transient climate change simulations with a coupled atmosphere-ocean GCM including the tropospheric sulfur cycle. *Journal of the Climate*, in press.

Roeckner E., Oberhuber J.M., Bacher A., Christoph M. & Kirchner I., 1996, ENSO variability and atmospheric response in a global atmosphere-ocean GCM. *Climate Dynamics* 12: 737-754.

Salvat B., 1991, Blanchissement et mortalite des scleractiniaries sur les recifs de Moorea archipel de la Societe en CR Academy of Science Paris 314: 353391.

Shulman M. & Robertson D. R., 1996, Changes in the coral reefs of San Blas, Caribbean Panama: 1983-1990. *Coral Reefs* 15: 231-236.

Spencer T., Teleki T.A., Bradshaw C. & Spalding M., 2000, Coral Bleaching in the Southern Seychelles During the 1997±1998 IndianOcean Warm Event. Mar. Pollut. Bull. 40: 569-586.

Streets D.G., Tsai N.Y., Akimoto H. & Oka K., 2000, Sulfur dioxide emissions in Asia in the period 1985-1997. *Atmos. Environ.* 34: 4413-4424

Strong A.E., Gjovig K.K. & Kearns E., 2000, Sea Surface Temperature Signals from Satellites - An Update. *Geophys. Res. Lett.* 2711: 1667-1670.

Strong A.E., Toscano M.A. & Guch I.C., 2000, Extent and Severity of the 1998 Mass Bleaching Event from Satellite SST "Hotspot" Mapping. *Proceedings of the 9th International Coral Reef Symposium*, Bali October 2000.

Swanson R. & Hoegh-Guldberg O., 1998, The assimilation of ammonium by the symbiotic sea anemone *Aiptasia pulchella. Marine Biology* 131: 83-93.

Thunnell R., Anderson D., Gellar D. & Miao Q., 1994, Sea-Surface Temperature Estimates for the Tropical Western Pacific during the Last Glaciation and their Implications for the Pacific Warm Pool. *Quaternary Research* 41: 255-264.

Timmermann A., Latif M., Bacher A., Oberhuber J.M. & Roeckner E., 1999, Increased El Niño frequency in a climate model forced by future greenhouse warming. *Nature* 398: 694-696.

Trench R.K., 1979, The Cell Biology of Plant-Animal Symbiosis. *Annu. Rev. Plant Physiol.* 30: 485-531

Veron J.E.N., 1986, *Corals of Australia and the Indo-Pacific*. Angus and Robertson, London, Sydney, 644pp.

Veron J.E.N., 2000, *Corals of the World*. Australian Institute of Marine Science 3 volumes.

Ward S., Jones R., Harrison P. & Hoegh-Guldberg O., 1998, *Changes in the reproduction, lipids and M.A.A.s of corals following the GBR mass bleaching event*. Abstract, Australian Coral Reef Society annual meeting in Port Douglas.

Westmacott S., Ngugi I. & Andersson J., 2000, Assessing the impacts of the 1998 coral reef bleaching on tourism in Tanzania and Kenya. In: S. Westmacott, H. Cesar, L. Pet Soede & J. De Schutter (eds.) *Assessing the socio-economic impacts of the coral reef bleaching in the Indian Ocean.* A CORDIO report submitted to the World Bank African Environment Department, Washington DC, USA.

Wilkerson C. R. (ed.), 2000, *Status of Coral reefs of the World: 2000.* Global Coral Reef Monitoring Network. Australian Institute of Marine Science publications, Townsville, Australia.

Fingerprints of climate changes on the photosynthetic apparatus' behaviour, monitored by the JIP-test
A case study on light and heat stress adaptation of the symbionts of temperate and coral reef foraminifers in hospite

MEROPE TSIMILLI-MICHAEL*,# & RETO J. STRASSER*
*Bioenergetics Laboratory, University of Geneva, CH-1254, Jussy-Geneva, Switzerland
#Cyprus Ministry of Education and Culture, CY-1434, Nicosia, Cyprus
Address for correspondence: M. Tsimilli-Michael, 3 Ath. Phylactou, CY-1100, Nicosia, Cyprus

Abstract: We here present an outline of the JIP-test, a screening test based on the quantitative analysis of the chlorophyll *a* fluorescence rise O-J-I-P transient exhibited by all photosynthetic organisms. The JIP-test leads to the calculation of several phenomenological and biophysical parameters quantifying the photosystem II (PSII) behaviour. The parameters express (a) the energy fluxes for absorption, trapping and electron transport, (b) the flux ratios or yields, (c) the concentration of reaction centres and (d) a performance index. The JIP-test was proved to be a very useful tool for the in vivo investigation of the adaptive behaviour of the photosynthetic apparatus to a wide variety and combination of stressors, e.g. high or low temperature, high light intensity, atmospheric CO_2 or ozone elevation etc. The successful "fingerprinting" of stressors on the behaviour/performance of the photosynthetic apparatus provides the basis for using this behaviour/performance as a bio-indicator of climate changes. As an example, a case study on light and heat stress adaptation of the symbionts of coral reef foraminifers in hospite is presented. This study tackles the question about the origin of the big environmental problem of massive bleaching of the reef ecosystem, which involves, besides corals, several other species among which large foraminifers, and corresponds to the loss of their photosynthetic symbionts and/or the symbionts' pigments. We used the JIP-test to investigate in three genera of large foraminifers the response of PSII of their symbionts in hospite upon light and heat stress (up to 32 °C). While low-light was found to offer a strong thermoprotection, heat stress in the dark was found to result in a wide decrease of the capacity for photosynthetic activity. Strong illumination induced even wider decreases of the photosynthetic activity, however highly reversible. The extent of the reversibility from the high-light stress was much bigger under low-light than in the dark. Since the symbionts' photosynthetic products are the major energy source for the symbiotic association, it is proposed that the reduction in their productivity and, concomitantly, of their delivery to the host, affects negatively the co-habitation and, hence, triggers bleaching. The results suggest that warm nights should be considered as a major factor provoking coral reef bleaching.

INTRODUCTION

The photosynthetic apparatus, and especially photosystem II (PSII), is well known to be very sensitive to different stresses. Stress and stress adaptation can therefore be monitored by following the behaviour of the photosynthetic apparatus. Chlorophyll (Chl) *a* fluorescence, though corresponding to a very small fraction of the dissipated energy from the photosynthetic apparatus, has been proven to be a very useful, non-invasive tool for the investigation of its structure and function. At ambient temperature Chl *a* fluorescence is basically emitted by PSII. The fluorescence transient, known as the Kautsky transient, consists of a rise completed in less than 1 s and a subsequent slower decline towards a steady state. The rise reflects the accumulation of the reduced form of the primary quinone acceptor Q_A, otherwise the closure of the reaction centres (RCs), which is the net result of Q_A reduction due to PSII activity and Q_A^- reoxidation due to photosystem I (PSI) activity. When the photosynthetic sample is kept for few minutes in the dark, Q_A is fully oxidised, hence the RCs are all open, and the fluorescence yield at the onset of illumination is denoted as F_0. The maximum yield F_P at the end of the fast rise, depending on the achieved reduction-oxidation balance, acquires its maximum possible value - denoted as F_M - if the illumination is strong enough to ensure the closure of all RCs. A lot of information has been driven during the last sixty years from the fluorescence transient (see e.g. Krause & Weis 1991, Govindjee 1995).

Transients recorded with high time-resolution fluorimeters, e.g. with the PEA-instrument (data acquisition every 10 μs for the first 2 ms and 1 ms thereafter), have provided additional and/or more accurate information (Strasser & Govindjee 1992, Strasser et al. 1995). It was shown that the fluorescence rise kinetics is polyphasic exhibiting clearly, when plotted on a logarithmic time scale, the steps J (at 2 ms) and I (30 ms) between the initial O (F_0) and maximum P level (F_P); moreover, a much more precise detection of F_0, as well of the initial slope which offers a link to the maximum rate of photochemical reaction, is succeeded.

All oxygenic photosynthetic material investigated so far using this method show this polyphasic rise, labelled as O-J-I-P. The shape of the O-J-I-P fluorescence transient has been found to be very sensitive to stress caused by changes in different environmental conditions, such as light intensity, temperature, drought, atmospheric CO_2 or ozone elevation and chemical influences (see e.g. Srivastava & Strasser 1996, 1997, Tsimilli-Michael et al. 1996, 1999, Van Rensburg et al. 1996, Krüger et al. 1997, Ouzounidou et al. 1997, Clark et al. 1998, 2000).

A quantitative analysis of the O-J-I-P transient has been introduced (Strasser & Strasser 1995) and further developed, named as the "JIP-test" after the steps of the transient, by which several phenomenological and biophysical - structural and functional - parameters quantifying the PSII behaviour are calculated (for a review see Strasser et al. 2000). The JIP-test was proven to be a very useful tool for the in vivo investigation of the adaptive behaviour of the photosynthetic apparatus and, especially, of PSII to a wide variety and combination of stressors, as it translates the shape changes of the O-J-I-P transient to quantitative changes of the several parameters. Hence the deviation of the constellation of these parameters from that of the non-stressed condition expresses quantitatively evaluated fingerprints of the stressors on the photosynthetic organism. Fingerprints of several kinds of stressors on the behaviour and performance of the photosynthetic apparatus have been so far detected, identified, analysed and "mapped" via the JIP-test (Srivastava & Strasser 1996, 1997, Tsimilli-Michael et al. 1996, 1999, 2000, Krüger et al. 1997, Clark et al. 1998, 2000). These research activities provide the basis for using the behaviour of the photosynthetic apparatus as a bio-indicator of climate changes.

We here present an outline of the JIP-test and an application of this approach in a case study on light and heat stress adaptation of the symbionts of coral reef foraminifers in hospite.

THE JIP-TEST

Chl a fluorescence transients exhibited by any photosynthetic material are measured by a PEA fluorimeter (Plant Efficiency Analyser, built by Hansatech Instruments Ltd. King's Lynn Norfolk, PE 30 4NE, GB). The transients are induced by a red light (peak at 650 nm) of 600 W m^{-2} (3200 µE m^{-2} s^{-1}) provided by an array of six light-emitting diodes, and recorded for 1 s with 12 bit resolution; the data acquisition is every 10 µs for the first 2 ms and every 1 ms thereafter (for details see Strasser et al. 1995).

A typical Chl a fluorescence transient O-J-I-P is shown in Figure 1, plotted on a logarithmic time scale so that the intermediate steps are clearly revealed. The following original data (see Figure 1) are utilised by the JIP-test: the maximal measured fluorescence intensity, F_P, equal here to F_M since the excitation intensity is high enough to ensure the closure of all RCs of PSII; the fluorescence intensity at 50 µs considered as the intensity F_0 when all RCs are open; the fluorescence intensity at 300 µs ($F_{300µs}$) or 150 µs ($F_{150µs}$) required for the calculation of the initial slope $M_0 = (dV/dt)_0 \cong (\Delta V/\Delta t)_0$ of the relative variable fluorescence (V) kinetics (see insert in

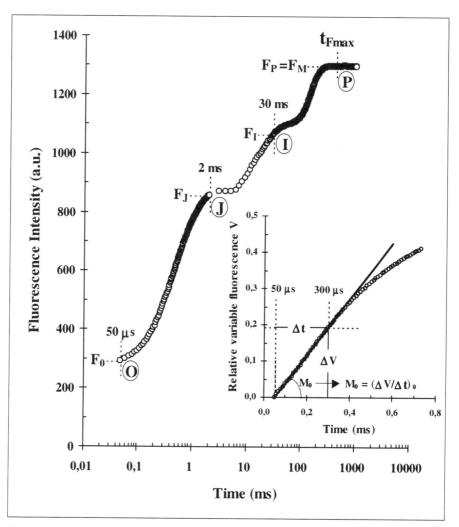

Figure 1. A typical Chl *a* polyphasic fluorescence rise O-J-I-P, exhibited by higher plants. The transient is plotted on a logarithmic time scale from 50 μs to 1 s. The marks refer to the selected fluorescence data used by the JIP-test for the calculation of structural and functional parameters. The signals are: the fluorescence intensity F_0 (at 50 μs); the fluorescence intensities F_J (at 2 ms) and F_I (at 30 ms); the maximal fluorescence intensity, $F_P = F_M$ (at t_{Fmax}). The insert presents the transient expressed as the relative variable fluorescence $V = (F-F_0)/(F_M-F_0)$ vs. time, from 50 μs to 1 ms on a linear time scale, demonstrating how the initial slope, also used by the JIP-test, is calculated: $M_0 = (dV/dt)_0 \cong (\Delta V/\Delta t)_0 = (V_{300\mu s})/(0{,}25\ ms)$.

Figure 1 and Table 1); the fluorescence intensities at 2 ms (J step) denoted as F_J, and at 30 ms (I-step) denoted as F_I (for a review see Strasser et al. 2000).

The JIP-test refers to a translation, through the formulae of Table 1, of the original data to the following biophysical parameters, all referring to time zero (onset of fluorescence induction), that quantify the PSII behaviour:

Fingerprints of climate changes on the photosynthetic apparatus 233

(a) the specific energy fluxes (per reaction centre) for absorption (ABS/RC), trapping (TR_0/RC), dissipation (DI_0/RC) and electron transport (ET_0/RC); (b) the flux ratios or yields, i.e. the maximum quantum yield of primary photochemistry ($\varphi_{P_o} = TR_0/ABS$), the efficiency ($\psi_0 = ET_0/TR_0$) with which a trapped exciton can move an electron into the electron transport chain further than Q_A^-, and the quantum yield of electron transport ($\varphi_{E_o} = ET_0/ABS = \varphi_{P_o} \bullet \psi_0$); the phenomenological energy fluxes (per excited cross section, CS) for absorption (ABS/CS), trapping (TR_0/CS), dissipation (DI_0/CS) and electron transport (ET_0/CS). The concentration of active PSII reaction centres per excited cross section (RC/CS) is also calculated. The set of formulae used for this translation is presented in Table 1.

Table 1. The formulae of the JIP-test

Specific fluxes		**Phenomenological fluxes**	
ABS/RC	= $(M_0/V_J) / [1 - (F_0/F_M)]$	ABS/CS	≈ F_0
TR_0/RC	= (M_0/V_J)	TR_0/CS	≈ $F_0 \bullet [1 - (F_0/F_M)]$
DI_0/RC	= (ABS/RC) - (TR_0/RC)	DI_0/CS	≈ (ABS/CS) - (TR_0/CS)
ET_0/RC	= $(M_0/V_J) \bullet (1-V_J)$	ET_0/CS	≈ $F_0 \bullet [1 - (F_0/F_M)] \bullet (1-V_J)$
		Density of reaction centres	
		RC/CS	≈ $F_0 \bullet [1 - (F_0/F_M)] / (M_0/V_J)$
Yields		**Yields as ratios of fluxes**	
φ_{P_o}	= $[1 - (F_0/F_M)]$	φ_{P_o}	= $(TR_0/RC)/(ABS/RC)$
φ_{E_o}	= $[1 - (F_0/F_M)] \bullet (1-V_J)$	φ_{E_o}	= $(ET_0/RC)/(ABS/RC)$
ψ_0	= $(1-V_J)$	ψ_0	= $(ET_0/RC)/(TR_0/RC)$
where, $V_J = (F_J-F_0)/(F_M-F_0)$			
and $M_0 = (dV/dt)_0 \equiv (\Delta V/\Delta t)_0 \cong 4 \bullet (F_{300\mu s}-F_0)/(F_M-F_0) \cong 10 \bullet (F_{150\mu s}-F_0)/(F_M-F_0)$			

Recently (for a review see Strasser et al. 2000) the performance index PI was introduced as:

$$PI = \frac{RC}{ABS} \bullet \frac{\varphi_{P_o}}{1-\varphi_{P_o}} \bullet \frac{\psi_0}{1-\psi_0}$$

Substitution of the biophysical by the experimental parameters (see Table 1) results in:

$$PI = \frac{1-(F_0/F_M)}{M_0/V_J} \bullet \frac{F_M - F_0}{F_0} \bullet \frac{1-V_J}{V_J}$$

As defined, the performance index is a product of expressions of the form $[p_i/(1-p_i)]$, where the p_i (i = 1, 2, ..., n) stand for probabilities or fractions. Such expressions are well-known in chemistry, with p_i representing e.g. the fraction of the reduced and $(1-p_i)$ the fraction of the oxidised form of a compound, in which case $\log[p_i/(1-p_i)]$ expresses the potential or driving force for the corresponding oxido-reduction reaction (Nernst's equation).

Extrapolating this inference from chemistry, the log(PI) can be defined as the total driving force (DF) for photosynthesis of the observed system, created by summing the partial driving forces for each of the several energy bifurcations (all at the onset of the fluorescence rise O-J-I-P):

$$DF = \log(PI) = \log\left(\frac{RC}{ABS}\right) + \log\left(\frac{\varphi_{Po}}{1-\varphi_{Po}}\right) + \log\left(\frac{\psi_o}{1-\psi_o}\right)$$

It is worth pointing out that the JIP-test reveals changes in the PSII behaviour that cannot be detected by the commonly used $\varphi_{Po} = (F_M-F_0)/F_M$, which is the least sensitive of all parameters. Moreover, φ_{Eo} and PI are related to the productivity of photosynthetic metabolites and, hence, they offer a diagnostic tool for the biomass production capability.

The JIP-test has been widely and successfully used for the study of PSII behaviour in various photosynthetic organisms under different stress conditions, which result in the establishment of different physiological states. The JIP-test has also been used to study synergistic and antagonistic effects of different co-stressors. Moreover, it has been proven a useful tool for the investigation of the beneficial effects of the participants in rhizosphere/mycorrhizosphere systems on the plant, as well as of their interactions. The big advantages of the method are: (1) it provides an early diagnosis of primary stress effects on the photosynthetic organisms; (2) it is rapid – less than a few seconds are needed for each measurement; (3) it can be applied in vivo; (4) it can be carried out anywhere - in the field, in the green house or even in tissue cultures - and even on samples as small as 2 mm^2; (5) it is non-invasive; (6) it is very inexpensive.

THE CASE STUDY: LIGHT AND HEAT STRESS ADAPTATION OF THE SYMBIONTS OF TEMPERATE AND CORAL REEF FORAMINIFERS IN HOSPITE

Since the early 80's massive bleaching affects the reef ecosystem at different places of the world (see e.g. Williams & Bunkley-Williams 1990). Bleaching involves not only corals but also all other cnidarians, molluscs, sponges, ascidians and large foraminifers in symbiotic association with diatoms, dinoflagellates or cyanobacteria, and corresponds to the loss of the symbionts or their photosynthetic pigments, hence the discoloration. Subsequent mortality is highly variable. The phenomenon is still poorly understood, though temperature elevation and strong irradiation are considered as primary factors. It is even not yet clear whether the hosts or the symbionts are more susceptible and thus responsible for the symbiosis rupture.

From studies of land plant stress it is known that photosystem II (PSII) is one of the most sensitive components of the photosynthetic apparatus towards both heat and light stress. Therefore, there were several attempts to investigate the behaviour of PSII of the photosynthetic symbionts upon bleaching-like conditions. Some of these studies used also Chl a fluorescence to probe PSII behaviour in corals associations or their isolated symbionts (Iglesias-Prieto et al. 1992, Iglesias-Prieto 1995, Warner et al. 1996). The implication of PSII in coral reef bleaching has become a strong consideration, as there are evidences of both its thermolability and its sensitivity to high light intensities.

It was therefore of interest to investigate whether the JIP-test could also be applied to investigate the response of PSII of foraminifers' symbionts in hospite upon light and heat stress (up to 32 °C). We used three genera of large foraminifers: *Amphistegina lobifera*, harbouring as symbiont the diatom *Fragilaria* sp., and *Amphisorus heimprichii* and *Sorites variabilis* who carry the same symbiont as corals (*Symbiodinium* sp.). *Amphistegina* and *Amphisorus* are main components of the coral reef ecosystem producing a large amount of $CaCO_3$, while *Sorites* is a temperate foraminifer.

The experiment

Amphistegina and *Amphisorus* were collected in Mauritius and *Sorites* in the Mediterranean sea near Nice, at 1,5 meters depth. The foraminifers were selected, cleaned and distributed in glass-tubes with 5,5 ml of daily exchanged Mediterranean sea-water (pCO_2-controlled pH at about 8,2), kept in a thermostated water bath with a temperature gradually elevated from 25 to 32 °C, and exposed to light-dark cycles: 12 h light at 70 $\mu E\ m^{-2}\ s^{-1}$ (L) – 12 h dark (D).

The fluorescence measurements were conducted every 6 hours. Two parallel sets, of 10 tubes each, were used. For the one set, the fluorescence transients, denoted as S_1, were recorded for 5 s, preceded by 1 min dark. For the other set the following light regime was applied at each measuring period: The organisms were first kept in the dark for 1 min, then excited for 5 s exhibiting the fluorescence transient at state-1, denoted however here as S_{11} (see in Results and Discussion). They were then kept again in the dark for 1 min and excited for 2 min exhibiting a fast rise followed by a slow decline – which reflects the state-1 to state-2 transition; hence, this transient is denoted as S_{12}. At the achieved state-2, a fluorescence transient (S_2) was then recorded for 5 s, preceded again by a dark interval of 1 min. (For more details on the experimental protocol see Tsimilli-Michael et al. 1998). All transients were analysed according to the JIP-test. The statistical analysis of the basic parameters, i.e. φ_{Po}, V_J and M_0 (see Table 1), showed that their changes upon temperature elevation and the light-dark cycles, as well as upon the state changes, are statistically significant.

Results and discussion

The O-J-I-P fluorescence transient of all the three studied species exhibited a sequence of more than one steps between the J-step and the fluorescence maximum, as shown for example in Figure 2 for *Amphisorus* being at the S_1 state. Therefore, the polyphasic transient is here denoted as O-K-J-I-H-G, where the labelling follows an alphabetic order from the slower to the faster phases; depending on the experimental conditions, any step can be the highest, i.e. the P-step (Tsimilli-Michael et al. 1998). As no changes of the F_0 level were observed upon temperature elevation, we express the fluorescence values by the F_t/F_0 ratios, thus avoiding fluctuations due to the movement of the cells in the glass tubes.

The shape of the transient depends on the cultivation light conditions (light-dark) while the amplitude of the variable fluorescence decreases upon temperature elevation; however, the decrease of F_t/F_0 occurs mainly when the cultures had been in the dark phase, while in the light phase (70 µE m^{-2} s^{-1}) only minor decreases take place (Figure 2). This is an expression of a low-light thermoprotection (Havaux & Strasser 1990, Srivastava & Strasser 1996), here realised by reversing the deformations occurring in the dark. The transients do not show the K-step (at 300 µs), which appears when a reduction of the oxygen evolving capacity occurs (Srivastava et al. 1997, Strasser 1997).

The O-J part of a transient is considered as the photochemical phase, i.e. the phase reflecting single turn over events (Q_A reduction by PSII activity), while the J-P phase reflects multiple turn over events, i.e. Q_A^- accumulation as the net result of Q_A reduction by PSII and Q_A^- reoxidation via the electron transport chain (Strasser et al. 1995). The distinction is supported by the finding that the in vivo exhibited O-J phase coincides with the monophasic (O-J=I=P) fluorescence rise exhibited in the presence of an inhibitor of Q_A^- reoxidation (e.g. DCMU: -3-(3,4-dichlorophenyl)-1,1-dimethylurea), when both are normalised from 0 to 1 (Strasser & Strasser 1995). Hence, for studying single turn over events, it is not necessary to use the relative variable fluorescence $V = (F-F_0)/(F_M-F_0)$ displayed by DCMU-poisoned photosynthetic material, but simply the $W = (F-F_0)/(F_J-F_0) = V/V_J$ from in vivo measurements (see e.g. Stirbet et al. 1998). Re-plotting the fluorescence transients of Figure 2 as $W = f(t)$, it is clearly revealed that the cultivation

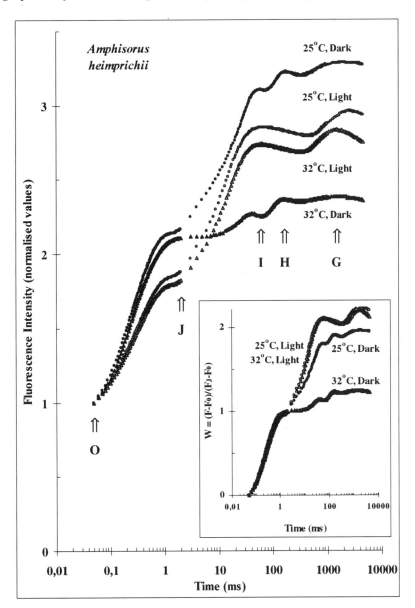

Figure 2. Polyphasic fluorescence transients exhibited by *Amphisorus heimprichii* at state S_1. The cultivation conditions were 12 hours in the dark or in the light of 70 µE m^{-2} s^{-1}, both at 25 °C and 32 °C. The transients were normalised on F_0 and presented on a logarithmic time scale from 50 µs to 5 s.

Compared to a typical O-J-I-P transient (Figure 1), a sequence of more steps is revealed after the I-step in the transients exhibited by *Amphisorus* (as well as by all tested foraminifers); hence the polyphasic transient is here denoted as O-K-J-I-H-G, where the labelling follows an alphabetic order from the slower to the faster phases; depending on the experimental conditions, any step can be the highest (P-step). The insert presents a re-plotting of the same transients expressed as W=f(t), where W=(F-F_0)/(F_J-F_0)=V/V_J. For other details see text.

conditions did not affect the single turn over events, as the O-J phase (0 to 1) is identical in all four transients (insert of Figure 2).

The transients presented in Figure 2 refer to a dark adapted or low-light adapted state, hence considered as state-1 (S_1), since they were exhibited by cells, which, throughout the experiment, were exposed every 6 h to the PEA-light for only 5 s, preceded by 1 min dark (mode A). Another excitation light regime (mode B) and a parallel set of biological samples were used to investigate, at each measuring period, the state changes under the strong red light of the PEA-fluorimeter (3200 µE m^{-2} s^{-1}). Three subsequent transients were recorded: the fast rise at the dark or low-light adapted state (state-1), the full Kautsky transient during the 2 min exposure, i.e. the fast rise followed by a slow decline – which reflects the state-1 to state-2 transition (hence the notation S_{12}), and finally the fast fluorescence rise (S_2) at the achieved state-2. Comparing the state-1 transients from the two sets, we observed that the transients recorded under mode B show a slightly lower variable fluorescence; the analysis according to the JIP-test revealed further differences concerning all calculated parameters (see below). This means that the changes upon the S_1 to S_2 transition are not completely reversible. Therefore, we denoted as S_{11} the state-1 recorded under mode B, which carries the "memory" of the previous S_1 to S_2 transitions, and kept the notation S_1 only for the state recorded under mode A, which has not a history of such transitions (for more details see Tsimilli-Michael et al. 1998). A set of the four transients exhibited by *Amphisorus* and recorded after a 6 hours light phase (70 µmol m^{-2} s^{-1}) at 30 °C is presented as an example in Figure 3. The S_1 transient was normalised on F_0. A slight and reversible decrease of F_0 appeared after exposure to the strong light of the PEA-instrument (3200 µE m^{-2} s^{-1}) inducing the S_{11} to S_2 transition, as also observed in higher plants when the photosynthetic apparatus is driven from the dark-adapted to a light-adapted state. In order to keep this information, the transients S_{11}, S_{12} and S_2 were normalised on the F_0 of S_{11}; the normalisation is well permitted since these three transients were recorded on the same sample.

The transients recorded throughout the experiment can be compared by plotting the ratios F_M/F_0, F_I/F_0 and F_J/F_0 for the whole sequence of measuring periods (Tsimilli-Michael et al. 1998), as e.g. shown in Figure 4 for the performance index. In order to proceed further than the phenomenological level and deduce biophysical information, we analysed all the transients according to the JIP-test, which leads to the calculation of the several structural and functional parameters quantifying the PSII behaviour (see Table 1; note that for M_0, the formula $M_0 = 10 \cdot (F_{150\mu s} - F_0)/(F_M - F_0)$ was used). Each single parameter, for each species and each state (S_1, S_{11}, S_{12} and S_2) can be presented for the whole sequence of measuring periods. Since the response of all three species to temperature elevation were found to show

Fingerprints of climate changes on the photosynthetic apparatus 239

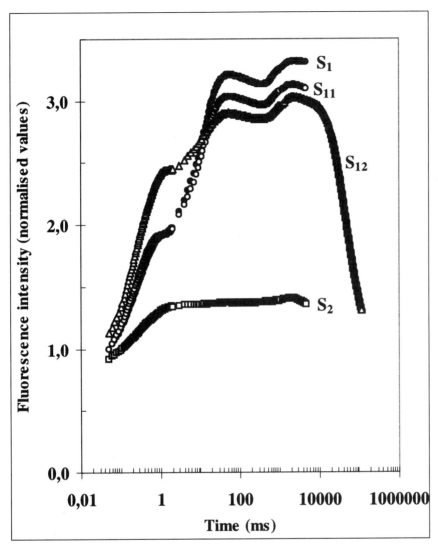

Figure 3. The fluorescence transients exhibited by *Amphisorus heimprichii* at the states S_1, S_{11}, S_{12} and S_2. The cultivation conditions were 6 hours in the light (70 μmol m^{-2} s^{-1}) at 30 °C. The transients are presented on a logarithmic time scale from 50 μs to 5 s, or up to 2 min (S_{12}). The S_1 transient was normalised on F_0, while the S_{11}, S_{12} and S_2 were normalised on the F_0 of S_{11}. For other details see text.

Figure 4 (next page). The response of *Amphisorus heimprichii* to the sequence of light-dark cycles (light of 70 μmol m^{-2} s^{-1}), under a gradually elevated cultivation temperature (25 °C, 30 °C and 32 °C). The PSII behaviour is described by the patterns of the total driving force log[PI] at all four states S_1, S_{11}, S_{12} and S_2. The patterns of the partial driving forces log[$\varphi_{Po}/(1-\varphi_{Po})$], log[$\psi_o/(1-\psi_o)$] and log[RC/ABS] are also presented, on the same scale, so that the contribution of each of them in building up the total driving force is visualised. For more details see text.

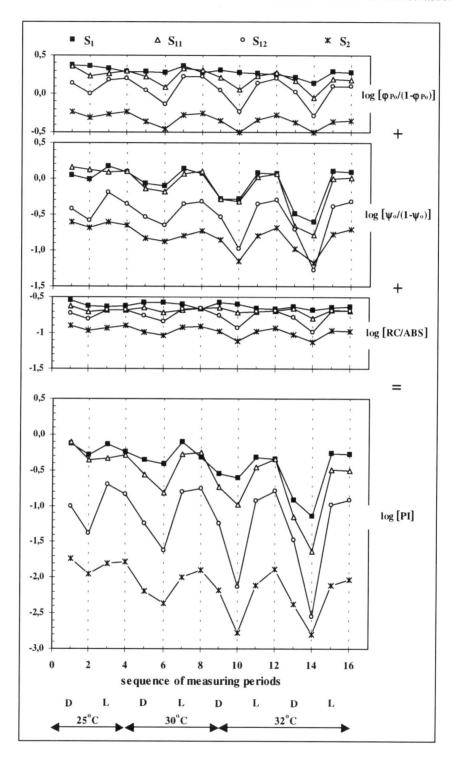

similar patterns, we present, as representative, the results from one of them, namely *Amphisorus*. However, it is worth pointing out that the three species showed different sensitivity. Upon temperature elevation, *Sorites* showed the most and *Amphistegina* the less wide changes of the parameters, both in the dark and in the light, indicating that the different species acquire different physiological sensitivities to heat. The difference between *Amphistegina* and *Amphisorus*, which are both coral reef foraminifers, reveals a different physiology of their symbionts. The difference between the two Soritids (*Amphisorus* and *Sorites*) hosting the same symbiont might be due to the long-term adaptation in their different natural habitats; the temperate foraminifer *Sorites*, adapted to a lower temperature range in the Mediterranean, show lower thermotolerance limits.

As mentioned in section 2, the comparison of the PSII behaviour by means of the different parameters of the JIP-test can be visualised by several types of plotting of the several calculated parameters. We here chose to present the driving forces. The logarithm of the performance index, defined as the total driving force (DF) for photosynthesis, DF = log[PI], can be written as the sum of three partial driving forces (see section 2). In Figure 4 we present the time course - throughout all the measuring periods - of the total driving force, together with its components. Note that the scale was kept the same for all of them, for a better visualisation of the contribution of each of the partial driving forces in building up the total driving force. The results refer to all the four states S_1, S_{11}, S_{12} and S_2 in *Amphisorus*. From the patterns of Figure 4 it is demonstrated that the partial driving forces undergo deformations of different extent, thus indicating that different survival strategies are employed in response to stress. Moreover, they show differences concerning the extent of low-light thermoprotection.

The quantum yield φ_{Po}, as experimentally determined by $[1-(F_0/F_M)]$, refers to the whole sample, averaging the yield of all photosynthetic units. The expression of φ_{Po} as ratio of fluxes (see Table 1) can be rewritten as $\varphi_{Po} = (TR_0/RC).(RC/ABS)$. It is worth pointing out that the trapping flux TR_0/RC is equal to the initial slope of the fluorescence transient when it is expressed as $W = f(t)$, as can be easily deduced from the formula $TR_0/RC = M_0/V_J$ given in Table 1. Hence, the coincidence of the O-J part of the transients $W = f(t)$ presented in the insert of Figure 2 reveals that the four cultivation conditions are characterised by the same TR_0/RC. As calculated from all the transients throughout the experiment, the trapping flux (TR_0/RC) was slightly higher when the cells were in the light phase of their cultivation, but it was not affected by temperature elevation, neither by the S_1 to S_2 transition. This stability is indirectly revealed in Figure 4, where it is shown that the patterns of $\log[\varphi_{Po}/(1-\varphi_{Po})]$ follow those of $\log[RC/ABS]$. The ratio RC/ABS refers to the ratio of active reaction centres - in the sense of Q_A

reducing - to the total absorption, i.e. the absorption by all photosynthetic units, including those with an inactive RC; inactivation is attributed to the transformation of RCs to quenching sinks (see Tsimilli-Michael et al. 1998). We can therefore conclude that a fraction of RCs is inactivated due to temperature elevation and a much bigger fraction due to the exposure to the strong light of the PEA-fluorimeter inducing the S_1 to S_2 transition, while the photochemical activity of the still active RCs is not affected.

The partial driving force $\log[\psi_o/(1-\psi_o)]$ appears much more sensitive to the cultivation conditions than $\log[\varphi_{Po}/(1-\varphi_{Po})]$. It should be noted that, though they are completely independent, their changes are in phase, thus resulting in even more pronounced changes of the total driving force $\log[PI]$.

The deviation of the patterns from S_{11} to S_2 through S_{12} appears unaffected by the cultivation conditions, which means that the applied heat stress did not alter the capability of the photosynthetic apparatus of the symbionts to undergo regulatory changes when exposed for 2 min to the strong light (3200 μE m^{-2} s^{-1}) of the PEA-fluorimeter. Moreover, the response of the driving forces to temperature elevation follows the same pattern at the S_2 state as at the S_{11}. For each plotted driving force, the deviation of the S_{11} from the S_1 values expresses the residual strain (i.e. the non-reversed part of a deformation) after the recovery from the S_{11} to S_2 transition. No residual strain of $\log[\psi_o/(1-\psi_o)]$ is observed (Figure 4), revealing that the deformations of this driving force are highly elastic. A partial plasticity is observed in the deformations of $\log[\varphi_{Po}/(1-\varphi_{Po})]$. The residual strain of $\log[\varphi_{Po}/(1-\varphi_{Po})]$ is due, as calculated, to an incomplete reactivation of the RCs that were inactivated at S_2. (For the definition of plasticity, elasticity and residual strain, see Krüger et al. 1997, Tsimilli-Michael & Strasser 2001). As shown in Figure 4, the residual strain is smaller when the cultures are in the light phase, which indicates that low light, beside offering a thermoprotection at all states, enhances also the reactivation of the RCs.

For each physiological state, established by the different temperature and light conditions, the energy fluxes can be visualised be means of the pipeline models (Strasser 1987, Krüger et al. 1997, Strasser et al. 2000, Tsimilli-Michael et al. 2000). In these dynamic models, which can be directly constructed from the recorded fluorescence transients (BIOLYSER program, created by Ronald M. Rodriguez, Bioenergetics Laboratory, University of Geneva, http://www.unige.ch/sciences/biologie/bioen), the value of each energy flux is expressed by the appropriately adjusted width of the corresponding arrow. As an example, Figure 5 demonstrates the pipeline models constructed from the four transients of Figure 2 exhibited by *Amphisorus heimprichii* (cultivation conditions; 12 hours in the dark or in the light of 70 μE m^{-2} s^{-1}, both at 25 °C and 32 °C) at state S_1. Two types of models are presented: the left one (membrane model) refers to the reaction centre and thus deals with the specific energy fluxes (per RC); the right one refers to the excited chlorophyll (chlorophyll model) and thus deals with the fluxes per Chl, which are proportional to the corresponding fluxes per absorption (ABS/Chl is constant, arbitrarily taken equal to 1). The membrane model includes also a demonstration of the average "antenna size", which is proportional to ABS/RC. This value expresses the total absorption of PSII antenna chlorophylls divided by the number of active (in the sense of Q_A reducing) RCs. The inactive RCs are presented in the chlorophyll model by dark circles.

Studies of the temperature effect on the photosynthetic apparatus of the symbionts under laboratory conditions are mainly attributing the initiation of bleaching to damages of PSII. A possible damage is the denaturation of PSII, well known from heat stress studies in land plants, which is revealed by a sharp increase of F_0 after a threshold temperature (Havaux 1993). As reported by Iglesias-Prieto (1995), this was indeed observed in isolated symbionts, however at temperatures higher than 34 °C. On the other hand, bleaching, either as loss of symbionts or their pigments, would result in lower F_0 values. Since throughout the full course of our experiments the F_0 level remained unaffected by the temperature elevation, we can safely conclude that, in the certain temperature range used, no such damage of PSII occurred. Neither any reduction of the oxygen evolving capacity, which is also considered as a possible damage, has been observed in our experiments, as above explained.

High levels of natural light were found to provoke coral bleaching in synergism with elevated temperatures (Coles & Jokiel 1978). However, in our experiments, the illumination for 2 min every 6 h by the strong light of the PEA-instrument (3200 µE m^{-2} s^{-1}) used to induce the S_{11} to S_2 transition did not result in any damage of PSII, even in combination with increased temperature (32 °C). There are indications that low salinity and seawater acidification (due to CO_2 rise) are also possible co-stressors in nature, the synergism of which we intend to investigate in future experiments.

We can therefore conclude that in our experiments the symbionts were exposed to a heat stress within the limits of their adaptability, which is in accordance with the fact that the cells, after the end of the experiment, were not bleached. However, the adaptation of the photosynthetic apparatus of the symbionts was realised by structural changes which resulted in a down regulation of their photosynthetic capacity expressed by a pronounced decrease of the driving force for photosynthesis DF = log[PI]. Though the light used to induce the transition to the S_2 state is an artificial light, it can be speculated that the modifications occurring upon the S_{11} - S_2 transition show also the trend of the effect of strong light in the natural habitat of foraminifers, while the behaviour at S_{11} shows the trend of the recovery that

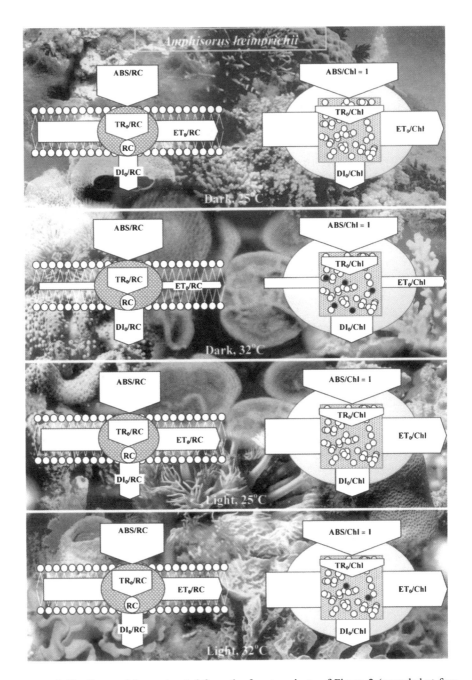

Figure 5. Pipeline models constructed from the four transients of Figure 2 (recorded at four different physiological states, see legend of Figure 2). For details see text.

follows. This speculation is based on the fact that, in the tropics at summer time, the photosynthetic photon flux density (PPFD) at the water surface during the diurnal changes is above 1500 µE m^{-2} s^{-1} for about 6 hours (Merino et al. 1995) and can be as high as 2000 µE m^{-2} s^{-1}. Moreover, the PPFD recorded in the water shows brief flashes with 60 % higher PPFD than that measured at the water surface, due to light focusing and defocusing by the waves (Merino et al. 1995). Taking into account the light absorption by a water layer of 1,5 m, we can estimate that the foraminifers receive in situ a basic PPFD up to 1600 µE m^{-2} s^{-1}, which can temporarily reach the 2500 µE m^{-2} s^{-1}.

It can be reasonably assumed, in agreement with a hypothesis proposed by Iglesias-Prieto (1995), that a reduction in the productivity of photosynthetic metabolites may cause a reduction of their delivery to the host. Since the symbionts' photosynthetic products are the major energy source for the symbiotic association, we speculate that, though for the symbionts as such the reduction of the productivity of metabolites is not harmful but represents an adaptive and protective procedure, it may have a negative effect on the co-habitation and may, thus, trigger bleaching by initiating the symbiosis rupture and the expulsion of symbionts. This working hypothesis has to be further tested, both in the laboratory and in the field.

It must be emphasised that the wide reduction of the photosynthetic capacity induced by heat stress, appeared when the cultures were kept in the dark prior to the measurements, while low light was found to offer thermoprotection; the reduction was even wider if the dark period was prolonged, as revealed by the comparison of the measurements after 12 h with those conducted after 6 h. On the other hand, though the decreases of the photosynthetic activity induced by strong illumination were even wider, they were highly reversible. The extent of the reversibility from the high-light stress was much bigger under low-light than in the dark. We can therefore speculate that a major factor triggering bleaching is the structural changes of PSII provoked by warm nights.

With the present work it is shown that the JIP-test can be considered as a convenient and powerful non-invasive tool for studying in vivo the vitality and adaptability of foraminifers. As we have successfully conducted fluorescence measurements in situ with the PEA-fluorimeter (unpublished results), we propose that the JIP-test could be used to screen the PSII behaviour in whole reef biotopes under stress caused by natural light and temperature changes, in an effort to get an access to the causes of bleaching.

ACKNOWLEDGEMENTS

We thank Dr. Martin Pêcheux, a specialist in foraminifers and coral reef bleaching, who provided the biological material and collaborated with us in the experiments on foraminifers. M. Tsimilli-Michael thanks the Ministry of Education and Culture of Cyprus for giving her the opportunity to carry out this work in Geneva. R. J. Strasser acknowledges financial support from the Swiss National Foundation (3100-052541.97 and 3100-057046.99).

REFERENCES

Clark A.J., Landolt W., Bucher J. & Strasser R.J., 1998, The response of *Fagus sylvatica* to elevated CO_2 and ozone probed by the JIP-test based on the chlorophyll fluorescence rise OJIP. In: L.J. De Kok & I. Stulen (eds.) *Responses of plant metabolism to air pollution and global change.* Backhuys Publishers, Leiden, The Netherlands, 283-286.

Clark A.J., Landolt W., Bucher J. & Strasser R.J., 2000, Beech (*Fagus sylvatica*) response to ozone exposure assessed with a chlorophyll a fluorescence performance index. *Environ. Pollut.* 109: 501-507.

Coles S.L & Jokiel P.L., 1978, Synergistic effects of temperature, salinity and light on the hermatypic coral *Montipora verrucosa. Mar. Biol.* 49: 187-195.

Govindjee, 1995, Sixty-three years since Kautsky: Chlorophyll a fluorescence. *Aust. J. Plant Physiol.* 22: 131-160.

Havaux M., 1993, Rapid photosynthetic adaptation to heat stress triggered in potato leaves by moderately elevated temperatures. *Plant Cell Environ.* 16: 461-467.

Havaux M. & Strasser R.J., 1990, Protection of photosystem II by light in heat-stressed pea leaves. *Z. Naturforsch.* 45C: 1133-1141.

Iglesias-Prieto R., 1995, The effects of elevated temperature on the photosynthetic responses of symbiotic dinoflagellates. In: P. Mathis (ed.) *Photosynthesis: From Light to Biosphere.* Kluwer Academic Publishers, The Netherlands, Vol V, 793-796.

Iglesias-Prieto R., Mata J.L., Robins W.A. & Trench R.K., 1992, Photosynthetic response to elevated temperature in the symbiotic dinoflagellate *Symbiodinium microadriaticun* in culture. *Proc. Natl. Acad. Sci. USA* 89: 10302-10305.

Krause G.H. & Weiss E., 1991, Chlorophyll fluorescence and photosynthesis: the basics. *Annu. Rev. Plant Physiol. Plant Mol. Biol.* 42: 313-349.

Krüger G.H.J., Tsimilli-Michael M. & Strasser R.J., 1997, Light stress provokes plastic and elastic modifications in structure and function of photosystem II in camellia leaves. *Physiol. Plant* 101: 265-277.

Merino M., Enríquez S., Strasser R.J. & Iglesias-Prieto R., 1995, Spatial and diurnal variations in the photosynthetic activity of the tropical seagrass *Thalassia testudinum*. In: P. Mathis (ed.) *Photosynthesis: From Light to Biosphere.* Kluwer Academic Publishers, The Netherlands, Vol V, 889-892.

Ouzounidou G., Moustakas M. & Strasser, R.J., 1997, Sites of action of copper in the photosynthetic apparatus of maize leaves: Kinetic analysis of chlorophyll fluorescence, oxygen evolution, absorption changes and thermal dissipation as monitored by photoacoustic signals. *Aust. J. Plant Physiol.* 24: 81-90.

Srivastava A. & Strasser R.J., 1996, Stress and stress management of land plants during a regular day. *J. Plant Physiol.* 148: 445-455.

Srivastava A. & Strasser, R.J., 1997, Constructive and destructive actions of light on the photosynthetic apparatus. *J. Sci. Ind. Res.* 56: 133-148.

Srivastava A., Guisse B., Greppin H. & Strasser R.J., 1997, Regulation of antenna structure and electron transport in PSII of *Pisum sativum* under elevated temperature probed by the fast polyphasic chlorophyll *a* fluorescence transient OKJIP. *Biochim. Biophys. Acta* 1320: 95-106.

Stirbet A., Srivastava A. & Stasser R.J., 1998, The energetic connectivity of PSII centres in higher plants probed in vivo by the fast fluorescence rise O-J-I-P and numerical simulations. In: G. Garab (ed.) *Photosynthesis: Mechanisms and Effects.* Kluwer Academic Publishers, The Netherlands, Vol V, 4317-4320.

Strasser B.J., 1997, Donor side capacity of photosystem II probed by chlorophyll *a* fluorescence transients. *Photosynth. Res.* 52: 147-155.

Strasser B.J. & Strasser R.J., 1995, Measuring fast fluorescence transients to address environmental questions: The JIP-test. In: P. Mathis (ed.) *Photosynthesis: From Light to Biosphere.* Kluwer Academic Publishers, The Netherlands, Vol V, 977-980.

Strasser R.J., 1987, Energy pipeline model of the photosynthetic apparatus. In: J. Biggins (ed.) *Progress in Photosynthesis Research.* Martinus Nijhoff, Dordrecht, Vol II, 717-720.

Strasser R.J. & Govindjee, 1992, The Fo and the O-J-I-P fluorescence rise in higher plants and algae. In: J.H. Argyroudi-Akoyunoglou (ed.) Regulation of Chloroplast Biogenesis, Plenum Press, New York, 423-426.

Strasser R.J., Srivastava A. & Govindjee, 1995, Polyphasic chlorophyll *a* fluorescence transient in plants and cyanobacteria. *Photochem. Photobiol.* 61: 32-42.

Strasser R.J., Srivastava A. & Tsimilli-Michael M., 2000, The fluorescence transient as a tool to characterize and screen photosynthetic samples. In: M. Yunus, U. Pathre & P. Mohanty (eds.) *Probing photosynthesis: mechanism, regulation and adaptation.* Taylor and Francis, London, UK, Chapter 25, 443-480.

Tsimilli-Michael M. & Strasser R.J., 2001, Mycorrhization as a stress adaptation procedure. In: S. Gianinazzi, H. Schüepp & K. Haselwandter (eds.) *Mycorrhizal Technology in Agriculture from Genes to Bioproducts.* Birkhauser, Basel (in press).

Tsimilli-Michael M., Krüger G.H.J. & Strasser R.J., 1996, About the perpetual state changes in plants approaching harmony with their environment. *Archs Sci. Genève* 49: 173-203.

Tsimilli-Michael M., Pêcheux M. & Strasser R.J., 1998, Vitality and stress adaptation of the symbionts of coral reef and temperate foraminifers probed *in hospite* by the fluorescence kinetics O-J-I-P. *Archs Sci. Genève* 51(2): 1-36.

Tsimilli-Michael M., Pêcheux M., & Strasser R.J., 1999, Light and heat stress adaptation of the symbionts of temperate and coral reef foraminifers probed *in hospite* by the chlorophyll *a* fluorescence kinetics O-J-I-P. *Z. Naturforsch.* 54C: 671-680.

Tsimilli-Michael M., Eggenberg P., Biro B., Köves-Pechy K., Vörös I. & Strasser R.J., 2000, Synergistic and antagonistic effects of arbuscular mycorrhizal fungi and *Azospirillum* and *Rhizobium* nitrogen-fixers on the photosynthetic activity of alfalfa, probed by the chlorophyll *a* polyphasic fluorescence transient O-J-I-P. *Appl. Soil Ecol.* 15: 169-182.

Van Rensburg, L., Krüger G.H.J., Eggenberg, P. & Strasser, R.J., 1996, Can screening criteria for drought resistance in *Nicotiana tabacum* L. be derived from the polyphasic rise of the chlorophyll *a* fluorescence transient (OJIP)? *S. Afr. J. Bot.* 62: 337-341.

Warner M.E., Fitt W.K. & Schmidt G.W., 1996, The effects of elevated temperature on the photosynthetic efficiency of zooxanthellae *in hospite* from four different species of reef coral: a novel approach. *Plant Cell Environ.* 19: 291-299.

Williams E.H. & Bunkley-Williams L., 1990, The world-wide coral reef bleaching cycle and related sources of coral mortality. *Atoll Res. Bull.* 335: 1-71.

Did recent climatic shifts affect productivity of grass-dominated vegetation in southern Switzerland?

Evidence from time series of two semi-natural grasslands and a maize field

ANDREAS STAMPFLI
Institute of Plant Sciences, University of Bern, Altenbergrain 21, CH-3013 Bern, Switzerland

Abstract: Time series of crop data from three grass-dominated ecosystems in southern Switzerland were selected for a comparison with climatic variables over the past decades. Standing crop from two permanent species-rich grasslands at Prugiasco and Salorino and the kernel yield of maize planted every year in monoculture at Cadenazzo were measured using standardised methods under controlled conditions in experimental areas over 13 (1988-2000) or 28 (1972-1999) years, respectively. The maize series was corrected for effects of sowing density and genetic variation due to the change of varieties over time. The sensitivity of crop variables to climatic variables, temperature, precipitation, relative humidity and duration of sunshine, recorded at Locarno-Monti, was calculated for 3-month and half-year intervals in 1-yr periods previous to harvest dates. Yields in semi-natural grasslands and in the maize field significantly responded to climatic variables, which were not subject to long-term trends, relative humidity or sunshine, respectively, during the growth period. A series of extremely dry summers negatively affected yields of the maize field and triggered a lagged shift towards a reduced grass-forb ratio in the harvest of the more intensively mown meadow. Dry summers were explained by a lee-effect of upper-level winds blowing from a more northward direction over the Alps in 1989-1991. It is concluded that grass-dominated vegetation in the Southern Alps was more sensitive to changing precipitation patterns than increased temperatures. The vegetation response to climate depended on the methods of human interference.

INTRODUCTION

In temperate areas an increase in summer or winter temperature is likely to alter the habitat conditions of living organisms and ecological processes such as migration or interactions among species in communities. Expected consequences include shifts of the geographical ranges of plant species and of the species composition in plant communities (community structure), and changes in ecosystem functioning, e.g. productivity (Westoby & Leishman 1997, Grime et al. 2000). Many aspects of the climate system have been observed and give a global picture of change (IPCC 2001). However, climatic shifts are not expected to be equally perceptible in all the regions of the world. In the Alps precipitation patterns are expected to respond sensitively to small changes in the large-scale atmospheric circulation (Gyalistras 2000). Shifts in precipitation patterns or drought frequency may alter the frequency or intensity of disturbances to which water-limited ecosystems are usually adapted. However, many other factors, which vary in relative importance from site to site, are also involved. In this context, long-term data sets are essential for improving our fragmentary understanding of how species-rich communities respond to variations in disturbance regimes.

Shifting patterns may also result from many other ecological processes driven by land-use change, a force which still has an increasing influence. Management systems and agricultural techniques have been subject to substantial changes over the past decades and progress in breeding techniques and varieties has resulted in a general linear increase in productivity of agricultural crops over the past 40 years (Weilenmann 1994). Data sets of biological variables over large spatial scales covering the full period of climatic change would ideally be needed to study the biological "fingerprints". Moreover, studies of species' and ecosystem responses to climate change must try to control for other superimposing driving forces. Opportunities for such studies, based on data series of standardised sampling quality, over many years or decades, are rare, however.

Over the past decades mean temperatures in Switzerland have shifted upwards. At low elevations in Ticino, this warming trend coincides with an unprecedented increase of exotic evergreen shrubs and trees in deciduous forests adjacent to gardens (Walther 2001). The phenomenon has been explained by a decreasing probability of lethal low temperatures in winter. Several authors in this volume have described the potential effects of climate warming in temperature-limited ecosystems.

In three grass-dominated plant communities in Ticino standardised sampling methods at yearly intervals, under the controlled conditions of field experiments, have been used. I measured standing crop in two species-rich meadows over 13 years, 1988-2000, one mown once, and the other twice, a year. Kernel yield was sampled in a maize monoculture over 28 years, 1972-1999, by few collaborators of the Swiss Federal Research Station for Agronomy. I compare the time series from these ecosystems with climatic variables calculated on daily measurements over different seasonal periods to answer the questions: (1) Are yields sensitive to particular climatic variables and do these variables show long-term trends? (2) Did differences in human impact such as replanting every year or irrigation affect the vegetation response to climate? A short summary of precise observations in one of the species-rich meadows will further elucidate the climate-productivity relationship in this type of vegetation in southern Switzerland.

DATA SERIES AND ANALYSIS

For many centuries semi-natural meadows and pastures have played a central role in the economy of the local people in the valleys of the Southern Alps. Over the past three to five decades they have strongly decreased in area because of economic changes (Antognoli et al. 1995). In Switzerland the few species-rich meadows which still exist today have become important habitats for conservation. Some of them are subject of long-term ecological studies (e.g. meadows at Negrentino or Pree).

In the valley bottoms of the Canton of Ticino maize (*Zea mays*) is the most important agricultural crop. As in other parts of Switzerland crop production has been characterised by a state-controlled crop market and a high turnover of varieties. Varieties have been tested continuously in field experiments by the Swiss Federal Research Station for Agronomy at Zürich.

In this chapter I combine yield data of two different studies sampled at three sites located in the 'lower montane' and 'lowland' zone of southern Switzerland: (a) slope of Monte Generoso, 960 m a.s.l., municipality of Salorino (a meadow at Pree, mown once every July, dominated by *Festuca tenuifolia, Anthoxanthum odoratum* and *Carex caryophyllea*), (b) slope of Blenio Valley, 850 m a.s.l., municipality of Prugiasco (a meadow at Negrentino, mown twice every June and September, dominated by *Bromus erectus, Festuca tenuifolia* and *Brachypodium pinnatum*), (c) valley bottom of Magadino, 210 m a.s.l., municipality of Cadenazzo (a maize field). The maize field is situated at a distance of 12 km to the east, and the meadows of Prugiasco and Salorino at distances of 33 km NNE and 35 km SSE of Locarno-Monti (371 m a.s.l.) where the climate data which I use in the analysis presented here, have been recorded by the Swiss Meteorological Institute at Zürich since 1965. From a data bank of monthly values, temperature, duration of sunshine, relative humidity and precipitation were calculated for seasonal periods of 3 months and half years in a 1-yr interval before harvesting, starting from closest to harvest dates. With this grouping of periods I intended to roughly divide the season of main growth (April-September, May-October) from the cold season. This resulted in 16 variables calculated over 3-month intervals corresponding to hay yields in June or July and eight variables calculated over half-year intervals corresponding to each of annual hay yield and kernel yield. I selected records of only one weather station as a reference because I expect annual series of climatic variables from different lowland stations in the Canton of Ticino to be strongly correlated.

Hay yield of meadows 1988-2000

Standing crop of the semi-natural grasslands was sampled along with other, population-level variables in the control plots of a succession experiment in species-rich meadows described earlier (Antognoli et al. 1995, Stampfli & Zeiter 2001, for sampling schemes see Stampfli 1993), and which has now run for 13 years, 1988-2000. At both sites fences put up in 1988 have excluded incidental grazers. In the control plots of these experiments the former mowing regimes have been maintained at traditional dates of hay-making (early July at Salorino, end of June and mid-September at Prugiasco). Yearly yield data are means of dried samples of phytomass (Salorino: five strips of 9 cm × 100 cm in each of four plots of 20 m^2; Prugiasco: one strip of 9 cm × 110 cm in each of nine plots of 4.4 m^2). Samples include all herbs in a strip cut with an electric lawn mower at peak standing crop, immediately before plots were mown with a scythe, and dried in a drying oven at temperatures of 80 °C for 24 h.

Kernel yield of maize fields 1972-1999

Kernel yield and other variables of a large number of maize varieties have been recorded in annually repeated experiments at the Stazioni Federali di Ricerche Agrarie, Sottostazione di Cadenazzo. Between 1972 and 1999 these experiments have been coordinated by Mathias Menzi and collaborators Flavio Lanini and Paolo Bassetti, and performed under standardised conditions. Every year, several varieties of maize were sown at the end of April or beginning of May in three replicates in a lattice-square design in unit plots of 10 m^2. The monocultures were fertilised with the minimum amounts of nitrogen, this based on the plants' requirements and soil analyses. They were normally irrigated if necessary, irrigation was not sufficient in 1991. No fungicides, insecticides or 'stalk shorteners' were applied. The plants were normally harvested in late October or early November. Kernel yields were standardised to a water content of 14 %.

Before 1987 the plant density was 6 m^{-2} when harvested, between 1987 and 1996 it was 7 m^{-2}, and after 1996 it was 6 - 6.5 m^{-2}. The effect of different harvest densities on kernel crop yield is known from other field experiments at different densities. An increase from 6 to 7 plants m^{-2} results in an 11 % increase in yield. I, therefore, corrected those kernel yields produced in years with a density of 7 m^{-2} by multiplication by 0.9, and in years with a density of 6.5 m^{-2} by 0.95.

Among the many varieties selected for experimental tests every year a few were maintained as standards for cross-reference between successive years. Before a standard variety was replaced it was grown in parallel with a new standard variety for several years. This offered the possibility of extending the time range of a standard variety beyond the period it was grown in the field, and to construct a time series for maize which was also corrected for the changes in varieties. From the archives of the Swiss Federal Research Station for Agronomy at Zürich I selected the variety *'orla 312'* because this had the longest unbroken time series, 1972-1992 (Figure 1). The varieties *'eva'* and *'orla 312'* overlapped for 9 years (1984-1992), and *'furio G-4207'*, which has been grown since 1990, overlapped with *'eva'* for 7 years (1990-1995, 1997). In overlapping years *'orla 312'* produced between 0.86- and 0.93-times (0.9006-times on average) smaller yields than *'eva'*, and *'eva'* between 0.89- and 0.99-times (0.9523-times on average) smaller yields than *'furio G-4207'* (results of 1991 excluded for both comparisons). Values for *'orla 312'* after 1992 were thus reconstructed by multiplication with 0.8577 of yields of *'furio G-4207'*.

Data analysis

The three time series of hay and kernel yield were analysed for temporal autocorrelation. For the corresponding time periods climatic trends were analysed by linear regressions of the climatic variables on year. Multiple linear regressions with an automatic stepwise-forward-selection procedure (with alpha-to enter and alpha-to-remove probability levels of 0.15) was used to find the best fitting set of climatic variables. Climatic variables, which contributed significantly ($p < 0.05$) were included in full models of the three yield series. Multiple linear regressions were repeated with climatic variables lagged 1 yr. All calculations were performed with the computer programme SYSTAT (version 8.0).

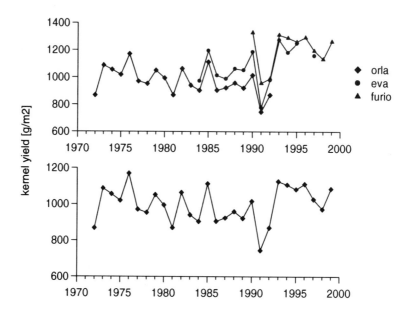

Figure 1. Kernel yield in maize field at Cadenazzo, varieties *orla, eva,* and *furio,* and extended 28-year series for *orla.*

RESULTS

Time series of hay and kernel crop yields showed considerable change form year to year, but no linear trends at the scale of decades, nor temporal autocorrelation. Climatic variables corresponding to the seasonal intervals of the June or July hay-yields series (1988-2000) did not show trends except for temperature in April-June, which was positive (0.12 K yr^{-1}, $p < 0.01$). Climatic variables corresponding to the seasonal intervals of the annual-yield series (1988-2000) did not show any trend, precipitation in April-September, however, tended to increase (34 mm yr^{-1}, $p = 0.07$). Among the eight climatic variables corresponding to the seasonal intervals of the kernel-yield series (1972-1999), temperatures in Mai-October and in November-April both increased (0.052 K yr^{-1}, $p < 0.001$; 0.034 K yr^{-1}, $p < 0.05$), relative humidity in November-April decreased (-0.21 % yr^{-1}, $p < 0.05$), and duration of sunshine tended to increase (4.3 h yr^{-1}, $p = 0.05$).

Forward selection found precipitation, relative humidity and duration of sunshine during the periods of main growth to be the best single variables to explain hay yields of June and annual harvests (Table 1a, b). Precipitation

Climatic shifts and productivity of grass-dominated vegetation 255

and duration of sunshine were also important variables to explain kernel crop, however, the climate-maize relationships have opposite signs (Table 1c). Sunshine in summer only shows a significant relation to kernel yields when values of 1991 with insufficient irrigation in the maize field are excluded from the analysis; with all values included only a negative relation to precipitation in the cold season remains.

Table 1. Correlation coefficients, r, of yields and best fitting climatic variables measured at Locarno-Monti and calculated for 1-yr periods previous to harvesting (significant $p < 0.05$ ones bold faced), at each site the two best fitting variables are at least included: (a) hay yield of first harvest (Prugiasco value 1991 estimated or missing) and climatic variables in 3-month periods, (b) hay yield of first and second harvest (Prugiasco value 1991 estimated or missing) and climatic variables in half-year periods, (c) kernel yield (Cadenazzo value 1991 included or excluded) and climatic variables in half-year periods.

a) Hay yield, 1st harvest	$r_{Salorino}$	$r_{Prugiasco}$	
	$df = 11$	$df = 11$	$df = 10$
April – June, sunshine	- .41	**- .70**	**- .67**
April – June, relative humidity	.39	**.67**	**.60**
July – Sept., precipitation	.48	**.71**	**.65**
July – Sept., sunshine	- .32	**- .63**	- .55
July – Sept., relative humidity	.33	**.63**	.54
Oct. – Dec., relative humidity	- .43	.27	.33

b) Hay yield, 1st and 2nd harvest	$r_{Prugiasco}$	
	$df = 11$	$df = 10$
April – Sept., relative humidity	**.74**	**.71**
April – Sept., precipitation	**.63**	**.60**
April – Sept., sunshine	**- .66**	- .58
Oct. - March, relative humidity	- .29	**- .61**
Oct. - March, temperature	**- .56**	- .46

c) Kernel yield	$r_{Cadenazzo}$	
	$df = 26$	$df = 25$
May – Oct., sunshine	.24	**.43**
Nov. – April, precipitation	**- .40**	**- .39**

In linear multiple regression models of hay yield from the meadows (Figures 2, 3) water-related climatic factors played a dominant and significant role: In the meadow mown once in July, yield was positively related to relative humidity in the summer months after harvesting of the year before, negatively related to relative humidity in late fall and positively related to precipitation in late winter (Table 2). In the meadow mown twice, annual yield was positively related to relative humidity in the main growth season and negatively related to relative humidity in the cold season. In the maize field, kernel yield was positively related to duration of sunshine in the growth season and negatively related to precipitation in the cold season before planting (Figure 4, Table 2).

Table 2. Models for yield (y) at three study sites based on multiple linear regressions on climatic variables h (relative humidity), p (precipitation), s (sunshine) in different 3-months or half-year periods before harvesting (subscripts indicate months), r^2, F-ratio, degrees of freedom *(df)* and significance level of full model *(p)* and variable within the model (* $p < 0.05$, ** $p < 0.01$, *** $p < 0.001$).

Model	r^2	F	df	p
Salorino				
$y = 11.99 \times h_{07-09}{}^{**} - 8.34 \times h_{10-12}{}^{**} + 0.202 \times p_{01-03}{}^{*} - 46.3$	0.71	7.2	3,9	0.01
Prugiasco				
$y = 28.29 \times h_{04-09}{}^{***} - 13.16 \times h_{10-03}{}^{*} - 738.8$	0.73	13.8	2,10	0.01
Cadenazzo				
$y = 0.524 \times s_{05-10}{}^{***} - 0.207 \times p_{11-04}{}^{**} + 488.2$	0.48	11.2	2,24	0.001

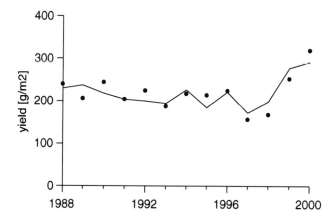

Figure 2. Hay yield of July harvest in meadow at Pree (Salorino), measured (closed circles) and modelled (line) based on climatic variables measured at Locarno-Monti (model *S*, Table 2): mean relative humidity in third (July-September) and forth (October-December) 3-month period of previous, and precipitation in first 3-month period (January-March) of current year. The model explains 71 % of the variation in July yield.

Climatic shifts and productivity of grass-dominated vegetation

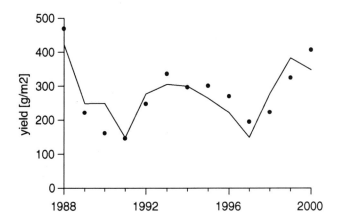

Figure 3. Annual hay yield of first and second harvest in meadow at Negrentino (Prugiasco), measured (closed circles) and modelled (line) based on summer and winter means of relative humidity measured at Locarno-Monti (model P, Table 2). Hay yield of first harvest in 1991 estimated (missing value). The model explains 73 % of the variation in annual yield.

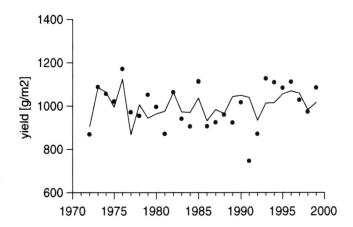

Figure 4. Kernel yield of maize field at Cadenazzo, measured (closed circles) and modelled (line) based on total hours of sunshine in summer (May-October) and total of precipitation in winter (November-April) measured at Locarno-Monti (model C, Table 2). The model, with 1991 value (outlier) excluded, explains 48 % of the variation in kernel yield.

Kernel yield was not significantly correlated with 1-yr lagged series of climatic variables. Annual hay yield at the Prugiasco site, however, showed an increased significant correlation with one of the eight 1-yr lagged series, precipitation in the interval April-September ($r = 0.75$, $df = 11$; $r = 0.69$, $df = 10$; cf. Table 1b). At the Salorino site hay yields of the first harvest did not show significant correlations with climatic variables when these were lagged 1 yr but the correlation coefficients increased with relative humidity (interval October-December) and precipitation (interval April-September).

Correlations among yields at the three sites were not significant. The correlation coefficient was largest between June hay-yield at Prugiasco and July hay-yield at Salorino ($r = 0.55$, $df = 10$, 1991 value missing). Hay yields were not correlated with kernel yields when the latter were lagged 1 yr.

DISCUSSION

In three time series of grass-dominated vegetation from the southern valleys of the Swiss Alps there was no sign of shifting productivity due to an increase of temperature (ca. 1 K) over the past three decades. Nevertheless, inter-annual variation of productivity strongly reflected inter-annual climatic variation during the growth period. Productivity was most sensitive to humidity-related climatic variables, including duration of sunshine, which is an inverse measure of cloudiness. The amount and frequency of humidity transported to the area much depends upon wind direction. Over the Alps westerly winds normally prevail and slight shifts in direction towards S or N may decide between wet weather on the windward or dry weather on the lee side of the mountains. In the semi-natural grasslands productivity positively responded to relative humidity, this latter, an average of continuous measurements over several months, being a surrogate for water availability in the soil. Lag-effects also stress the high importance of water-related variables for permanent grasslands in which populations of perennial plants "remember" favourable growth periods and droughts over more than one growing season (Dunnett et al. 1998). Productivity responded in quite the opposite manner in the maize field in which irrigation during dry periods reduces the negative effects of limited water. A lot of sunshine and few clouds were favourable conditions as long as there was sufficient water from irrigation. This was not the case in the extremely dry summer 1991. The negative response of maize productivity to precipitation in the cold season is probably linked with human interference. In spring the necessity to irrigate might have been underestimated after a cold season with relatively high precipitation. No lag-effects were found in the maize field which is annually replanted. To a much smaller extent, human interference probably also modulated the climate-productivity relationship between the two meadows. Under a 2-yr^{-1} mowing regime, dry conditions over the years 1989 to 1991 had a more severe negative effect, resulting in a much stronger variation at the Prugiasco site. Alternatively, a high small-scale variability of summer thunderstorms with scattered rains at Salorino but not at Prugiasco might also explain these differences.

Since water availability is a limiting factor in species-rich grasslands, disturbance is expected to exceed normal levels during years with extreme droughts. The variables, measured at Locarno-Monti and calculated over 36 years for intervals in the growing season, showed that a series of extremely dry summers was concentrated between 1989 and 1991 with absolute minima of relative humidity in 1991 (intervals April-June and April-September) and precipitation in 1990 (April-September and July-September) and absolute maxima of temperature in 1991 (July-September and April-September). This remarkable series was clearly linked with more northerly upper-level winds over the Alps. There is no direct coincidence with the most persistent and extreme positive phase of the North Atlantic Oscillation (NAO) observed in the late 1980s/early 1990s (Wanner & Schmutz 2000, Ottersen et al. 2001), because this pattern appears during winter months.

At the Prugiasco site the vapour pressure deficit (VPD) exceeded assumed thresholds of drought stress caused by the atmosphere (1 kPa) in more than 50 % of all daytime-hours in summer 1991. These values are not based on measurements of the microclimate at the site but on records in a weather station at a distance of 1 km from the Prugiasco site (Figure 5). Here, precise observations of the inter-annual shifts in the species composition revealed that drought-induced gaps were recovered between 1991 and 1995 by regenerating populations of several forb species. This resulted in a shift of the graminoid-forb ratio in the yield, which lagged 1-3 years behind. In semi-natural grasslands the climate-productivity relationship is thus more complex than in annually replanted grain fields. Direct effects of climatic variables, which limit plant growth are superimposed by long-lasting effects of extreme events, which are able to trigger shifts in species composition and community structure. For species-rich grassland in southern Switzerland patterns of atmospheric circulation seem to be of primary concern in scenarios to predict their future response to climatic change.

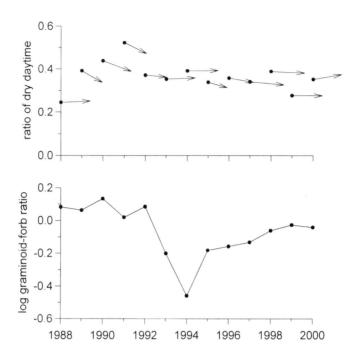

Figure 5. Graminoid-forb ratio in yield of second harvest in meadow at Negrentino (Prugiasco), ratio of dry daytime and resultant vector of added daily upper-level wind directions (mainly West-East) over the Alps in a 155-day summer period (14 April-15 September). Ratio of dry daytime calculated as proportion of daytime hours with a vapour pressure deficit VPD > 1 kPa, VPD based on measurements of temperature and relative humidity at Comprovasco (SMI data), upper-level winds measured at Payerne, ca. 9200 m a.s.l. (SMI data), vector length is proportional to 'concentration', a negative linear transformation of 'circular variance'.

ACKNOWLEDGEMENTS

I thank the Swiss NSF and the Museo cantonale di storia naturale of the Canton of Ticino for their support of my annual sampling in the meadows of Negrentino and Pree over many years (NSF grant nr 31-9096.87, 31-30055.90, 31-39431.93, 31-41922.94, 31-55917.98), R. Hunt for coordinating a series of stimulating meetings on "East-West Variation in North-Atlantic Impacts on Ecosystem Processes in Europe", and M. Menzi for his kind introduction to the Swiss maize experiments and his excellent data archive. The Swiss Meteorological Institute provided all the climatic data, M. Zeiter helped with data transformations, and D.M. Newbery kindly read the manuscript.

REFERENCES

Antognoli C., Guggisberg F., Lörtscher M., Häfelfinger S. & Stampfli A., 1995, Prati magri ticinesi tra passato e futuro, *Memorie* 5, Poncioni, Losone.
Dunnett N.P., Willis A.J., Hunt R. & Grime J.P., 1998, A thirty-eight year study of relations between weather and vegetation dynamics in road verges near Bibury, Gloucestershire. *J. Ecol.* 86: 610-623.
Grime J.P., Brown V.K., Thompson K., Masters G.J., Hillier S.H., Clarke I.P., Askew A.P., Corker D. & Kielty J.P., 2000, The response of two contrasting limestone grasslands to simulated climate change. *Science* 289: 762-765.
Gyalistras D., 2000, Zur zukünftigen Entwicklung des Klimas im Alpenraum und in der Schweiz. In: H. Wanner, D. Gyalistras, J. Luterbacher, R. Rickli, E. Salvisberg & C. Schmutz (eds.) *Klimawandel im Schweizer Alpenraum*. vdf, Zürich.
IPPC, 2001, *Summary for Policymakers, A Report of the Working Group I of the Intergovernmental Panel on Climatic Change*. www.ipcc.ch.
Ottersen G., Planque B., Belgrano A., Post E., Reid P.C. & Stenseth N.C., 2001, Ecological effects of the North Atlantic Oscillation. *Oecologia*, in press.
Stampfli A., 1993, *Veränderungen in Tessiner Magerwiesen: Experimentelle Untersuchungen auf Dauerflächen*, Inauguraldissertation Universität Bern.
Stampfli A. & Zeiter M., 2001, Species responses to climatic variation and land-use change in grasslands of southern Switzerland. In: C.A. Burga & A. Kratochwil (eds.) *Biomonitoring: General and Applied Aspects on Regional and Global Scales*. Kluwer Academic Publishers, Dordrecht.
Walther G.-R., 2001, Laurophyllisation – a sign of changing climate? In: C.A. Burga & A. Kratochwil (eds.) *Biomonitoring: General and Applied Aspects on Regional and Global Scales*. Kluwer Academic Publishers, Dordrecht.
Wanner H. & Schmutz C., 2000, Dynamische Überlegungen zur geographischen und synoptischen Kontrolle des Alpenklimas. In: H. Wanner, D. Gyalistras, J. Luterbacher, R. Rickli, E. Salvisberg & C. Schmutz (eds.) *Klimawandel im Schweizer Alpenraum*. vdf, Zürich.
Weilenmann F., 1994, *Sortenentwicklung und Sortenstrategie bei Getreide in der Schweiz*. Bericht 45. Arbeitstagung 1994 Arbeitsgemeinschaft Saatzuchtleiter 'Vereinigung österreichischer Pflanzenzüchter', BAL Gumpenstein.
Westoby M. & Leishman M., 1997, Categorizing plant species into functional types. In: T.M. Smith, H.H. Shugart & F.I. Woodward (eds.) *Plant Functional Types, their relevance to ecosystem properties and global change*. University Press, Cambridge.

Responses of some Austrian glacier foreland plants to experimentally changed microclimatic conditions

BRIGITTA ERSCHBAMER
Institute of Botany, University of Innsbruck, Sternwartestr. 15, A-6020 Innsbruck, Austria

Abstract: A modified International Tundra Experiment (ITEX) was carried out on the glacier foreland of the Rotmoosferner (Ötztal, Tyrol, Austria) to study the effects of changing the microclimate. Two early successional species (*Trifolium pallescens, Poa alpina* f. *vivipara*) and one late successional species (*Carex curvula*) were transplanted to open-top chambers (OTCs) and to control plots in 1996 and in 1998, respectively. The OTCs were left on the field site also during the winter. As a consequence, the snowmelt occurred at least one week earlier in the OTCs. Growth dynamics were followed over five and three year periods, respectively. The *Trifolium* and *Poa* plants were harvested in August 2000.

The growth responses of the two early successional species differed. *Trifolium pallescens* showed a significantly higher leaf development and a significantly higher total phytomass in the OTCs than in the control plots. *Poa alpina* f. *vivipara* was negatively affected by the artificial warming.

A considerable transplantation shock was observed for the *Carex curvula* plants. The shoot numbers decreased in the OTCs as well as in the control plots. The experiment with this species remained for further monitoring.

INTRODUCTION

Global temperatures are predicted to increase by 1 to 4.5 °C by the middle of the 21st century (IPCC = Intergovernmental Panel on Climate

Change 1996) with the highest increases expected in the Arctic. Considerable efforts have been made to investigate the effects of experimental warming on the growth of plants (Chapin & Shaver 1985, Welker et al. 1993, Wookey et al. 1993, Parsons et al. 1994, Walker et al. 1994, Wookey et al. 1994, Chapin et al. 1995, Wookey et al. 1995, Michelsen et al. 1996, Henry & Molau 1997, Jones et al. 1997, Molau 1997, Molau & Shaver 1997, Welker et al. 1997, Stenström & Jónsdóttir 1997, Suzuki & Kudo 1997, 2000, Totland 1997, Robinson et al. 1998, Press et al. 1998, Arft et al. 1999, Nyléhn & Totland 1999), although few studies been carried out in the Alps so far (Körner et al. 1996, Stenström et al. 1997).

Guisan et al. (1995) have suggested that the greatest effects of climatic change will probably be indirect, through their influence on cryological features such as glacier melting, permafrost degradation, and changes in the duration of the snow cover. In fact, glacier melting is one of the most visible signs of fast climatic change on high mountain ecosystems in the Alps during the 20th century (Haeberli 1995). Areas that have recently become free of ice would therefore seem to be ideal sites on which to study the effects of changed microclimate (with a concentration on enhanced temperatures) for several reasons:

- reactions of plants to changed conditions can be expected especially in those communities where competition is unimportant, i.e. in open communities (Körner 1992, Arft et al. 1999). On the young morainic sites of a glacier foreland, pioneer communities with scattered plant individuals and a very low plant cover can be found;

- revegetation of the terrain following deglaciation is slow under the harsh climatic conditions in front of a glacier. A change in the microclimate could thus enhance both the vegetative and generative growth;

- greater reproductive success and increased vegetative growth on the part of the colonisers could accelerate the colonisation process.

Therefore, an artificial environmental warming experiment was installed on a glacier foreland of the Central Alps, using open-top chambers (OTCs), in accord with the ITEX (International Tundra Experiment) programme (Molau & Mølgaard 1996, Henry & Molau 1997). The OTCs should induce a change in the microclimatic conditions and an increase in the ambient temperatures by at least 1 °C throughout the snow-free season (= minimum IPCC scenario). The OTCs were left on the field site also during the winter.

The main aim of the project was to study the growth dynamics and phytomass production of two early and one late successional species under ambient conditions and within the OTCs. The selected species were transplanted to the OTCs and to control plots. It was hypothesised that the vegetative and generative growth of the species should increase due to the experimental warming, as it has been found from the results of the International Tundra Experiments (Arft et al. 1999). Attention was focused on the question as to how this may affect the primary succession on the glacier foreland.

MATERIAL AND METHODS

Study area and studied species

The study area was located at 2400 m a.s.l. on the glacier foreland of the Rotmoosferner (Obergurgl, Ötztal, Tyrol, Austria). The area is characterised by quartzofeldspathic and metapelitic rocks intercalated by orthogneisses. Metacarbonates, metamarls and amphibolites are folded into these rocks (Frank et al. 1987, Hoinkes & Thöni 1993).

On a morainic site, ice-free for 30 years, a research plot of 20 x 30 m was fenced off in 1996 by an electric fence, in order to exclude large herbivores (chamois, sheep, horses). The area was covered by a sparse pioneer vegetation (plant cover 15 %). The most common species were *Saxifraga aizoides, S. oppositifolia, Artemisia genipi, Linaria alpina, Poa alpina* f. *vivipara,* and *Trifolium pallescens.*

On the oldest moraines (ice-free for 142 years), an initial alpine grassland can be found (*Kobresia myosuroides* community), with scattered tussocks of *Carex curvula.* This species is highly dominant on the adjacent slopes of the glacier valley and it may be hypothesised that it could become dominant in the glacier foreland vegetation in a late successional stage on acidified soils. *Carex curvula* was selected as a late-successional species for the present experiment, *Trifolium pallescens* and *Poa alpina* f. *vivipara* were selected as representatives of the early successional species.

In 1996, ten pairs of plots were randomly selected. 10 OTCs (tetragonal open top chambers, height = 30 cm, lower diameter = 84.6 cm, upper diameter = 50 cm) and 10 control plots (1 m^2) were established in pairs, another 2 two pairs (2 OTCs and 2 controls) were added in 1998.

Plant material

In July 1996, seedlings of *Trifolium pallescens* were collected from an older moraine (ice-free for 44 years) and immediately transplanted to five OTCs and to five control plots (10 seedlings per OTC and per control plot, respectively) on the 30-year ice-free moraine. Tussocks of *Carex curvula* were dug up at 2300 m a.s.l. in a typical alpine grassland on the slopes surrounding the glacier foreland and divided into ramet groups of 5-14 connected shoots. Ten ramet groups were planted on each OTC and control plot, respectively.

In October 1997, viviparous units of *Poa alpina* were collected and planted in the Botanical Garden in Innsbruck. In July 1998, they were transplanted to the two new OTCs and control plots and also to those plots and OTCs in which the *Trifolium pallescens* seedlings had died off.

The number of shoots and flowers of *Carex curvula*, and the number of leaves of *Trifolium pallescens* were counted each year in August. Leaves, flowers, and shoots of *Poa alpina* were counted only in July 2000.

In August 2000, all the *Trifolium pallescens* and *Poa alpina* plants were harvested. Leaves, flowers, shoots and roots were separated, washed and dried at 80 °C for 48 hours. Leaf areas were measured using a Delta-T Image Analysis System (Delta-T Devices LTD, Cambridge). The specific leaf area (SLA = ratio leaf area : leaf dry weight) and the leaf weight ratio (LWR = ratio leaf dry weight : total dry weight) were calculated.

The experiment with *Carex curvula* remained for further monitoring.

Microclimate

Temperatures at the soil surface and at 3 cm soil depth were measured at hourly intervals in 1996 in one OTC and in one control plot, using a Grant Squirrel 1250 data logger. From 1997 until 2000, surface temperatures were recorded at 15-minute intervals in four OTCs and four control plots, and soil temperatures at 3 cm depth in two OTCs and two control plots, using Optic Stowaway loggers (Onset Computer Corporation, Pocasset, MA). The differences in soil moisture (volumetric water content) in two OTCs and two control plots (at hourly intervals) were recorded from 1997 to 2000, using CYCLOBIOS sensors (Dr. R. Kaufmann, Institute of Zoology and Limnology, University of Innsbruck).

C/N-Analysis

Soil samples were collected from 0 – 5 cm soil depth immediately after the plants had been harvested and were then air-dried. The samples were sieved (2 mm mesh) and ground in an analysis mill. The dried above- and below-ground parts were also ground. Fifteen soil samples and six plant samples were analysed: approximately 2 mg of soil and 1 mg of plant material were weighted into tin capsulas and analysed for carbon and nitrogen on an elemental analyser (EA 1110, CE Instruments, Milan, Italy) coupled to a gas isotope ratio mass spectrometer (Delta^{PLUS}, Finnigan MAT).

The analyses were carried out by Dr. A. Richter, Institute of Ecology and Conservation Biology, University of Vienna.

Data analysis

Data analyses were performed using the programme SPSS. To assess the effects of treatment (= increased ambient temperature) on the number of shoots and flowers of *Carex curvula*, a multivariate ANOVA was performed, with 'treatment' and 'year' as fixed factors and 'blocks' (= pairs of OTC/control plot) as covariate. The values of the shoot numbers were log-transformed. A cross-table was used to test the survival of *Trifolium pallescens* plants (Chi-Quadrat-test). A univariate ANOVA was performed to test the effects of treatment on leaf numbers of *Trifolium pallescens* (the values were log-transformed) with 'treatment' and 'year' as fixed factors and 'blocks' as random factor.

The effects of treatment on leaf, flower and shoot numbers of *Poa alpina* f. *vivipara* were analysed by a multivariate ANOVA, with 'treatment' as fixed factor and 'blocks' as covariate. An ANOVA was also performed to compare the effects on dry weights, leaf areas, SLAs, LWRs, root/shoot ratios, C and N contents of *Trifolium pallescens* and *Poa alpina* f. *vivipara*.

RESULTS

Microclimate

Within the OTCs, the snow melted at least one week earlier compared to the control plots. The differences of the surface temperatures during the growing seasons (June to September) from 1996 to 2000 are shown in Table 1. Within the OTCs, daytime temperatures generally rose above the ambient value, whereas night-time cooling below ambient temperature was occasionally recorded. The soil temperatures were also higher in the OTCs. However, the increase in soil temperature within the OTCs was not as pronounced as was the surface temperature.

Soil moisture values were slightly higher in the OTCs than in the control plots (Figure 1) with exception of 1997, though the differences were not statistically significant.

Table 1. Differences of mean, maximum and minimum surface temperature between OTCs and the control plots during the growing seasons (July – September) from 1996 till 2000.

Year	Differences in surface temperature		
	Mean	Maximum	Minimum
1996	0.94	1.15	1.33
1997	1.05	3.96	0.97
1998	1.01	2.53	0.61
1999	0.56	1.35	0.28
2000	1.27	3.40	0.45

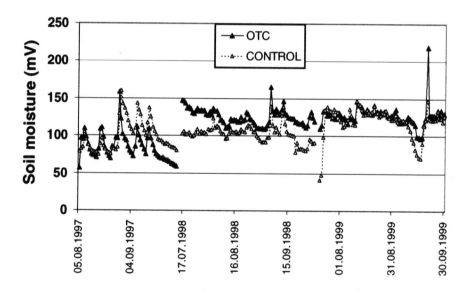

Figure 1. Soil moisture curves (measured as volumetric water content) in one OTC and in one control plot during the growing season (July – September) from 1997 till 1999.

Plant responses

Phenological development (production of leaves and flowering buds, opening of flowers) occurred from a few days up to one week earlier in the OTCs compared to the control plots.

Carex curvula showed a considerable decrease in shoot number between 1996 and 2000 (Figure 2). A greater decrease was detectable in the control plots.

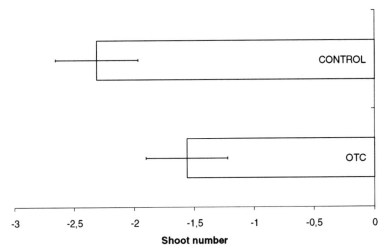

Figure 2. Mean decrease in shoot numbers and standard errors of *Carex curvula* in the OTCs and in the control plots between July 1996 and August 2000.

In the ANOVA, 'treatment', 'year' and 'blocks' showed significant effects on the decrease of shoots and on the development of flowers (Table 2). Flower production of *Carex curvula* was significantly higher in the OTCs compared to the controls.

Table 2. Response of shoot and flower numbers of *Carex curvula* to increased ambient temperature: results of the multivariate ANOVA with 'treatment' and 'year' as fixed factors, and 'blocks' as a covariate. d.f. = degrees of freedom; MS = mean squares; F = F-ratio; Sign. = significance.

Variable	Source of Variation	d.f.	MS	F	Sign.
Shoots	Treatment	1	0.212	9.716	0.002
	Year	4	0.259	11.903	<0.001
	Blocks	1	0.524	24.065	<0.001
	Treatment x Year	4	0.0125	0.572	0.683
Flowers	Treatment	1	1.685	9.888	0.002
	Year	4	1.617	9.488	<0.001
	Blocks	1	0.729	4.279	0.039
	Treatment x Year	4	0.276	1.618	0.169

Significantly more plants of *Trifolium pallescens* died off in the control plots (60 %) compared to the OTCs (40 %, Chi-square = 0.046).

In August 2000, *Trifolium pallescens* in the OTCs showed a significantly higher increase in leaf number (+12.7 leaves, standard error = 1.73) compared to the control plots (+6.35 leaves, standard error = 1.75). The fixed factors 'treatment' and 'year' resulted in a significant effect on the development of leaves (Table 3).

Table 3. Response of leaf numbers of *Trifolium pallescens* to increased ambient temperature: results of the univariate ANOVA with 'treatment' and 'year' as fixed factors, and 'blocks' as a random factor. d.f. = degrees of freedom; MS = mean squares; F = F-ratio; Sign. = significance.

Source of Variation	d.f.	MS	F	Sign.
Treatment	1	1.597	16.357	0.014
Year	4	4.204	103.441	<0.001
Blocks	4	0.056	0.509	0.734
Treatment x Year	4	0.268	8.771	<0.001
Year x Blocks	16	0.041	1.353	0.276
Treatment x Blocks	4	0.101	3.296	0.034
Treatment x Year x Blocks	16	0.030	0.797	0.689

The interactions 'treatment x year' and 'treatment x blocks' were also significant (Table 3). Leaf area, leaf dry weight, and total phytomass were significantly lower in the control plots (Table 4). Flower production in the OTCs increased nearly fivefold (P = 0.003). The dry weights of the flowering shoots, however, did not differ significantly from those of the controls. SLA, LWR, and root/shoot ratio did not change significantly (Table 4).

Table 4. Response of leaf, shoot and flower numbers of *Poa alpina* f. *vivipara* to increased ambient temperature: results of the multivariate ANOVA with 'treatment' as fixed factor, and 'blocks' as a covariate. d.f. = degrees of freedom; MS = mean squares; F = F-ratio; Sign. = significance.

Variable	Source of Variation	d.f.	MS	F	Sign.
Leaves	Treatment	1	74.23	10.301	0.004
	Blocks	1	41.831	5.805	0.024
Shoots	Treatment	1	4.041	3.239	0.084
	Blocks	1	5.217	4.182	0.052
Flowers	Treatment	1	0.710	7.437	0.012
	Blocks	1	0.769	8.053	0.009

In the control plots, *Poa alpina* f. *vivipara* produced a higher number of leaves, vegetative shoots, and flowering shoots (Figure 3). The differences in leaf and flower numbers were statistically significant, but not the values for the vegetative shoots (Table 5). The covariate 'blocks' showed significant effects on all the three variables (Table 5).

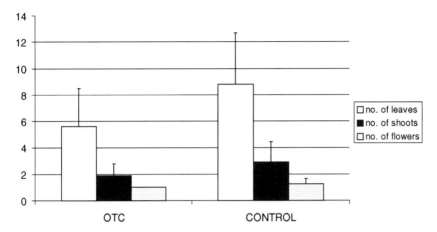

Figure 3. Mean numbers of leaves (+ standard errors), shoots and flowers of *Poa alpina* f. *vivipara* in the OTCs and in the control plots in July 2000.

Table 5. Measured plant variables (means + se = standard errors) for *Trifolium pallescens* and *Poa alpina* f. *vivipara* from the OTCs and the control plots. P = significance.

Trifolium pallescens	OTC		CONTROL		P
	mean	se	mean	se	
Leaf area (mm^2)	564.02	93.80	301.85	64.54	0.038
Leaf dry weight (mg)	54.25	8.65	28.64	6.32	0.030
SLA	10.77	0.27	10.96	0.24	0.599
LWR	0.40	0.02	0.42	0.03	0.559
Shoot dry weight (mg)	19.46	3.23	8.68	2.40	0.016
Root dry weight (mg)	59.84	7.31	29.47	6.47	0.004
Flowering shoots (mg)	57.76	7.62	35.05	11.55	0.184
Total vegetative phytomass (mg)	131.61	17.47	66.01	14.41	0.008
Root/shoot ratio	1.03	0.083	1.22	0.32	0.510

Poa alpina f. *vivipara*	OTC		CONTROL		P
	mean	se	mean	se	
Leaf area (mm^2)	189.15	24.23	302.56	36.09	0.010
Leaf dry weight (mg)	27.20	3.78	47.30	6.26	0.007
SLA	7.12	0.16	6.85	0.24	0.338
LWR	0.49	0.03	0.42	0.03	0.094
Root dry weight (mg)	22.40	3.23	41.10	5.09	0.003
Flowering shoots (mg)	39.50	5.18	59.90	9.27	0.100
Total vegetative phytomass (mg)	49.65	6.88	88.42	10.99	0.003
Root/shoot ratio	0.84	0.07	0.96	0.08	0.257

In the OTCs, significantly smaller leaf areas, leaf dry weights and a lower total phytomass were detected (Table 4). No significant differences were observed with regard to the SLA, LWR, and root/shoot ratio.

C and N content

The abundance of N in the soil samples was low for both the OTCs and the control plots (Table 6). *Trifolium* plants exhibited higher N contents compared to *Poa* plants. No significant differences were detected between any of the means from the OTCs and the control plots.

Table 6. Mean values (with standard deviation) for C and N (in %) in the soil samples and the shoots and roots of *Trifolium pallescens* and *Poa alpina* f. *vivipara* in the OTCs and control plots in August 2000.

C and N natural abundance %			*Trifolium pallescens*				*Poa alpina* f. *vivipara*			
	Soil		Shoots		Roots		Shoots		Roots	
	C	N	C	N	C	N	C	N	C	N
OTC	2.44 ± 0.10	0.03 ± 0.001	38.87 ± 2.01	2.22 ± 0.26	42.34 ± 0.93	2.29 ± 0.14	42.11 ± 0.35	1.06 ± 0.24	42.54 ± 2.88	0.69 ± 0.03
CONTROL	2.36 ± 0.07	0.03 ± 0.001	38.11 ± 0.69	2.65 ± 0.12	44.65 ± 3.36	1.82 ± 0.22	42.81 ± 0.99	1.15 ± 0.13	44.73 ± 0.69	0.80 ± 0.05

DISCUSSION

Microclimate is one of the main external factors controlling plant colonisation on the glacier foreland (Matthews 1992). The OTCs provide effective shelter against the strong glacier winds; they mainly change the length of the growing season by inducing an earlier snowmelt and they alter the summer temperature regimes (Henry & Molau 1997, Marion et al. 1997). The changes in other microclimatic conditions (light, moisture, gas concentration) are less pronounced (Marion et al. 1997), wherefore the effects of the OTCs can be interpreted mainly as being one on air and soil temperature. OTCs have been regarded as useful manipulative tools for increasing temperatures in the field (Kennedy 1995, Hollister & Webber 2000). However, „they are not perfect surrogates of natural variation" wherefore the results should be validated carefully considering natural temperature gradients and plant biology (Hollister & Webber 2000). Kennedy (1995) stressed the importance of „a priori testing" of greenhouse treatment effects.

The earlier onset of the snowmelt in the OTCs resulted in an earlier start of the growing season, as also confirmed for the ITEX sites (Arft et al. 1999). The observed changes in the phenology of the studied plants are well documented also for several arctic species (Welker et al. 1993, Wookey et al. 1993, 1994, Parsons et al. 1994, Alatalo & Totland 1997, Jones et al. 1997, Molau 1997, Mølgaard & Christensen 1997, Stenström & Jonsdottir 1997, Totland 1997, Arft et al. 1999). Although, some species, such as *Polygonum bistorta* (Starr et al. 2000) seem to be unaffected by soil warming and prolongation of the growing season. Divergent reactions within the same dwarf shrub species growing on different sites were also observed (Havström et al. 1993).

The positive effect of increased temperatures on the vegetative growth of *Trifolium pallescens* is consistent with the findings made for forbs on the ITEX sites (Arft et al. 1999). This species is one of the most successful colonisers on the 30-year ice-free moraine onwards to the oldest moraines. The capability of symbiotic N_2-fixation by this plant is certainly highly important in this context. The N-content was clearly higher in *Trifolium* plants compared to *Poa* plants (Table 6). Due to the lack of N in the soils of the glacier foreland, N_2-fixation may play an essential role in structuring the pioneer communities (Chapin et al. 1994), by facilitating invasion by late successional species and accelerating the speed of primary succession. From the results of Thomas & Bowmann (1998) we know that N_2-fixation by *Trifolium* species in alpine tundra communities led to a doubling of the above-ground production compared to tundra sites without *Trifolium*. It may be hypothesised that, in a longterm view, the increase in growth of *Trifolium pallescens* due to potential environmental warming may have a positive effect on the whole colonisation process, by accelerating the growth of mid and late successional species as well.

Surprisingly, *Trifolium alpinum*, a key leguminous species in the Caricetum curvulae in the Swiss Alps, was not favoured by elevated CO_2 concentration (Schäppi & Körner 1996). However, the ambient temperature was not enhanced in this experiment.

Poa alpina f. *vivipara* in the present experiment seemed to be negatively affected by enhanced temperatures (-56 % shoot growth). This contrasts strikingly with the findings of Schäppi & Körner (1996), who reported a pronounced growth stimulation (+47 % shoot growth) of *Poa alpina* under elevated CO_2 values. However, the authors also pointed out that the air and soil temperatures and soil moisture values did not differ significantly between the chambered and unchambered plots. From the results of the ITEX sites, a positive mean effect of artificial warming on plant size was reported for graminoids with regard to vegetative growth, with significant changes in the second year of treatment (Arft et al. 1999). Further experiments, however, are needed to clarify the responses of grasses to changed temperatures.

The viviparous form of *Poa alpina* has a highly efficient dispersal mechanism and colonisation ability on the glacier foreland. The species occurs constantly in the pioneer communities as well as on the oldest morainic sites. Although seed producing plants are also present, they seem to play a minor role in the colonisation process. It may be hypothesised that higher temperatures on the glacier foreland induce a shift of the vegetative reproduction of *Poa alpina* towards seed production. An increase in reproduction and a shift from clonal to sexual colonisation has already been predicted for the High Arctic (Arft et al. 1999). Totland & Nylehn (1998), however, found neither an increase of flower nor bulbil number in plants of *Bistorta vivipara*, whereas Gugerli & Bauert (2001) detected a lower number of bubils in plants of the OTCs. Longterm experiments are needed to clarify the effects of warming on both the vegetative and generative reproduction of grasses and herbs.

In the Arctic, striking effects of temperature enhancement were found on the seed-setting capability and on the seed weight of certain species (Wookey et al. 1993, Molau & Shaver 1997, Totland 1997, 1999). Körner et al. (1997) found that the seed mass of *Carex curvula* under elevated CO_2 levels only increased in the final experimental year. In the present experiment, a higher reproductive effort was perceived for *Trifolium* and *Carex curvula* within the OTCs, where more flowers were developed compared to the control plots. A higher allocation of dry weight to femal organs due to increased ambient warming was also found for the alpine dwarf shrub *Sieversia pentapetala* in the Japanese mountains (Wada 2000). The same results were obtained for *Dryas octopetala* analysed in OTCs in Svalbard (Wada & Kanda 2000). In Southwestern Norway, *Saxifraga stellaris* showed shorter prefloration times and seed maturation was faster in OTCs compared to the control plots (Sandvik & Totland 2000).

Late successional species only establish on the oldest moraines in the study area where a shallow soil (Pararendzina) has already developed (Erschbamer et al. 1999). Therefore, it was not surprising to find that *Carex curvula* had difficulties in growing on the 30-years old morainic site, where no real soil so far had developed (only a Syrosem). Moreover, the species is greatly affected by transplantation having problems in subsequently becoming established even on good soils (Erschbamer et al. 1998). However, a potential for colonisation under poor soil conditions has to be assumed, since the uppermost closed swards of *Carex curvula* are found at 3480 m a.s.l. (Grabherr et al. 1995). This species has also migrated to high altitudes in the Ötztal Alps during the last 50 years (Gottfried et al. 1994). In general, *Carex curvula* with its clonal growth seems to react extremely slowly to environmental changes. For instance, in CO_2 enrichment plots in the Swiss Alps no changes in the biomass of this species were detected (Schäppi & Körner 1996). The present experiment with *Carex curvula* is still going on. Longterm observations will validate the results obtained in this study.

An increase in C/N ratios has been reported for dwarf shrubs following induced artificial warming (Michelsen et al. 1996). Welker et al. (1997) found that the leaf nitrogen values for *Dryas octopetala* were significantly lowered under warmer conditions, wherefore senescent leaves of plants in OTCs had a higher C/N ratio than those on the control plots. In the present study, a tendency to higher C/N ratios was found in the OTCs compared to the control plots, however, the differences were not significant.

CONCLUSIONS

The results showed that functional groups react differently to changed microclimatic conditions: the grass *Poa alpina* f. *vivipara* responded negatively, whereas the growth of the forb *Trifolium pallescens* was favoured. It remains to be tested whether the graminoids of the high mountain ecosystems in the Alps are in general adversely affected by enhanced temperature conditions and how other herbaceous species will react.

The consequences of the difference in the reaction of *Poa alpina* f. *vivipara* and of *Trifolium pallescens* to artificial warming could be that the plant succession on the glacier foreland will change. With increasing ambient temperatures, *Trifolium pallescens* could become even more efficient in colonising morainic areas and *Poa alpina* f. *vivipara* might be adversely affected by such a development. The present dominance of *Poa alpina* f. *vivipara* all over the glacier foreland could be brought to a halt by an improved growth of the herbaceous species. In total, the rate of change in the primary succession may be enhanced by the forecast increase in ambient temperatures and, in a longterm view, by the site amelioration due to the N_2-fixing ability of *Trifolium pallescens*.

ACKNOWLEDGEMENTS

I would like to thank all the people who have helped in establishing the fences and the experiment and in repairing the equipment on the glacier foreland of the Rotmoosferner. I am particularly grateful to Bertram Piest, Manuela Hunn, Josef Schlag, Ruth Niederfriniger Schlag, Elisabeth

Kneringer, Corinna Raffl, Rüdiger Kaufmann, Helmut Scherer, Klaus Vorhauser, Dirk Lederbogen, Martin Mallaun, Meini Strobl, Max Kirchmair, Walter Steger, Peter Unterluggauer, Verena Kuen, to the participants of the botanical course of the University of Innsbruck held at Obergurgl in 1996 and to the participants of the botanical course of the University of Essen (Prof. Dr. Maren Jochimsen) held at Obergurgl in 1997. Many thanks to Erich Schwienbacher for assistance during the harvest. Thanks also to Dr. P.A. Tallantire for linguistic help. I am indebted to Dr. Andreas Richter (University of Vienna) for the C and N analyses. The University of Innsbruck provided the financial support.

REFERENCES

Alatalo J.M. & Totland Ø., 1997, Response to simulated climatic change in an alpine and subarctic pollen-risk strategist, *Silene acaulis*. *Global Change Biol.* 3 (Supplement 1): 74-79.

Arft A.M., Walker M.D., Gurevitch J., Alatalo J.M., Bret-Harte M.S., Dale M., Diemer M., Gugerli F., Henry G.H.R., Jones M.H., Hollister R.D., Jónsdóttir I.S., Laine K., Lévesque E., Marion G.M., Molau U., Mølgaard P., Nordenhäll U., Raszhivin V., Robinson C.H., Starr G., Stenström A., Stenström M., Totland Ø., Turner P.L., Walker L.J., Webber P.J., Welker J.M. & Wookey P.A., 1999, Responses of tundra plants to experimental warming: meta-analysis of the International Tundra Experiment. *Ecol. Monogr.* 69: 491-511.

Chapin F.S., III & Shaver G.R., 1985, Individualistic growth response of tundra plant species to environmental manipulations in the field. *Ecology* 66: 564-576.

Chapin F.S., III, Walker L.R., Fastie C.L. & Sharman L.C., 1994, Mechanisms of primary succession following deglaciation at Glacier Bay, Alaska. *Ecol. Monogr.* 64: 149-175.

Chapin F.S., III, Shaver G.R., Giblin A.E., Nadelhoffer K.J. & Laundre J.A., 1995, Responses of arctic tundra to experimental and observed change in climate. *Ecology* 76: 694-711.

Erschbamer B., Buratti U. & Winkler U., 1998, Long-term population dynamics of two *Carex curvula* species in the Central Alps on native and alien soils. *Oecologia* 115: 114-119.

Erschbamer B., Bitterlich W. & Raffl C., 1999, Die Vegetation als Indikator für die Bodenbildung im Gletschervorfeld des Rotmoosferners (Obergurgl, Ötztal, Nordtirol). *Ber. Naturwiss.-Med. Ver. Innsbruck* 86: 107-122.

Frank W., Hoinkes G., Purtscheller F. & Thöni M., 1987, The Austroalpine Unit West of the Hohe Tauern: The Ötztal-Stubai Complex as an example for the eoalpine metamorphic evolution. In: H.W. Flügel & P. Faupl (eds.), *Geodynamics of the Eastern Alps*, Franz Deuticke, Vienna, pp. 179-225.

Grabherr G., Gottfried M., Gruber A. & Pauli H., 1995, Patterns and current changes in alpine plant diversity. In: F.S. Chapin III & C. Körner (eds.), *Arctic and alpine biodiversity*, Springer, Berlin, Heidelberg, New York, London, Paris, Tokyo, Hong Kong, Barcelona, Budapest, pp. 167-182.

Gottfried M., Pauli H. & Grabherr G., 1994, Die Alpen im „Treibhaus": Nachweise für das erwärmungsbedingte Höhersteigen der alpinen und nivalen Vegetation. *Jahrbuch des Vereins zum Schutz der Bergwelt*, 59. Jahrgang: 13-27.

Gugerli F. & Bauert M.R. (2001): Growth and reproduction of *Polygonum viviparum* show weak responses to experimentally increased temperature at a Swiss Alpine site (submitted).

Guisan A., Tessier L., Holten J.I., Haeberli W. & Baumgartner M., 1995, Understanding the impact of climate change on mountain ecosystems: an overview. In: A. Guisan, J.I. Holten, R. Spichiger & L. Tessier (eds.), *Potential ecological impacts of climate change in the Alps and Fennoscandian Mountains*, Publ. Hors-série n° 8 des Conservatoire et Jardin botaniques de la Ville de Genève, Genève, pp. 15-38.

Haeberli W., 1995, Climate change impacts on glaciers and permafrost. In: A. Guisan, J.I. Holten, R. Spichiger & L. Tessier (eds.), *Potential ecological impacts of climate change in the Alps and Fennoscandian Mountains*, Publ. Hors-série n° 8 des Conservatoire et Jardin botaniques de la Ville de Genève, Genève, pp. 97-103.

Havström M., Callaghan T.V. & Jonasson S., 1993, Differential growth responses of *Cassiope tetragona*, an arctic dwarf-shrub, to environmental perturbations among three contrasting high- and subarctic sites. *Oikos* 66: 389-402.

Hollister R.D. & Webber P.J., 2000, Biotic validation of small open-top chambers in a tundra ecosystem. *Global Change Biol.* 6: 835-842.

Henry G.H.R. & Molau U., 1997, Tundra plants and climate change: the International Tundra Experiment (ITEX). *Global Change Biol.* 3 (Supplement 1): 1-9.

Hoinkes G. & Thöni M., 1993, Evolution of the Ötztal-Stubai, Scarl-Campo and Ulten Basement Units. In: J.F. von Raumer & and F. Neubauer (eds.) *Pre-mesozoic geology in the Alps*, Springer Verlag, Berlin, pp. 485-494.

IPCC (=Intergovernmental Panel on Climate Change), 1996, Technical summary. In: Houghton J.T., Meria Filho L.G., Callander B.A., Harris N., Kattenberg A. & Maskell K. (eds.), *Climate change 1995: the science of climate change*. Press Syndicate of the University of Cambridge, Cambridge, UK, pp. 13-49.

Jones M.H., Bay C. & Nordenhäll U., 1997, Effects of experimental warming on arctic willows (*Salix* spp.): a comparison on responses from the Canadian High Arctic, Alaskan Arctic, and Swedish Subarctic. *Global Change Biol.* 3 (Supplement 1): 55-60.

Kennedy A.D., 1995, Simulated climate change: are passive greenhouses a valid microcosm for testing the biological effects of environmental perturbations? *Global Change Biol.* 1: 29-42.

Körner C., 1992, Response of alpine vegetation to global climate change. *Catena* Supplement 22: 85-96.

Körner C., Diemer M., Schäppi B. & Zimmermann L., 1996, Response of alpine vegetation to elevated CO_2. In: G.W. Koch & H.A. Mooney (eds.) *Carbon dioxide and terrestrial ecosystems*. Academic Press, San Diego, California, USA, pp. 177-196.

Körner C., Diemer M., Schäppi B., Niklaus P. & Arnone J., III, 1997, The response of alpine grassland to four seasons of CO_2 enrichment: a synthesis. *Acta Oecol.* 18: 165-175.

Marion G.M., Henry G.H.R., Freckman D.W., Johnstone J., Jones G., Jones M.H., Lévesque E., Molau U., Mølgaard P., Parsons A.N., Svoboda J. & Virginia R.A., 1997, Open-top designs for manipulating field temperature in high-latitude ecosystems. *Global Change Biol.* 3 (Supplement 1): 20-32.

Matthews J.A., 1992, *The ecology of recently-deglaciated terrain*. Cambridge University Books, Cambridge, New York, Port Chester, Melbourne, Sydney.

Michelsen A., Jonasson S., Sleep D., Havström M. & Callaghan T.V., 1996, Shoot biomass, $\delta^{13}C$, nitrogen and chlorophyll responses of two arctic dwarf shrubs to in situ shading, nutrient application and warming simulating climatic change. *Oecologia* 105: 1-12.

Molau U., 1997, Responses to natural climatic variation and experimental warming in two tundra plant species with contrasting life forms: *Cassiope tetragona* and *Ranunculus nivalis*. *Global Change Biol.* 3 (Supplement 1): 97-107.

Molau U. & Mølgaard P., 1996, *ITEX Manual* (2nd edn), Danish Polar Center, Copenhagen, Denmark.

Molau U. & Shaver G.R., 1997, Controls on seed production and seed germinability in *Eriophorum vaginatum*. *Global Change Biol.* 3 (Supplement 1): 80-88.

Mølgaard P. & Christensen, K., 1997, Response to experimental warming in a population of *Papaver radicatum* in Greenland. *Global Change Biol.* 3 (Supplement 1): 116-124.

Nyléhn J. & Totland Ø., 1999, Effects of temperature and natural disturbance on growth, reproduction, and population density in the alpine annual hemiparasite *Euphrasia frigida*. *Arct., Antarct. Alp. Res.* 31: 259-263.

Parsons A.N., Welker J.M., Wookey P.A., Press M.C., Callaghan T.V. & Lee J.A., 1994, Growth responses of four sub-arctic dwarf shrubs to simulated environmental change. *J. Ecol.* 82: 307-318.

Press M.C., Potter J.A., Burke M.J.W., Callaghan T.V. & Lee J.A., 1998, Responses of a subarctic dwarf shrub heath community to simulated environmental change. *J. Ecol.* 86: 315-327.

Robinson C.H., Wookey J.A., Lee T., Callaghan T.V. & Press M.C., 1998, Plant communitiy responses to simulated environmental change at a high arctic site. *Ecology* 79: 856-866.

Sandvik, S.M. & Totland Ø., 2000, Short-term effects of simulated environmental changes on phenology, reproduction, and growth in the late-flowering snowbed herb *Saxifraga stellaris* L. *Ecoscience* 7: 201-213.

Schäppi B. & Körner C., 1996, Growth responses of an alpine grassland to elevated CO_2. *Oecologia* 105: 43-52.

Starr G., Oberbauer S.F. & Pop E.W., 2000, Effects of lengthened growing season and soil warming on the phenology and physiology of *Polygonum bistorta*. *Global Change Biol.* 6: 357-369.

Stenström A. & Jónsdóttir I.S., 1997, Responses of the clonal sedge, *Carex bigelowii*, to two seasons of simulated climate change. *Global Change Biol.* 3 (Supplement 1): 89-96.

Stenström A., Gugerli F. & Henry G.H.R., 1997, Response of *Saxifraga oppositifolia* L. to simulated climate change at three contrasting latitudes. *Global Change Biol.* 3 (Supplement 1): 44-54.

Suzuki S. & Kudo G., 1997, Short-term effects to simulated environmental change on phenology, leaf traits, and shoot growth of alpine plants on a temperate mountain, northern Japan. *Global Change Biol.* 3 (Supplement 1): 108-115.

Suzuki S. & Kudo G., 2000, Responses of alpine shrubs to simulated environmental change during three years in the mid-latitude mountain, northern Japan. *Ecography* 23: 553-564.

Thomas B.D. & Bowman W.D., 1998, Influence of N_2-fixing *Trifolium* on plant species composition and biomass production in alpine tundra. *Oecologia* 115: 26-31.

Totland Ø., 1997, Effects of flowering time and temperature on growth and reproduction in *Leontodon autumnalis* var. *taraxici*, a late-flowering alpine plant. *Arct. Alp. Res.* 29: 285-290.

Totland Ø., 1999, Effects of temperature on performance and phenotypic selection on plant traits in alpine *Ranunculus acris*. *Oecologia* 120: 242-251.

Totland Ø. & Nyléhn J., 1998, Assessment of the effects of environmental change on the performance and density of *Bistorta vivipara*: the use of multivariate analysis and experimental manipulation. *J. Ecol.* 86: 989-998.

Wada N., 2000, Responses of floral traits and increase in female reproductive effort to a simulated environmental amelioration in a hermaphrodite alpine dwarf shrub, *Sieversia pentapetala* (Rosaceae). *Arct. Antarct. Alp. Res.* 32: 208-211.

Wada N. & Kanda H., 2000, Notes on floral traits and gender expression of *Dryas octopetala* under a simulated environmental change. *Polar Bioscience* 13: 147-151.

Walker M.D., Webber P.J., Arnold E.H. & Ebert-May D., 1994, Effects of interannual climate variation on aboveground phytomass in Alpine vegetation. *Ecology* 75: 393-408.

Welker J.M., Wookey P.A., Parsons A.N., Press M.C., Callaghan T.V. & Lee J.A., 1993, Leaf carbon isotope discrimination and vegetative responses of *Dryas octopetala* to temperature and water manipulations in a high arctic polar semi-desert, Svalbard. *Oecologia* 95: 463-469.

Welker J.M., Molau U., Parsons A.N., Robinson C.H. & Wookey P.A., 1997, Responses of *Dryas octopetala* to ITEX environmental manipulations: a synthesis with circumpolar comparisons. *Global Change Biol.* 3 (Supplement 1): 61-73.

Wookey P.A., Parsons A.N., Welker J.M., Potter J.A., Callaghan T.V., Lee J.A. & Press M.C., 1993, Comparative responses of phenology and reproductive development to simulated environmental change in sub-arctic and high arctic plants. *Oikos* 67: 490-502.

Wookey P.A., Welker J.M., Parsons A.N., Press M.C., Callaghan T.V. & Lee J.A., 1994, Differential growth, allocation and photosynthetic responses of *Polygonum viviparum* to simulated environmental change at a high arctic polar semi-desert. *Oikos* 70: 131-139.

Wookey P.A., Robinson C.H., Parsons A.N., Welker J.M., Press M.C., Callaghan T.V. & Lee J.A., 1995, Environmental constraints on the growth and performance of *Dryas octopetala* ssp. *octopetala* at a high arctic polar semi-desert. *Oeocologia* 104: 567-578.

Reliability and effectiveness of Ellenberg's indices in checking flora and vegetation changes induced by climatic variations

SANDRO PIGNATTI*, PIETRO BIANCO*, GIULIANO FANELLI*,
RICCARDO GUARINO[+], JÖRG PETERSEN[°] & PAOLO TESCAROLLO*
* *Dipartimento di Biologia Vegetale, Orto Botanico, Largo Cristina di Svezia 24, I-00165 Rome.*
[+] *Dipartimento di Botanica, Universitá di Catania, Via A. Longo 19, I-95125 Catania.*
[°] *Institut für Geobotanik, Universität Hannover, Nienburger Str. 17, D-30167 Hanover.*

Abstract: Two methods are proposed to assess the significance of time-spanned local variations of Ellenberg's indicator values in floristic and phytosociological data sets respectively. Both methods are based on frequencies and averages of the indicator values within and among data sets. Main goal of proposed methods is to determine the threshold above which the observed variations are to be considered significant. In order to relate the observed floristic variations to recent climatic changes, attention was focused on the Ellenberg's indices dealing with the main climatic variables: heliophany, temperature and precipitation. Study-cases from Zannone Island (Pontine Archipelago, Mediterranean region), Inferno Valley (Rome, Mediterranean region) and Braulio Valley (Stelvio National Park, Alpine region) are reported. For Mediterranean data- sets, integration and adjustments to the Ellenberg's indicator values have been adopted.

INTRODUCTION

Many scientific contributions of the last ten years deal with the responses of living organisms to the recent global warming caused by human activities. As to plants, it is quite difficult to claim that the observed variations are strictly related to the global warming, because distribution of species largely depends on non-linear interactions among different environmental variables. Moreover, plant communities often bear feedback mechanisms that buffer slight variations of environmental factors without radical changes in physiognomy and species composition (Wilson & Agnew, 1992).

An attempt to draw indications on the influence of each main environmental factor in determining flora and vegetation changes can be made with the help of Ellenberg's indices. There are large experimental evidence that ecological factors determine both composition and structure of plant communities. On this basis, Ellenberg (1974, 1996; Ellenberg et al., 1992) outlined the synecological requirements of each species belonging to the vascular flora of Central Europe by means of numerical indices referred to 7 main environmental factors. These can be divided in two subgroups of three and four indices respectively. The first three indices are linked to climatic variables: light regime (L), temperatures (T) and continentality of climate (C); the other four deal with edaphic conditions: moisture of soils (M), pH, nutrients availability (N) and salt concentration (S). All indices are arranged in ordinal scales ranging between 1 and 9 (only M ranges between 1 and 12 and S between 0 and 9), where only a nominal correlation with the physical/chemical parameters, measuring the environmental variables to which indices are referred, is given. Another important feature of Ellenberg's indices is that they are not related to the ecological optimum of a given species, but to its synecological optimum, expressing the optimal ecological requirements of a species when in competition with other species, which can vary from region to region.

Ellenberg's indices are defined by the author himself as a useful paradigm to summarise interactions between plants and environment, recognising to each species a role as biological indicator, which has been often neglected or disregarded in recent literature. Ellenberg's indices have been successfully applied to "fingerprint" plant communities and higher syntaxa according to their floristic composition, by means of synoptic representations of their ecological spectra (Van der Maarel 1993, Pignatti et al. 1996, Pignatti 1998, 1999, Petersen 2000, Bianco et al. 2001, Guarino & Bernardini 2001). Time-spanned variations of such spectra might also be useful in tracking the effects of global changes on local flora and vegetation. Methods for assessing the thresholds above which the observed variations can be considered significant will be the main subject of the present paper.

On Ellenberg's indices

From Central Europe, Ellenberg's model has been extended eastwards to Poland by Zarzycky (1984) and to Hungary by Borhidi (1995). This extension was not so problematic, thanks to the large number of species in common and to the relatively similar latitudinal distribution. Synecological requirements of a species may change over its range, especially when moving to different latitudes. In order to extend the model to the Italian flora, a data-base has been collected by Pignatti and collaborators since more than twenty years. In this data-base all the species of the Italian flora are reported, together with ecological and ecophysiological measurements for each species, if available. The complete list of indicator values referred to the Italian flora should be published in a relatively short time (Pignatti et al. 1996). From the methodological point of view, the only relevant problem was the necessity to enlarge the scale of T to 12 possible values, since, compared to central Europe, higher temperature may occur in Italy. Aim of the present study was also to check whether the integration proposed for the Italian flora fits the Ellenberg's model in terms of statistical homogeneity.

Some authors (Durwen 1982, Böcker et al. 1983, Möller 1992, Kowarik & Seidling 1989), even if sometimes recognising the usefulness of calculating average values of the Ellenberg's indices for florae or plant communities, believe that such use is incorrect from the mathematical point of view, since Ellenberg's indices are ordinal scales, without dimensional correlation with chemical/physical parameters. It is well-demonstrated (Figure 1) that when the number of data is sufficiently high, Ellenberg's variables fit the normal distribution (Gaussian curve), which is described by the parameters average and variance. In this case, in addition to non-parametric statistics, all statistical tests referring to the normal distribution can be used, since they are based on differences of each sample from the average. Calculating average and variance of data sets is, therefore, allowed as well. When only a small number of data is available, a parametric statistical approach can be tested anyway, because in the case of Ellenberg's scales it is always possible to transform data in their respective ranks, to

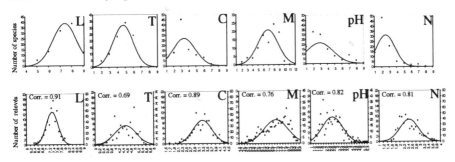

Figure 1. Distribution of Ellenberg's indicator values within a phytosociological table (449 relevés × 233 species) of Empetro-Ericetum from Wadden Sea Islands (Northern Sea). The first row of diagrams is referred to the distribution of values among the species occurring in the relevés, the second one to the distribution of average indicator values among the relevés.

better approximate the normal distribution and to create the conditions to use a parametric approach.

Moreover, several contributors demonstrated a fairly good correlation between M, pH, N, S indices and measured environmental variables (Degorski 1982, Seidling & Rohner 1993, Möller 1997, Petersen 2000, Schaffers & Sýkora 2000). From the theoretical point of view it is therefore possible, at least for the above-mentioned scales, to find out algorithms that describe linear or non-linear correlation between Ellenberg's scales and chemical concentrations, so to give a dimensional meaning to the Ellenberg's values. Similarly, this holds true also for the others indices (L, T and C).

Carrying out parametric statistical tests and using average indicator values referred to floras and plant communities should be thus considered admissible even from the epistemological viewpoint, as already accepted by many authors and by Ellenberg himself (Pignatti et al. 1996). This greatly broadens the usefulness of such indices.

METHODS

Variations of Ellenberg's indices over the past fifty years were analysed for floras of Inferno Valley (Rome, central Italy), of Zannone (an island belonging to the Pontine Archipelago, central Italy) and for vegetation of Braulio Valley (Stelvio National Park, central Alps) (Figure 2). In the case of Zannone, variations occurred over the first half of the last century have been considered as well. Analyses were performed either on floristic or phytosociological data sets depending on the availability of old data. For the first two localities, belonging to the Mediterranean region, integration and adjustments proposed by Pignatti to the Ellenberg's model were adopted.

In case of floristic data sets (Inferno Valley and Zannone Island), frequencies and average of the indicator values were evaluated for each of the Ellenberg's indices. Such estimates were calculated on total floras, on species in common and on appeared/disappeared species. A similar evaluation was performed for phytosociological data (Braulio Valley), but in this case results have been weighted in the frequency of each species within the phytosociological tables. Cover values of species have been neglected, since the considered time-span was too short to cause remarkable changes in the dominance of plant communities, while appeared or disappeared species were mostly found in few individuals per plot. Indeed, the use of presence/absence values in the calculation of site indicator values is recommended by several authors and by Ellenberg himself (Durwen 1982, Kowarik & Seidling 1989, Ellenberg 1991).

Numerical analyses have been based in each case on parametric statistics and combinatory calculations.

In order to relate the observed floristic variations to recent climatic changes, attention was focused on the Ellenberg's indices dealing with the main climatic variables: heliophany, temperature and precipitation. These are, respectively, the L, T, C and M indices. Owing to the provenance of data, the C index was disregarded, since values of continentality are always very low for Mediterranean species (that dominate the floras of Inferno Valley and of Zannone), while are poorly distinctive for Alpine species. On the other hand, the M index, even if refers to water availability in soil, has been considered because it can be affected temperature and precipitation trend and by variation in total rainfall. Moreover, for the alpine sites, soil moisture can be also influenced by an increased melting rate of glaciers during the vegetative period.

Figure 2. Approx. localisation of data sources (centre of the black circles). From south to north: Zannone Island, Inferno Valley and Braulio Valley.

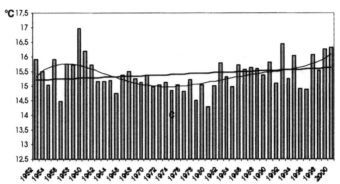

Figure 3. Infrared scanners reveal an average increase of 0.6 °C proceeding from non-urban to urban areas of Rome. The trend of the mean annual temperature (period: 1952-2000) at Monte Mario, representing the south-eastern boundary of the Inferno Valley, is plotted. Besides the regression line, a polinomial curve is presented, expressing the minimum squares within groups of 5 temperature values.

FLORISTIC DATA

Data sets

Inferno Valley has an extension of approx. 2.5 km^2 and it is located in Rome, only 2 km far from the centre of the town. Starting from 1930, it has been progressively included in the urban tissue, but it was managed in a traditional way ("Ager Romanus") until 20 years ago. By comparing the flora of 1954 (Montelucci) to the present one (Bianco & Fanelli, in press) a slight decrease of the total number of species (from 667 to 644) was noted due to the disappearance of 64 species (9.3 % of the 1954 flora) and to the appearance of 41 new ones (6.4 % of the actual flora). Disappeared species belong mostly to the Eurasiatic chorotype, while new ones are mostly adventitious synanthropic and ephemeral Mediterranean species (Table 1).

Observed variations can be related to the increase of human disturbance, to the local/global warming (Figure 3), to the ending of traditional farming and subsequent soil erosion of large areas within the park.

By considering the total floras, the above-mentioned floristic changes led to a slight decrease (-0.13) of the average L value and to an increase of T and M values (0.27 and 0.08, respectively). By comparing the average values of the 64 disappeared species to those of the 41 newly appeared (Figure 4), L and T increased (0.14 and 0.50, respectively) while H decreased (0.56).

Zannone Island has an extension of approx. 1.03 km^2 and it is located along the Thyrrenian coast, 7 miles off the Circeo headland. It was inhabited by Benedictine monks until the end of the 13th century, later only beeing periodically used as a refuge by Saracen pirates and successively as hunting and wood reserve by local people. At the end of the 19th century, few couples of mufflons were introduced on the island and around 1950 the population reached 200 individuals. Since 1979, Zannone Island belongs to the Circeo National Park; the mufflon-population has been reduced to the actual 30 individuals and fires are extremely rare, the island being permanently patrolled by foresters. Floras of the island were published in 1905 (Bèguinot), 1954 (Anzalone), 1997 (Anzalone et al.). A noteworthy decrease in species richness occurred over the years: from 272 species recorded in 1905, 241 were scored in 1954 and 193 in 1997. The highest turn-over took place in the time gap between the two old floras and the recent one: in addition to the 57 species found only in 1905 and to the 25 ones found only in 1954, a total of 60 species were recorded only in these old floras. Disappeared species are to be ascribed mostly to the Steno-Mediterranean chorotype, while among the new ones a large number are exotic synanthropic species (Table 2). Many of the disappeared species are therophytes, the spreading of which was probably due to the mufflon-grazing and to the higher frequency of fires. It appears likely that these mammals caused probably the extinction of the 57 species recorded only in 1905, represented mostly by perennial palatable grasses. Anyway, global warming can not be excluded a priori from the possible causes of such changes.

Observed variations of average L, T, and M values passing between the flora of 1905 and that of 1954 were: -0.02, -0.09 and +0.08, respectively.

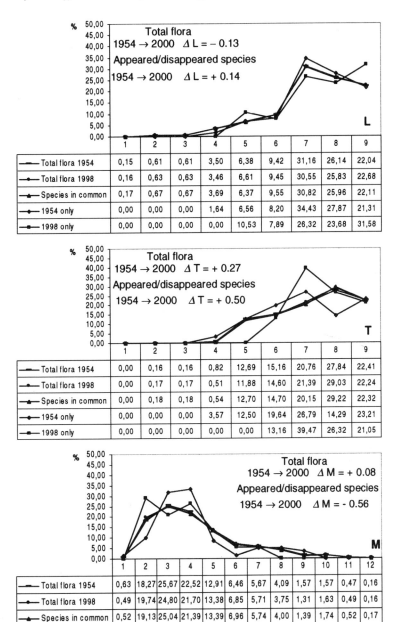

Figure 4. Frequency distribution (percentage) of Ellenberg's indices for light (L), temperature (T) and moisture (M, see next page), for the vascular flora of Inferno Valley. Differences between the average values of the total floras and of the appeared/disappeared species are also given.

observed variations of the same values between the flora of 1954 and that of 1997 were: +0.04, -0.05 and -0.05 respectively. Variations of the average values between the species found exclusively in 1905 and those found exclusively in 1954 or 1997 were the following: $\Delta L = +0.04$; $\Delta T = -0.32$ and $\Delta H = +0.11$. Variations of the average values passing from the species exclusively found in 1905 or 1954 to those exclusively found in 1997 were the following: $\Delta L = +0.19$; $\Delta T = -0.20$ and $\Delta H = +0.04$ (Figure 5).

Data analysis

Can the observed variations of L, T and M be related to variations of environmental factors or should they be considered not significant and therefore negligible?

To answer the question, it was considered that each flora can be divided in subgroups formed by n species, with n varying between 2 and the number of species forming the total flora (N). The number of possible combinations (C) for such groups is given by the following expression:

$$(1) \quad C = \sum_{n=2}^{N} N!/(N-n)!\,n!$$

According to the expression (1), when a flora is composed by more than 100 species, the total number of possible combinations can be approximated to $+\infty$ (Figure 6). Theoretically, it is possible to calculate the average value of a given Ellenberg's index for each of the aforesaid combinations. For each combination it is possible, as well, to calculate the mean absolute deviation D_m from the average indicator value of the total flora μ, according to the following expression:

$$(2) \quad D_m = \sum_{i=1}^{n} |x_i - \mu| / n$$

where x_i is the average indicator value of a given combination and n the total number of possible combinations. Variance of D_m calculated for each rank of the possible combinations has the theoretical trend represented in Figure 7, with a maximum for combinations including 15 % of the total flora.

Observed variations in Ellenberg's mean values have been considered significant only when both the absolute differences between the mean values referred to the total floras and to the appeared/disappeared species exceeded the highest D_m, measured within at least 50 subgroups, randomly chosen among the possible combinations including 15 % of the species belonging to each flora. D_m was preferred to other methods to assess the variability of data sets (such as mean square and standard deviation) because it is directly comparable to the above-mentioned differences.

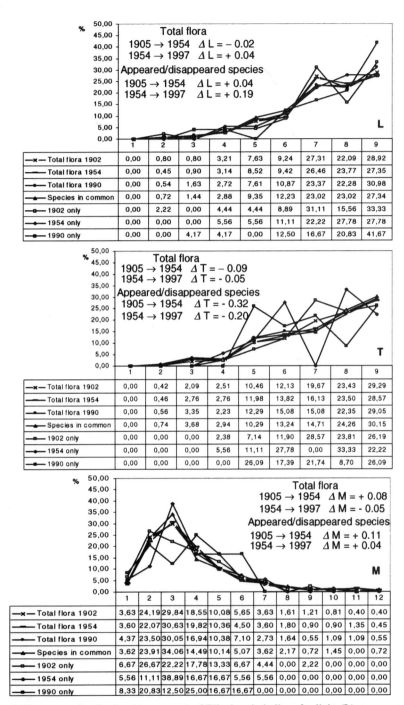

Figure 5. Frequency distribution (percentage) of Ellenberg's indices for light (L), temperature (T) and moisture (M), for the vascular flora of Zannone Island. Differences between the average values of the total floras and of the appeared/disappeared species are also given.

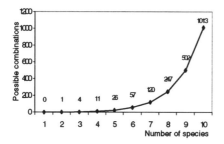
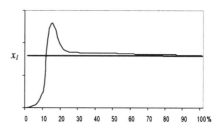

Figure 6. Curve given by the expression (1) for hypothetical floras composed by 1 to 10 species.

Figure 7. Theoretical curve of the variance of D_m (see expression 2), calculated on subgroups including from 0 to 100 % of the total flora. x_1 represents the variance of D_m referred to the total flora.

Results

In the case of <u>Inferno Valley</u>, the mean absolute deviations of the average L, T, and M values, measured among 50 subgroups of 100 and 128 species (corresponding to the 15 % of the flora of 1954 and of the present one, respectively) gave the following figures:

	D_m (L)	D_m (T)	D_m (M)
Flora 1954	0.201	0.256	0.243
Actual flora	0.187	0.219	0.195

According to the proposed approach, it was concluded that the observed variation of average T indicator values over the last fifty years could be related to the combined effect of local and global thermal increase occurred in the Inferno Valley over the same period (Figure 3). Observed variations of L and M values should be neglected, even if the variation of M (-0.56) occurred within the appeared/disappeared species gives an important indication about a possible evolution towards drier environmental conditions in the next future. These conclusions are resumed by the following expressions:

$|\Delta L_{total}|$ [0.13] $\wedge$ $|\Delta L_{app./disapp.}|$ [0.14] < 0.201 $\Rightarrow$ negligible
$|\Delta T_{total}|$ [0.27] $\wedge$ $|\Delta T_{app./disapp.}|$ [0.50] > 0.256 $\Rightarrow$ meaningful
$|\Delta M_{total}|$ [0.08] < 0.195 < $|\Delta M_{app./disapp.}|$ [0.56] $\Rightarrow$ negligible

Variations of Ellenberg's indices reflected quite well that the more thermophilous character of the new entries, mostly belonging to the Mediterranean chorotype, compared to the disappeared ones, mostly belonging to the Eurasiatic one. The higher xerophily of the new entries can be accounted for in the same way.

In the case of <u>Zannone Island</u>, the mean absolute deviations of the average L, T, and M values, measured among 50 subgroups of 41, 36 and 29 species (corresponding to the 15 % of the flora of 1905, 1954 and 1997, respectively) gave the following results:

	D_m (L)	D_m (T)	D_m (M)
Flora 1905	0.084	0.127	0.083
Flora 1954	0.160	0.119	0.192
Flora 1997	0.133	0.141	0.108

According to the proposed approach, all the observed variation of the average indicators referred to the flora of Zannone Island should be neglected, as shown by the following expressions:

1905 → 1954

$|\Delta L_{total}|$ [0.02] ∧ $|\Delta L_{app./disapp.}|$ [0.04] < 0.160 ⇒ negligible
$|\Delta T_{total}|$ [0.09] < 0.127 < $|\Delta T_{app./disapp.}|$ [0.32] ⇒ negligible
$|\Delta M_{total}|$ [0.08] ∧ $|\Delta M_{app./disapp.}|$ [0.11] < 0.192 ⇒ negligible

1954 → 1997

$|\Delta L_{total}|$ [0.04] < 0.160 < $|\Delta L_{app./disapp.}|$ [0.19] ⇒ negligible
$|\Delta T_{total}|$ [0.05] < 0.141 < $|\Delta T_{app./disapp.}|$ [0.20] ⇒ negligible
$|\Delta M_{total}|$ [0.05] ∧ $|\Delta M_{app./disapp.}|$ [0.04] < 0.192 ⇒ negligible

In the case of Zannone Island, in spite of a total turnover of species higher than 37 % (21 % higher than in Inferno Valley), all the observed variation of the considered Ellenberg's indices are negligible. This supports the hypothesis that floristic variation on the island should be mostly ascribed to the presence of mufflons and to changes in fire frequency. The decreasing trend of ΔT within the appeared/disappeared species (-0.32 and -0.20, respectively) can be explained if we consider that up to 1939 the southern slope of the island was exploited for charcoal production. It is likely that markedly xerothermophilous species belonging to the class Thero-Brachypodietea (Lygeo-Stipetea syn. synt.), forming perennial post-fire grasslands widespread on the island in 1905, were heavily grazed by the 200 mufflons occurring at that time. Therophytic, less palatable, grasses belonging to the class Tuberarietea guttatae, and characterised by a lower T value (as only their seed survive to the driest period of the year) replaced the overgrazed perennial species. Present vegetation is rapidly evolving toward more complex and stable wooden types, whith T values that reflect the relatively mild suboceanic climatic conditions given by the Thyrrenian Sea. This explanation is well supported by differences in ecological spectra.

PHYTOSOCIOLOGICAL DATA

Data sets

Braulio Valley ranges between 2,250 and 3,000 m a.s.l. and has an extension of approx. 9.5 km^2. It belongs to the Stelvio National Park and witnessed a progressive decrease in traditional land use ("alpeggio") and an increase of tourism during the last fifty years. An exhaustive monograph about flora (including bryophytes and lichens) and vegetation of the valley, including a phytosociological map of the area, was published in 1955 (Giacomini & Pignatti). At present, a working group co-ordinated by Pignatti himself is carrying out a similar research to check possible changes occurred meanwhile. The complete results of such research are supposed to be published within March 2002. According to the preliminary results, no meaningful variation has been recognised within the local flora during the last fifty years. This would agree with the hypothesis, supported by some authors (e.g. Körner 1994, Theurillat 1995), that a slight increase of temperature would not affect significantly the floristic diversity of the Alpine chain.

As to vegetation, 28 different plant communities belonging to 13 phytosociological classes have been investigated up to now by means of the Braun-Blanquet's approach (Braun-Blanquet 1964, Westhoff & Van der Maarel 1976). Small variations have been noted in structure and composition of some phytocoenoses, as well as in the extensions covered by them, as already observed in Braulio Valley by Faifer & Pirola (1995). Among the surveyed associations, Caricetum curvulae bears the highest variation in Ellenberg's L, T and M indexes. This is largely due to the enrichment in species of Nardetea strictae (and lower syntaxa), that are spreading from the subalpine to the alpine belt (Figure 8). Even if these species are commonly favoured by a high grazing pressure; it is interesting to note that within the Festucetum halleri, a plant association widespread in the subalpine belt and supporting the highest grazing pressure in Braulio Valley, such species are clearly decreasing. This observation is in accordance to the progressive reduction of grazers caused by the vanishing traditional way of farming, locally called "alpeggio". This consideration suggests that upwards spreading of Nardetea-species might be related to global warming.

Observed variations of average L, T, and M values, weighted in frequency of species within the phytosociological tables, were: -0.37, +0.28 and +0.44 respectively (Figure 9 and Table 3). In the case of Braulio Valley, no integration neither adjustment of Ellenberg's indicator values has been adopted. The few species not occurring in the original list of indicator values have been ignored.

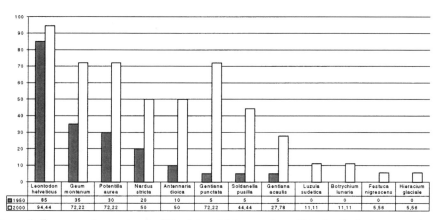

Figure 8. Frequency (percentage) of Nardetea strictae-species within the Caricetum curvulae in Braulio Valley in 1955 and in 2000.

Data analysis

1) Was the number of relevés sufficient to render significant the observed variations?

All relevés were taken along 4 transects approx. 100 m wide. The total estimated area covered by Caricetum curvulae along the transects was 36,000 m². In 1955, 20 relevés of this vegetation have been published (Giacomini & Pignatti 1955a, b). The mean plot size was 68.2 m², corresponding to an investigated area of 1365 m² ($\cong$ 3.8 % of the total estimated area along the transects). In 2000, 18 relevés of Caricetum curvulae have been taken, with mean plot size of 62,3 m², corresponding to an investigated area of 1121 m² ($\cong$ 3.1 % of the total estimated area along the transects). In order to assess whether the number of relevés was sufficient to include the whole range of variation of Ellenberg's indices within Caricetum curvulae of Braulio Valley, the n possible plots along the transects (528 and 745 respectively) were ascribed to two different classes, A and B, with relative frequencies p and q. The probability P to assign a sample i times to A (and therefore n - i to B) is given by the following binomial distribution:

$$(3) \quad P = [n!/i!(n-i)!] \, p^i \, q^{n-i}$$

By means of the expression (3), it is possible to describe, for example, the probability P that 75 % of the 20 samples of 1950 was among those having T values lower than the average (and therefore belonging to the class A) and that 75 % of the 18 samples of 2000 was among those having T values higher than the average (and therefore belonging to the class B),

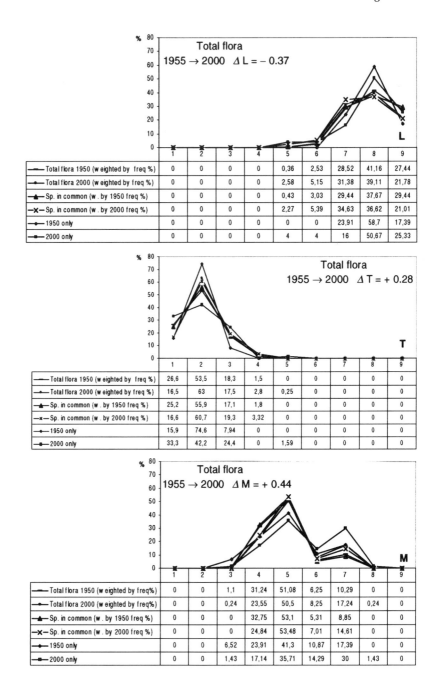

Figure 9. Frequency distribution (percentage) of Ellenberg's indices for light (L), temperature (T) and moisture (M, see next page), referred to the Caricetum curvulae of Braulio Valley. Frequencies have been weighted in the frequency of species within the phytosociological tables. Differences between the average values of the total floras are also given.

according to the variations of p and q (Figure 10). By considering the trend of the curves of iso-probability in Figure 10, it is likely that the considered relevés were quite homogeneously distributed along the gradient of possible variations of Ellenberg's indices.

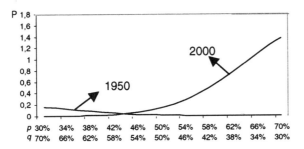

Figure 10. Curve given by the expression (3), describing the probability P that 75 % of the 20 samples of 1950 was among those having T values lower than the average (and therefore belonging to the class A), and that 75 % of the 18 samples of 2000 was among those having T values higher than the average (and therefore belonging to the class B), according to the variations of frequencies of the two classes. Probability is expressed in percentage.

2) Can the observed variations of L, T and M be related to variations of environmental factors or should they be considered not significant and therefore negligible?

The proposed approach used to answer the question will be illustrated taking the L index as example. The same procedure was followed to assess the significance of the observed variations of T and M.

The first step was to calculate the mean L value for each of the 38 available relevés. It was assumed that indicator values of Caricetum curvulae in Braulio Valley were normally distributed among the n possible plots within the valley. The present approach loses its meaning for other distribution models, as those described by bimodal or sigmoid curves. If the normal distribution of average indicator values is accepted, the distribution of L values for each sampling area of 1955 can be described by a Gaussian (Figure 11, C_1) settled by the average L of the 20 relevés of 1955 (L_1) and by the standard deviation estimated among the L values of the same set of relevés (Table 4). Analogously, the distribution of L values for each sampling area of 2000 can be described with a Gaussian (Figure 11, C_2) settled by the average L of the 18 relevés of 2000 (L_2) and by the standard deviation estimated among the L values of the same set of relevés (Table 4).

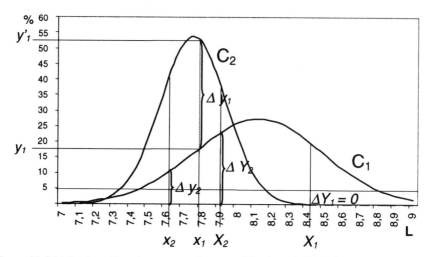

Figure 11. Distribution of L values among the n possible plots of 1955 (C_1) and of 2000 (C_2). See text.

With reference to Figure 11, given:
- x_1 as the average of L values lower than L_1 among the relevés of 1950;
- y_1 as the probability (in %) expressed by C_1 in x_1;
- y'_1 as the probability expressed by C_2 in x_1;
- Δy_1 as the difference between y_1 and y'_1 (i.e. the difference between the probability expressed by C_1 and the probability expressed by C_2 in x_1);
- x_2 as the average of L values lower than L_2 among the relevés of 2000;
- Δy_2 as the difference between the probability expressed by C_2 and the probability expressed by C_1 in x_2;
- X_1 as the average of L values higher than L_1 among the relevés of 1950;
- ΔY_1 as the difference between the probability expressed by C_1 and the probability expressed by C_2 in X_1;
- X_2 as the average of L values higher than L_1 among the relevés of 2000;
- ΔY_2 as the difference between the probability expressed by C_2 and the probability expressed by C_1 in X_2);

it was assumed that the observed variation of L was significant and not given by chance only when:

$$0 < \Delta y_1 \vee \Delta y_2 \vee \Delta Y_1 \vee \Delta Y_2 < 5\%$$

i.e. when at least one of the averages of L values above or below L_1 and L_2 were confined in the portion C_2 or C_1 expressing a probability lower than 5 %.

Results

According to the proposed approach, variations observed in the average indicator value for light (L_{av}) can be considered significant. The same holds for temperature and moisture indicator values (Figures 12&13). A first observation in Figures 11-13 is the different width of C_1 and C_2: data of 1955 are more dispersed than those of 2000 (Table 3). Around 1955, pastures of Braulio Valley were intensively grazed by 268 cows and 600 sheep on average (Giacomini & Pignatti 1955b). By excluding rocks, screes and glaciers surrounding the valley, all grazers were concentrated on a surface smaller than 8 km^2. Nowadays the situation is quite different, as no sheep and less than 100 cows are bred in Braulio Valley during the summer. It is likely that relevés of Caricetum curvulae of 1955 were more heterogeneous than those of 2000 because of the more intense grazing pressure. This could also explain the higher L_{av} of 1955 relevés: disturbance opened new niches in Caricetum curvulae, suitable for pioneer species dispersing from contiguous plant communities and characterised by higher L values. Besides, L decrease might be partially related to the decreased albedo caused by the recent retreat of glaciers surrounding the valley. The more intense melting rate of glaciers during the vegetative period (Orombelli et al., 1999) is probably responsible for the higher soil moisture suggested by the increase of M_{av} indicator value. This conclusion is supported by the similar trend observed in the variation of T_{av} indicator value, the recent increase of which might be directly related to the global warming and to the shifting upwards of species normally occurring in the subalpine belt (Table 4).

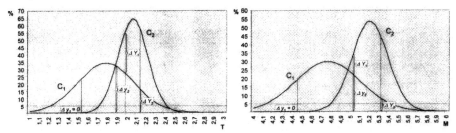

Figures 12 & 13. Distribution of T and M values among the n possible plots of 1955 (C_1) and of 2000 (C_2). See text.

CONCLUSIONS

Ellenberg's indices seem to be a consistent system to relate observed variations in flora and vegetation of a given site to variations of environmental factors, and to compare data from different sites. Up to now, the attention of scientists was mostly concentrated in testing the consistency of the system for M, pH, N and S indicator values, by referring to instrumental measurements on a topographic sequence. In the present study, the consistency of Ellenberg's indices was tested against the effects of climate change on a chronological sequence for the first time. Since no local instrumental measurements were available to support the observed variations of indicator values, an attempt to validate the hypotheses per analogy, by means of statistical inferences, was carried out. On this basis, a tentative approach to assess the significance-thresholds of the observed variations has been proposed. A similar approach, with different calibration, might be also used for indices similar to the Ellenberg's ones but adopting scales of different amplitude, such as those proposed by Landolt (1977) for the Swiss flora.

The proposed approach can be used even in case of a hypothetical incongruity of ecological preferences of a given species from Central to Southern Europe, since in time-spanned observations, variations are referred to the same locality. Therefore, the ordinal position of the indicator value of a given species is irrelevant, meaning only that the species has the same indicator value in different data-sets. Anyway, integration and adjustments proposed by Pignatti to adapt the Ellenberg's model to the Italian flora resulted quite homogeneous in the latter, in terms of amplitude and distribution of variance.

Usefulness of Ellenberg's indices in environmental analyses mostly derives from the simplicity of the model. This model does not consider interactions among environmental variables and operates in the frame of a reductionistic paradigm. Any calibration or integration including such interactions would unavoidably complicate the model and render it to less general application, as already noted by Ter Braak & Gremmen (1987). Ellenberg's indices achieve their maximum significance in synthetic-comparative studies on large temporal and/or spatial scales, so any complication of the model should be avoided. Averaging indicator values can be recommended as the simplest way to present such values in a synthetic way, worthy to be statistically compared on macroscales. By studying plants at the small scale, one can observe many differences even among contiguous populations of the same species. Ellenberg's indices certainly are not useful to check such variations, but "duc in altum" It is impossible to appreciate paintings keeping the nose upon the canvas!

"Generalisations about vegetation are urgently required to solve pressing problems created by modern land use, climate change and pollution. If one approaches vegetation with the tools of the watchmaker, there is no limit to the dissection which can be achieved. But if, like Heinz Ellenberg, one raises one's eyes to the broader canvas, the generalisations are there to be discovered" (after Thompson et al., 1993).

ACKNOWLEDGEMENTS

We are grateful to Dr. Annette Menzel and to Prof. Conradin A. Burga, Peter J. Edwards, Frank Klötzli, Christian Körner, Hans Möller, Michael Richter, and Reto J. Strasser for their helpful comments and suggestions.

REFERENCES

Anzalone B.,1954, Flora e vegetazione dell'Isola di Zannone (Arcip. Pontino). *Rend. Accad. Naz. Quaranta* 4-5: 23-72.

Anzalone B., Lattanzi E., Lucchese F. & Padula M., 1997, Flora vascolare del Parco Nazionale del Circeo (Lazio). *Webbia* 51(2): 67-105.

Bèguinot A., 1905, La vegetazione delle Isole Ponziane e Napoletane. *Annali di Botanica* 3: 181-453.

Bianco P. & Fanelli G., 2001, Flora e vegetazione della Valle dell'Inferno (Roma). *Annali di Botanica* (in press).

Bianco P., Tescarollo, P., Pignatti, S., 2001, La vegetazione di Castelporziano (Roma). *Quad. Accad. Scienze* (in press).

Borhidi A., 1995, Social behaviour types, the naturalness and relative ecological indicator values of the higher plants in the Hungarian flora. *Acta Bot. Hungar.* 39: 97-181.

Böcker R., Kowarik I. & Bornkamm R., 1983, Untersuchungen zur Anwendung der Zeigerwerte nach Ellenberg. *Verh. Ges. Ökol.* 11: 35-56.

Braun-Blanquet J., 1964, *Pflanzensoziologie* (3rd edition). Springer, Wien.

Degorski M., 1982, Usefullness of Ellenberg bioindicators in characteristic plant communities and forest habitats on the basis of data from the range Grabowy in Kampinos Forest. *Ekol. Pol. (Warsaw)* 30: 453-477.

Durwen K.J., 1982, Zur Nutzung von Zeigerwerten und artspezifischen Merkmalen der Gefäβpflanzen Mitteleuropas für Zwecke der Landschaftökologie und –planung mit Hilfe der EDV. *Arbeitsber. Lehrst. Landschaftökol. Münster* 5: 1-138.

Ellenberg H., 1974, Zeigerwerte der Gefäβpflanzen Mitteleuropas. *Scripta Geobot.* 9: 1-97.

Ellenberg H., 1996, *Vegetation Mitteleuropas mit den Alpen* (5th edition). Ulmer, Stuttgart

Ellenberg H., Weber H.E., Düll R., Wirth V., Werner W., Pauliβen D., 1992, Zeigerwerte von Pflanzen in Mitteleuropa *Scripta Geobot.* 18(2): 1-258.

Faifer D. & Pirola A., 1995, Nota preliminare sui cambiamenti nella vegetazione dell'alta valle del Braulio (Parco Nazionale dello Stelvio, Lombardia). *Arch. Geobot.* 1(2): 189-190.

Giacomini V. & Pignatti S., 1955a: Flora e vegetazione dell'alta valle del Braulio con particolare riguardo ai pascoli d'altitudine. *Mem. Soc. Ital. Sc. Nat. Mus. Civ. St. Nat. Milano* 11(2-3): 1-194.

Giacomini V. & Pignatti S., 1955b: I pascoli dell'Alpe dello Stelvio (alta Valtellina), saggio di fitosociologia applicata e di cartografia fitosociologica. *Ann. Sperim. Agraria* (n.s.) 9: 1-49.

Guarino R. & Bernardini A., 2001, *Indagine sulla diversità floro-vegetazionale del Comprensorio del Cuoio (Toscana centro-settentrionale)*. Pubblicazioni dell'Amministrazione Provinciale, Pisa.

Körner C, 1994, Impact of atmospheric changes on high mountain vegetation. In: Benistorm M. (ed.): *Mountain environments in changing climates*. Routledge, London, pp. 155-166.

Kowarik I. & Seidling W., 1989, Zeigerwertberechnungen nach Ellenberg, zur Problemen und Einschränkungen einer sinnvollen Methode. *Landschaft u. Stadt* 21(4): 132-143.

Landolt E., 1977, Ökologische Zeigerwerte zur Schweizer Flora. *Veröff. Geobot. Inst. ETH*, Stiftung Rübel, Zürich 64: 1-208.

Möller H., 1992, Zur Verwendung des Medians bei Zeigerwertberechnungen nach Ellenberg. *Tuexenia* 12: 25-28.

Möller H., 1997, Reaktions- und Stickstoffzahlen nach Ellenberg als Indikatoren für die Humusform in terrestrischen Waldökosystemen im Raum Hannover. *Tuexenia* 17: 349-365.

Montelucci G., 1954: Flora e vegetazione della Valle dell'Inferno. *Annali di Botanica* 24(2): 241-289.

Petersen J., 2000, *Die Dünentalvegetation der Wattenmeer-Inseln in der südlichen Nordsee*. Husum Verlag, Husum.

Pignatti S., 1982, *Flora d'Italia* (3 volumes). Edagricole, Bologna.

Pignatti S., 1998, *I boschi d'Italia, sinecologia e biodiversità*. UTET, Torino.

Pignatti S., 1999, La phytosociologie Braun-Blanquetiste et ses perspectives. *Coll. Phytosoc.* 27: 1-15.

Pignatti S., Ellenberg H. & Pietrosanti S., 1996, Ecograms for phytosociological tables based on Ellenberg's Zeigerwerte. *Annali di Botanica* 54: 5-14.

Seidling W. & Rohner M.S., 1993, Zusammenhänge zwischen Reaktions-Zeigerwerten und bodenchemischen Parametern am Beispiel von Waldbodenvegetation. *Phytocoenologia* 23: 301-317.

Schaffers A.P. & Sýkora K.V., 2000, Reliability of Ellenberg indicator values for moisture, nitrogen and soil reaction: a comparison with field measurements. *J. Veg. Sci.* 11: 225-244.

Ter Braak C.J. & Gremmen N.J.M., 1987, Ecological amplitudes of plant species and the internal consistency of Ellenberg's indicator values for moisture. *Vegetatio* 69: 79-87.

Theurillat J.P., 1995, Climate change and the alpine flora: some perspectives. In: Guisan A., Holten J.I., Spichinger R. & Tessier L. (eds.): *Potential ecological impacts of climate change in the Alps and Fennoscandian mountains*. Ed. Conserv Jard. Bot., Genève, pp. 121-127.

Thompson K., Hodgson J.G., Grime P., Rorison I.H., Band S.R. & Spencer R.E., 1993, Ellenberg numbers revisited. *Phytocoenologia* 23: 277-289.

Van der Maarel E., 1993, Relations between sociological-ecological species groups and Ellenberg indicator values. *Phytocoenologia* 23: 343-362.

Westhoff V. & Van der Maarel E., 1978, The Braun-Blanquet approach. In: Whittaker R.H. (ed.): *Classification of plant communities*. Junk, The Hague, pp. 287-399.

Wilson J.B. & Agnew A.D.Q., 1992, Positive-feedback switches in plant communities. *Adv. Ecol. Res.* 23: 263-336.

Zarzycky K., 1984, *Indicator values of vascular plants in Poland* (in polish). Krakow Inst. Bot. Polska Akad. Nauk.

TABLES

Table 1. List of appeared and disappeared species in the present flora of Inferno Valley, compared to the flora of 1954. L, T, and M indicator values are reported. Nomenclature of species follows Pignatti (1982).

2000 only	L T M			
Allium caepa	8 8 6	Setaria verticillata	7 8 4	Geranium colombinum	7 9 2
Alopecurus myosuroides	6 6 6	Silene bellidifolia	7 8 2	Hieracium boreale	× × ×
Amaranthus blitoides	9 7 3	Stipa caudata	× × ×	Hieracium sabaudum	5 6 4
Anagallis parviflora	7 8 5	Vicia atropurpurea	9 9 2	Inula conyza	6 6 4
Anthemis cotula	7 7 4	Vicia peregrina	7 7 4	Inula graveolens	9 8 3
Antirrhinum majus	9 8 2	Viola odorata	5 6 5	Inula salicina	7 5 4
Artemisia verlotiorum	7 7 5	**1954 only**		Lactuca saligna	9 7 4
Biserrula pelecinus	9 9 2	Allium oleraceum	7 8 3	Legousia speculum-veneris	7 7 4
Brachypodium phoenicoides	8 8 3	Allium paniculatum	7 7 3	Linaria pellisseriana	9 9 3
Bromus inermis	8 × 4	Allium vineale	8 7 4	Lysimachia vulgaris	7 × 9
Carduus micropterus	7 6 3	Allium chamaemoly	8 9 2	Melilotus altissima	8 6 7
Carex hallerana	5 7 3	Alopecurus utriculatus	8 7 8	Melittis melissophyllum	5 6 4
Centaurea napifolia	8 9 4	Amaranthus albus	9 9 3	Mentha longifolia	7 5 8
Chenopodium multifidum	8 7 2	Apium graveolens	7 7 7	Misopates orontium	7 7 5
Conyza albida	8 8 3	Arabis sagittata	7 6 4	Nigella damascena	8 9 3
Cuscuta cesatiana	8 7 ×	Asperula laevigata	6 6 4	Odontites lutea	7 7 2
Eleagnus angustifolia	9 7 3	Aster linosyris	8 7 3	Ophrys fuciflora	9 9 3
Eleusine indica	9 8 2	Brachypodium distachyum	9 9 1	Ophrys tenthredinifera	8 9 3
Galinsoga ciliata	7 6 7	Bromus erectus	8 5 3	Pastinaca sativa	8 6 4
Limodorum abortivum	× 7 4	Bromus rubens	8 9 2	Phalaris coerulescens	7 6 5
Lycium chinense	9 7 3	Buglossoides arvensis	5 × ×	Phalaris minor	7 7 4
Medicago litoralis	9 9 2	Bupleurum praealtum	6 7 4	Polygala vulgaris	7 4 5
Mirabilis jalapa	6 7 4	Bupleurum tenuissimum	9 8 4	Reseda lutea	7 6 3
Nardurus halleri	8 9 2	Campanula erinus	7 8 2	Satureja alpina	9 × 5
Onobrychis caput-galli	9 9 2	Carduncellus coeruleus	9 9 3	Satureja graeca	8 8 2
Oxalis articulata	× × ×	Carex pendula	5 5 8	Satureja vulgaris	7 5 4
Oxalis dilleni	7 7 5	Catabrosa acquatica	8 4 9	Scandix pecten-veneris	7 7 3
Papaver dubium	6 6 4	Cephalaria transsylvanica	7 6 3	Sisymbrium irio.	8 8 3
Phytolacca americana	9 8 5	Cerinthe major.	7 8 4	Thymus serpyllum	× × ×
Plantago coronopus	8 7 7	Consolida regalis	9 7 4	Thlaspi perfoliatum	8 6 4
Pyracantha coccinea	5 8 3	Convolvolus cantabrica	9 8 3	Trifolium tomentosum	9 9 4
Robinia pseudoacacia	5 7 4	Delphinium halteratum	8 9 3	Trifolium vesiculosum	8 9 3
Senecio leucanthemifolius	9 9 2	Euphorbia falcata	9 7 4	Viola hirta.	6 5 3
Serapias parviflora	9 9 2	Festuca fenas	8 7 6	Viola tricolor	7 X 5
Setaria ambigua	7 7 4	Festuca gigantea	4 5 7	Zizyphus sativus	× × ×
		Fragaria vesca	6 × 4		

Table 2. List of appeared and disappeared species in the present flora of Zannone Island, compared to the floras of 1905 and 1954. L, T, and M indicator values are reported. Nomenclature of species follows Pignatti (1982).

1905 only	L	T	M
Allium triquetrum	6	9	4
Ambrosia maritima	8	7	5
Anemone hortensis	×	×	×
Avellinia michelii	9	8	1
Beta vulgaris	9	7	6
Brachypodium sylvaticum	4	5	5
Brassica fruticosa	×	×	×
Bromus hordeaceus	7	6	2
Buglossoides purpurocaer	7	×	5
Carex sylvatica	2	5	5
Cercis siliquastrum	×	×	×
Cistus crispus	9	9	2
Digitaria sanguinalis	7	7	3
Echium plantagineum	9	8	3
Erodium chium	×	×	×
Fumaria parviflora	×	×	×
Galium divaricatum	7	7	5
Galium verrucosum	9	9	1
Geranium columbinum	7	9	2
Hainardia cylindrica	×	×	×
Amaranthus deflexus	8	8	4
Bromus madritensis	8	7	3
Bromus rigidus rigidus	7	7	4
Calamintha nepeta	5	7	3
Carduus pycnocephalus	7	8	3
Carex otrubae	9	5	9
Carlina lanata	9	×	6
Chenopodium vulvaria	7	7	4
Cirsium vulgare	8	5	5
Convolvulus althaeoides	9	9	3
Conyza bonariensis	8	8	3
Cynoglossum creticum	9	9	3
Cyperus longus	9	8	10
Erodium cicutarium	8	7	3
Euphorbia helioscopia	6	7	4
Fumana thymifolia	7	7	8
Fumaria flabellata	×	×	×
Geranium dissectum	7	9	2
Geranium purpureum	7	8	3
Geranium rotundifolium	7	8	3
Hedypnois cretica	9	9	2
Holcus lanatus	7	5	6
Inula graveolens	9	8	3
Lathyrus annuus	8	8	3
Lavatera cretica	8	9	2
Lolium perenne	8	5	5
Luzula forsteri	4	7	4
Lythrum hyssopifolia	8	7	7
Melica arrecta	8	9	2
Moehringia pentandra	×	×	×
Myosotis arvensis	5	4	7
Myosotis ramosissima	9	6	2
Narcissus tazetta	7	8	4
Neotinea intacta	4	5	3
Oryzopsis miliacea	5	7	4
Petrorhagia velutina	×	×	×
Phytolacca americana	9	×	×
Polycarpon alsinifolium	9	9	2
Rumex pulcher	8	9	2
Sedum hispanicum	7	6	4
Senecio leucanthemifolius	7	7	4
Senecio lividus	7	×	5
Inula conyza	6	6	4
Juncus bufonius	4	7	6
Juncus pygmaeus	7	4	9
Lactuca serriola	9	7	4
Lathyrus aphaca	6	6	3
Lathyrus cicera	×	×	×
Lathyrus sphaericus	9	9	2
Lolium multiflorum	7	7	4
Medicago arabica	8	7	3
Medicago praecox	8	6	7
Osyris alba	7	8	3
Paronychia echinulata	9	×	4
Periballia minuta	8	9	2
Phagnalon saxatile	×	×	×
Phalaris minor	7	7	4
Plantago lanceolata	6	8	2
Rhagadiolus edulis	×	×	×
Rumex bucephalophorus	8	9	2
Rumex conglomeratus	8	7	7
Sedum andegavense	9	9	1
Senecio vulgaris	7	9	5
Sisymbrium officinale	8	6	4
Sonchus asper	7	5	4
Sonchus oleraceus	7	5	4
Spartium junceum	7	7	4
Teucrium flavum	6	6	4
Torilis arvensis	7	8	4
Trachynia distachya	×	×	×
Trifolium angustifolium	9	9	2
Trifolium ligusticum	8	9	2
Trifolium tomentosum	9	9	4
Urospermum picroides	9	9	2
Veronica arvensis	5	5	5
Veronica cymbalaria	7	7	4
Vicia pubescens	×	×	×
Vulpia bromoides	8	9	2
1954 only			
Adonis microcarpa	6	6	5
Avena barbata	8	8	3
Brassica oleracea v. sylv.	7	9	2
Crepis foetida	8	8	2
Crepis leontodontoides	7	6	4
Cyclamen hederifolium	5	5	5
Epipactis microphylla	4	4	6
Fumaria densiflora	8	6	3
Juniperus oxycedrus	8	8	3
Leontodon tuberosus	9	8	3
Lycopersicon esculentum	8	9	5
Medicago lupulina	7	5	4
Medicago truncatula	9	8	1
Melilotus elegans	×	×	×
Mentha piperita	×	×	×
Pallenis spinosa	9	9	4
Panicum miliaceum	6	8	7
Polypodium australe	×	×	×
Radiola linoides	×	×	×
Ranunculus bulbosus	8	6	3
Raphanus sativus	×	×	×
Reseda luteola	7	6	3
Serapias vomeracea	9	8	3
Smyrnium olusatrum	×	×	×
Sedum stellatum	9	8	2
Silene alba	7	5	6
Tolpis umbellata	9	×	4
Trifolium bocconei	×	×	×
Trifolium stellatum	9	9	2
Trifolium subterraneum	9	9	2
Umbilicus horizontalis	5	8	3
Verbascum thapsus	7	7	2
Vicia bithynica	7	7	3
Vicia hirsuta	×	×	×
Vicia hybrida	7	8	3
Vicia lutea ssp. vestita	9	8	3
Vicia peregrina	9	8	3
Vicia tenuissima	7	8	4
Viola odorata	5	6	5
Vulpia ciliata	8	9	2
1905 and 1954 only			
Aegilops geniculata	9	9	2
Allium ampeloprasum	7	7	3
1954 and 1997 only			
Anthemis arvensis incr.	7	6	4
Centaurium pulchellum	9	×	4
Cynosurus echinatus	9	9	2
Lemna gibba	7	6	11
Lemna minor	7	6	11
Limonium pontium	8	8	1
Lotus angustissimus	9	8	4
Matthiola incana	9	9	3
Petroselinum sativum	7	5	2
Sagina apetala	8	7	4
Valantia muralis	5	5	5
1997 only			
Anagallis parviflora	9	×	2
Brachypodium distachyum	9	9	1
Bromus fasciculatus	6	5	6
Bromus rubens	6	5	6
Capparis spinosa	9	6	5
Carex hallerana	3	5	4
Daucus carota	8	6	4
Daucus gingidium	4	5	5
Fumaria officinalis	7	7	4
Galium tricornutum	6	5	5
Hedypnois rhagadioloides	×	×	×
Hordeum leporinum	9	9	3
Linum strictum	9	9	2
Lotus parviflorus	8	6	6
Malva nicaeensis	9	8	3
Matricaria chamomilla	×	×	×
Parapholis incurva	9	7	5
Phalaris bulbosa	7	7	4
Plantago psyllium	9	6	3
Polycarpon diphyllum	9	8	2
Polycarpon tetraphyllum	7	7	4
Rumex crispus	7	5	6
Senecio cineraria	8	7	4
Vicia disperma	×	×	×
Vulpia geniculata	9	9	1
Vulpia ligustica	8	9	2
Vulpia muralis	8	9	2

Table 3. List of species occurring in the phytosociological tables of Caricetum curvulae in Braulio Valley. In the first two columns frequencies (in percent) of each species in 1955 and in 2000 are reported. L, T, and M indicator values are reported as well. The few species not occurring in the Ellenberg's list of indicator values have been ignored. Nomenclature of species follows Pignatti (1982).

v%$_1$	v%$_2$	Species	L	T	M
45	0	*Luzula spicata*	8	2	4
35	0	*Soldanella alpina*	7	1	7
30	0	*Euphrasia drosocalyx*	8	3	5
20	0	*Erigeron uniflorus*	9	1	5
15	0	*Cardamine resedifolia*	8	2	5
15	0	*Sagina saginoides*	7	3	6
10	0	*Saxifraga bryoides*	9	1	5
10	0	*Sedum alpestre*	8	2	5
5	0	*Potentilla grandiflora*	8	3	3
5	0	*Cerastium arvense strictum*	9	3	3
5	0	*Draba aizoides*	8	×	3
5	0	*Saxifraga exarata*	8	1	4
5	0	*Potentilla frigida*	9	1	4
5	0	*Ranunculus parnassifolium*	8	2	5
5	0	*Festuca pumila*	8	2	5
5	0	*Salix retusa*	7	2	6
5	0	*Taraxacum alpinum*	8	2	6
5	0	*Saxifraga seguieri*	8	2	7
0	33,3	*Juncus jacquinii*	9	2	5
0	27,8	*Silene acaulis*	9	1	4
0	22,2	*Salix reticulata*	8	2	6
0	16,7	*Antennaria carpatica*	8	2	5
0	16,7	*Campanula scheuchzeri*	8	2	5
0	16,7	*Cirsium spinosissimum*	7	2	6
0	16,7	*Veronica alpina*	7	2	6
0	16,7	*Vaccinium myrtillus*	5	×	×
0	11,1	*Hieracium pilosella*	7	×	4
0	11,1	*Botrychium lunaria*	8	×	4
0	11,1	*Lloydia serotina*	9	1	5
0	11,1	*Sedum atratum*	9	2	5
0	11,1	*Luzula sudetica*	8	3	5
0	11,1	*Carex parviflora*	9	2	7
0	5,6	*Festuca alpina*	8	1	3
0	5,6	*Trifolium alpinum*	8	2	4
0	5,6	*Gentiana germanica*	7	5	4
0	5,6	*Carex sempervirens*	7	×	4
0	5,6	*Hieracium glaciale*	8	1	5
0	5,6	*Trisetum spicatum*	9	1	5
0	5,6	*Leontodon montanus*	8	2	5
0	5,6	*Carex atrata*	9	2	5
0	5,6	*Cerastium fontanum fontanum*	6	3	5
0	5,6	*Aster bellidiastrum*	7	3	5
0	5,6	*Trifolium repens. prostratum*	8	×	5
0	5,6	*Gnaphalium hoppeanum*	8	2	7
0	5,6	*Dechampsia caespitosa*	6	×	7
0	5,6	*Bartsia alpina*	8	3	8
0	5,6	*Deschampsia flexuosa*	6	×	×
0	5,6	*Festuca nigrescens*	7	×	×
100	100	*Carex curvula*	9	1	4
90	94,4	*Leucanthemopsis alpina*	8	2	7
85	94,4	*Leontodon helveticus*	8	3	5
65	88,9	*Agrostis rupestris*	8	2	4
60	88,9	*Poa alpina* (incl. *f.ma vivipara*)	7	3	5
45	100,0	*Euphrasia minima*	7	2	5
55	77,8	*Polygonum viviparum*	7	2	5
55	77,8	*Phyteuma hemisphaericum*	8	2	5
60	66,7	*Senecio incanus carniolicus*	8	2	5
30	88,9	*Salix herbacea*	7	2	7
40	72,2	*Ligusticum mutellina*	7	2	6
35	72,2	*Geum montanum*	7	2	5
40	66,7	*Avenula versicolor*	9	2	5
55	50,0	*Oreochloa disticha*	9	1	5
30	72,2	*Potentilla aurea*	8	3	4
45	50,0	*Festuca halleri*	9	1	4
55	38,9	*Veronica bellidioides*	8	2	4
50	33,3	*Primula daonensis*	8	2	6
20	61,1	*Homogyne alpina*	6	4	6
5	72,2	*Gentiana punctata*	8	2	5
5	66,7	*Alchemilla pentaphyllea*	7	2	7
20	50,0	*Nardus stricta*	8	×	×
30	38,9	*Hieracium glanduliferum*	9	1	5
10	55,6	*Luzula alpino-pilosa*	7	2	7
15	50,0	*Anthoxanthum alpinum*	7	3	2
10	50,0	*Antennaria dioica*	8	×	4
25	33,3	*Loiseleuria procumbens*	9	2	5
5	44,4	*Soldanella pusilla*	5	3	5
25	22,2	*Minuartia sedoides*	9	1	4
25	22,2	*Sibbaldia procumbens*	7	2	7
25	22,2	*Gnaphalium supinum*	7	2	7
5	33,3	*Ranunculus montanus*	6	3	5
20	16,7	*Juncus trifidus*	8	2	4
5	27,8	*Gentiana acaulis*	8	2	5
10	22,2	*Elyna myosuroides*	9	2	4
15	16,7	*Phyteuma globulariifolium ped.*	8	1	5
5	22,2	*Gentiana nivalis*	9	1	5
5	22,2	*Agrostis alpina*	8	2	5
5	22,2	*Selaginella selaginoides*	8	3	7
10	11,1	*Pulsatilla vernalis*	7	×	4
5	5,6	*Gentiana tenella*	9	1	5
5	5,6	*Agrostis schraderiana*	6	2	5
5	5,6	*Vaccinium uliginosum*	6	×	×
Mosses					
35	66,7	*Polytrichum piliferum*	9	2	5
25	50,0	*Racomitrium canescens*	9	3	1
20	16,7	*Sanionia uncinata*	×	×	7
20	22,2	*Paraleucobryum enerve*	9	1	7
15	38,9	*Polytrichum juniperinum*	8	2	4
5	5,6	*Dicranum elongatum*	7	1	6
5	22,2	*Dicranum fuscescens*	7	2	6
5	44,4	*Pohlia cruda*	4	×	5
5	5,6	*Sphenolobus minutus*	7	2	5
Lichens					
100	94,4	*Cetraria islandica*	8	×	5
70	66,7	*Cladonia rangiferina*	6	4	5
60	66,7	*Cladonia pyxidata*	7	×	×
55	22,2	*Cetraria nivalis*	9	1	8
25	16,7	*Thamnolia vermicularis*	8	1	8
20	11,1	*Alectoria ochroleuca*	8	1	8
15	5,6	*Cetraria cucullata*	9	2	8
15	11,1	*Cladonia uncialis*	8	4	5
15	0,0	*Peltigera venosa*	5	2	6

Table 4. Average L, T and M values of the species occurring in each of 20 relevés of Caricetum curvulae of 1955 (left columns) and of the 18 relevés of 2000 (right columns). μ and σ are averages and standard deviations, respectively μ_1 and μ_2 are averages of the indicator values lower and higher than μ, respectively. Curves in fig. 11-13 are settled by μ and σ, while μ_1 and μ_2 were used to assess the significance of the observed variations of L, T and M values in Caricetum curvulae of Braulio Valley over the last 45 years.

	L_{1955}	L_{2000}	T_{1955}	T_{2000}	M_{1955}	M_{2000}
	7,63	7,50	1,33	1,71	4,00	4,77
	7,67	7,57	1,33	1,79	4,25	4,96
	7,67	7,60	1,33	1,88	4,40	5,00
	7,75	7,60	1,50	1,95	4,50	5,00
	7,78	7,63	1,57	2,00	4,50	5,04
	7,83	7,63	1,60	2,00	4,57	5,12
	7,86	7,66	1,60	2,00	4,60	5,17
	7,89	7,67	1,71	2,06	4,67	5,19
	8,00	7,67	1,75	2,08	4,67	5,19
	8,00	7,73	1,75	2,11	4,70	5,24
	8,20	7,73	1,80	2,11	4,89	5,32
	8,22	7,76	1,83	2,14	4,89	5,33
	8,25	7,81	1,89	2,14	5,00	5,34
	8,30	7,81	1,89	2,17	5,00	5,38
	8,33	7,82	1,90	2,18	5,00	5,38
	8,50	7,95	2,00	2,19	5,00	5,38
	8,50	8,14	2,00	2,26	5,00	5,40
	8,67	8,18	2,22	2,27	5,11	5,41
	8,67		2,25		5,20	
	8,71		2,29		5,33	
μ	8,12	7,75	1,78	2,06	4,76	5,20
σ	0,36	0,19	0,29	0,15	0,34	0,19
μ_1	7,81	7,63	1,55	1,90	4,49	5,05
μ_2	8,44	7,93	2,01	2,16	5,04	5,35

Correspondence and request for materials should be addressed to Pignatti (e-mail: pignatti@axrma.uniroma1.it) if concerning the data-base of Ellenberg's indices for the Italian flora, to Guarino (e-mail: guarinotro@hotmail.com) if concerning the statistical treatment of data proposed in this paper.

Fingerprints of climate change – concluding remarks

CHRISTIAN KÖRNER* & GIAN-RETO WALTHER[#]
*Institute of Botany, University of Basel, Schoenbeinstr. 6, CH-4056 Basel, Switzerland
[#]Institute of Geobotany, University of Hannover, Nienburger Str. 17, DE-30167 Hannover, Germany

Abstract: This volume documents that ongoing climate change affects organisms. Climate change emerges as an important biodiversity issue with functional implications on the (i) species, (ii) population and community, and (iii) ecosystem level. The climatic regime of the last three decades induced changes in species ranges and behaviour more than might have been expected from the purely physical magnitude of the change, thus, leaving clear "fingerprints" of change, which might impress those, who find the 0.6 K mean global warming itself negligible.

INTRODUCTION

Global change issues address mainly three different aspects: (i) change in atmospheric chemistry, (ii) climate change and (iii) land use change (Figure 1). All these three aspects of global change rose public concern and scientific interest. The contributions in the present volume mainly deal with the second aspect, i.e. climatic warming and its consequences on biological systems, the two other components always interfere with any climatic impacts.

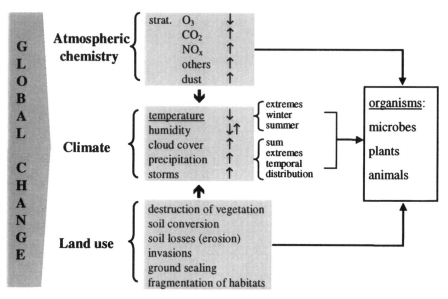

Figure 1. Aspects of global change and their interactions.

Often it is hard to disentangle these three aspects and to attribute just one particular cause to an observed effect. There will always remain some uncertainty. Too much of the political debate is mislead by the believe in certainty, in 'yes' or 'no' scenarios, when in fact we deal with probability. In this regard, the Third Assessment Report of the Intergovernmental Panel on Climate Change (IPCC 2001a, b) provides a good example on how to deal with uncertainties and confidence.

When does weather become climate? What length of a period is needed for weather anomalies to become "climate change"? Does it need 10, 20, 30 or even 50 years of a continued anomaly for the observed change to be considered a trend rather than a fluctuation? Singularities such as El Niño-Southern Oscillation (ENSO) or extreme events such as heavy floods or storms such as "Lothar", which occurred in December 1999 in Central Europe (cf. WSL & BUWAL 2001) may become 'trends' when they occur more regularly as series of singularities, e.g. many ENSOs or many "Lothars".

In the case of temperature it is unclear to which degree the change in means in minima or maxima, in the amplitude or frequency or duration (e.g. season length) matter. Figure 2 illustrates various scenarios. One for instance shows a case with little change in the mean, but a significant reduction in minimum temperatures, similar to what happened over the past decades around Ascona, where this meeting was held, and which appears to have opened terrain to laurophyllous vegetation.

Fingerprints of climate change – concluding remarks

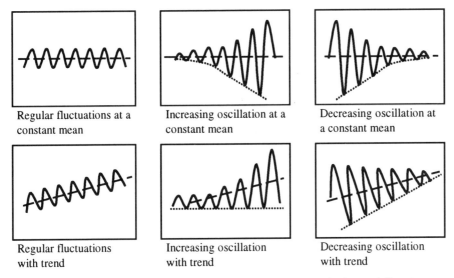

Figure 2. Idealised hypothetical changes of a fluctuating parameter in time and direction.

In evolutionary time scales, climate has experienced many periods of change and organisms adjusted themselves or died out (great extinctions). For the biota, the rate of change may be more important than the degree of change. The 1990s represent the warmest decade of the millennium, with 1998 the warmest year in that period (IPCC 2001a). A statistical analysis of thermal oscillations suggests, that "the warming over the past 100 years is very unlikely to be due to internal variability alone" (IPCC 2001a). But, in geological time scales, fossil records suggest periods with a humid and subtropical climate for regions experiencing a temperate climate today (e.g. Mai 1995).

This meeting aimed at collecting biological "fingerprints" of recent climate warming. Some of these fingerprints are rather subtle, such as changed fecundity in populations of birds, dragonflies or walruses, others materialise as solid environmental 'memory' such as treerings. For instance, Paulsen et al. (2000) tested climatic influences on trees along elevational transects at and below the upper treeline in the European Alps. In the uppermost zone of the treeline ecotone, radial stem increments were measured at three different ranges of the elevational distance from the outpost treeline over the past 170 years (Figure 3). These results show that the former elevational differences in vigour of trees had gradually disappeared.

This volume documents a number of similarly dramatic responses of plants and animals form various latitudes. Taken together, these signals are much more pronounced as could be expected from the climatic changes we have seen in recent years. They illustrate the sensitivity of organisms to environmental alterations which, by their purely physical magnitude, hardly affect humans.

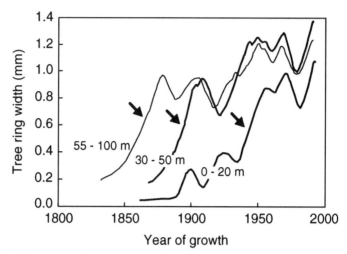

Figure 3. Radial tree stem increment (age detrended) of the past 170 years in the uppermost zone of the treeline ecotone. The three lines represent tree rings from nine transects cored at three different ranges of elevational distance from the outpost treeline. Arrows indicate periods of major change (modified from Paulsen et al. 2000).

BIOLOGICAL CONSEQUENCES OF CLIMATIC CHANGE

For a given type (species) of organisms climate change impacts can be divided into two ecologically different phenomena:
- gradual phenomenon, e.g. inducing faster or slower growth
- threshold phenomenon; i.e. "yes" or "no" in terms of the (im-)possibility to exist.

For instance, tree height gradually decreases towards the treeline in contrast to radial tree growth, which shows rather abrupt changes, implying threshold effects of temperature and a wave-like upslope movement of the signal as temperatures increased over the past 170 years (Figure 3).

The interpretation of "fingerprints" becomes very difficult when biological invasions, for what ever reasons, overlap with climatic changes. For indigenous species this means a combined alteration in competition with exotics and climate. An example illustrated in Figure 4 is the laurophyllous invasion seen along the lower slopes at the southern Alps in the so-called Insubrian climate (Klötzli et al. 1996).

Fingerprints of climate change – concluding remarks 309

Figure 4. Picture from a forest, heavily invaded by exotic evergreen broad-leaved species, in the vicinity of Locarno, Ticino, Switzerland.

Changes in biota, the vegetation in particular, will influence other ecosystem properties such as soils. Soils are only as stable and secure as the vegetation is; the vegetation is only as stable and secure as its components resists extreme events, which in turn, is a function of internal functional redundancies. The arrival of new species or the disappearance of keystone species may, at least periodically, reduce the strength of the vegetation to protect soils from erosion. A risky situation on steep slopes.

CLIMATIC CHANGE: A BIODIVERSITY ISSUE

Differential sensitivity of organisms causes climate change to become a biodiversity issue. On the one hand, there may be a loss of species, as was exemplified for pinnipeds (Kelly, chapter 3), corals (Hoegh-Guldberg, chapter 13) and species on mountain tops (Pauli et al., chapter 9), on the other hand, there may be a gain in species, e.g. butterflies (Parmesan and Hill et al., chapters 4 & 5), dragonflies (Ott, chapter 6), laurophyllous species (Walther et al., chapter 10) and spreading species (Convey and Vesperinas et al., chapters 2 & 11).

The examples presented in this volume suggest that common species become more abundant and 'win', whereas rare species tend to 'sit and loose' (cf. also IPCC 2001b, chapter 19.3.3).

Climate change may act as driving factor for migration processes. These can either lead to local re-arrangements of species or to a "moving front" of migrating synusia. Space for time experimental or modelling approaches are useful tools for testing climate change scenarios (see e.g. Erschbamer, chapter 16). They offer the possibility to investigate whether a species, taken from its present range and moved to its expected future habitat may survive. However, different migration rates and migration routes, as well as varying availability of species specific sites in future habitats may set limits to the interpretation of space for time models.

Functional measures of change

Species matter. Sometimes their functional significance is not obvious, but often their presence or abundance affects the functioning of the whole ecosystem. Several of the presentations illustrated cases where functional implications could be documented. These examples fall in three different groups and refer to functional changes (1) at the species level (species or indicators), (2) of life processes and (3) changes at the ecosystem level.

- The use of indicator values or species attributes:
  - "Mediterranean" species (see e.g. Ott, chapter 6)
  - Ellenberg indicator values (Pignatti et al., chapter 17)
  - Frost-indicators resp. thermophilous species (Walther et al. and Vesperinas et al., chapters 10 & 11)
- Processes, organ / individual:
  - Photoinhibition, f(heat) (Hoegh-Guldberg, chapter 13)
  - Fluorescence signals (Tsimilli-Michael & Strasser, chapter 14)
  - Growth / development (e.g. Paulsen et al. 2000; this chapter Figure 3)
  - Phenology (Defila & Clot and Menzel & Estrella, chapters 7 & 8)
- Ecosystem level
  - C-release (e.g. Möller et al. 2001)
  - N-release/trapping (see case study on laurophyllisation, below)

In any case, environmental change results in some sort of adjustment in terms of acclimation and/or modulation. In the long run, selection may lead to evolutionary adaptations, i.e. to greater abundance of formerly rare or new genotypes (ecotypes). Species replacement and re-arrangements strongly depend on the regional availability of species and migration corridors, and finally, new (exotic) species may seep in.

Implications of laurophyllisation – a case study

The spread and establishment of many laurophyllous plants in a former deciduous habitat, called laurophyllisation (see Klötzli & Walther 1999), was the subject of a student field course, with ecophysiological and population ecological studies carried out in a forest, heavily invaded by exotic evergreen broad-leaved species, in the vicinity of Locarno (Körner & Stöcklin 1999). In the following, two examples will be shown to demonstrate some of the consequences of interactions of climate change and invasion processes.

Structural change due to the arrival of new elements

The new laurophyllous species germinate in protected niches within an established deciduous chestnut forest. Their survival is likely due to their ability to photosynthesise (with their evergreen broad-leaved foliage) in winter. This might be a crucial advantage, which itself depends on the relatively mild winter temperatures. This also explains the unexpected high leaf area index (LAI) of 8: the laurophyllous species have established as a second layer of foliage underneath the deciduous layer, active in winter and just surviving the 6 month summer season (Figure 6). In fact, it is the longevity of the leaves which permits survival in deep shade. This longevity of leaves also traps nutrients, no longer available to the overstorey.

Population dynamics

Once a fertile adult neophytic individual has established in the forest, there is a new source for seed dispersal which may serve as 'stepping stone' for the next stage of the invasion process. It was observed, that the density of young plants decreases with increased distance from a source plant (Figure 7). In the vicinity of a reproductive individual, the number of seedlings was ten times that around non-reproductive individuals of the palm *Trachycarpus fortunei*. Given the observed inverted hyperbolic decrease of the number of juveniles around the adult, the source effect is apparent up to 5 m distance, i.e. one adult mother plant has a direct influence on the colonisation of the surrounding area of ca. 80 m^2. Thus, 130 fertile adult neophytic palms could ensure the supply of saplings to cover at least 1 ha (disregarding long distance transport by birds).

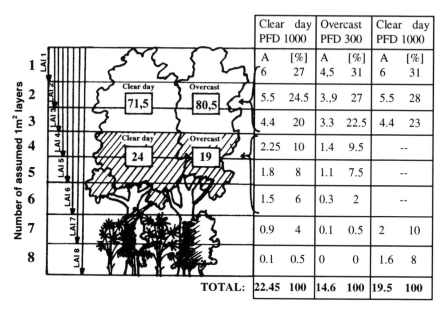

	Clear day PFD 1000		Overcast PFD 300		Clear day PFD 1000	
	A	[%]	A	[%]	A	[%]
1	6	27	4,5	31	6	31
2	5.5	24.5	3.,9	27	5.5	28
3	4.4	20	3.3	22.5	4.4	23
4	2.25	10	1.4	9.5	--	
5	1.8	8	1.1	7.5	--	
6	1.5	6	0.3	2	--	
7	0.9	4	0.1	0.5	2	10
8	0.1	0.5	0	0	1.6	8
TOTAL:	22.45	100	14.6	100	19.5	100

Figure 6. Contribution of the different layers to the total photosynthetic yield of a forest stand under different meteorological situations (averaged values over the time span from 8-16 h). Clear day conditions: extinction coefficient k = 0.5, PFD = 1000 μmol photons $m^{-2}s^{-1}$
Overcast conditions: extinction coefficient k = 0.4, PFD = 300 μmol photons $m^{-2}s^{-1}$
The layer specific photosynthetic rates (A in μmol CO_2 $m^{-2}s^{-1}$) sum up to 100 % (= total photosynthetic rate of the forest stand). The third column illustrates a situation with LAI = 3 (instead of 6) for the canopy layer, which results in an increased gain for the laurophyllous understorey. (Körner & Stöcklin 1999)

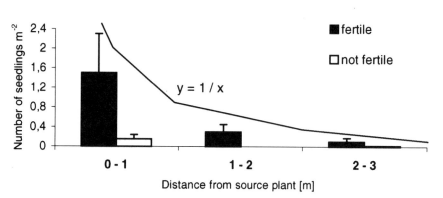

Figure 7. Density of young plants in the surroundings of an adult source plant (the palm *Trachycarpus fortunei*, see Figure 4) compared to central areas. Values are given in means ± 1SD (Körner & Stöcklin 1999).

Consequences for ecosystem processes

With the change from a mainly deciduous forest type to one dominated by evergreen broad-leaved species, another hypothesis arises. With the decelerated turnover due to the longer life time of evergreen leaves, one might assume that nutrients become immobilised. In the end, this would mean, that indigenous species become increasingly outcompeted due to their enhanced nutrient demand, compared with the exotic species.

'Warming the alpine' – a case study

Another example is the question of whether alpine plants will take any advantage of a warmer climate and associated earlier snowmelt. A recent test, in which plants were experimentally allowed to emerge from winter dormancy to spring growth and flowering at different photoperiods, revealed (1) a number of opportunists (flowering always when released), but (2) a number of species which 'waited' for 'their photoperiod' to arrive (F. Keller & Ch. Körner, pers. comm.). Hence these species took no advantage of earlier snowmelt. About half of all 23 alpine species which flowered under the experimental spring conditions were photoperiod sensitive, which means, they had a preferred photoperiod for flowering (time of snow melt). Surprisingly, temperature per se (experimental warming) had a smaller affect on flowering in these species than photoperiod. This example illustrates that secondary effects of climatic warming such as time of snowmelt may be of greater impact than warming as such. The diverse responses of these alpine species complicate predictions and will affect species distribution and abundance in such cold environments.

THE EFFECT OF WARMING ON ECOSYSTEM CARBON FLUXES

It is often assumed that warming enhances respiratory losses, causing ecosystem C-stocks to deplete. However, this is an oversimplistic assumption derived from well known instantaneous temperature responses of active tissue. Global warming must not be confused with warming of leaf in a measurement cuvette. Given sufficient time such initial effects will disappear (thermal acclimation). It is known since the 1930s that in situ respiratory losses of willow leaves in Greenland and tropical tree leaves in Java do not differ (Stocker 1935).

The same is true for soil CO_2-release. Labile C-compounds may "burn off" faster if one heats soil, but in the long run, soil CO_2-release equals soil

C-input, as was nicely documented in a large scale data comparison (Figure 8).

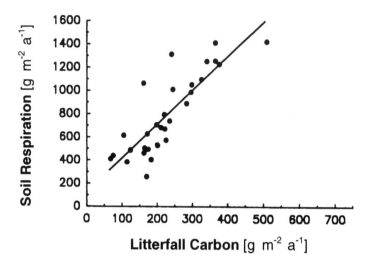

Figure 8. Carbon input controls carbon output. The relationship between soil respiration (measured as CO_2 carbon) and litterfall in forest ecosystems of the world is linear, with surprisingly little scatter, given the wide range of site conditions (from Raich & Nadelhoffer 1989, modified).

Figure 8 illustrates that no matter whether the data come from the Amazonian rainforest or from Alaska, input and output are strongly coupled, suggesting that above- and below-ground production are controlled by the same factors. Hence any warming effect would jointly change input and output, making output-only-oriented scenarios irrelevant (the often assumed carbon loss in the case of warming rests on this illusion).

Hence, unlike climate change induced changes in biodiversity or keystone species presence, overall ecosystem processes per se are not necessarily affected. Too much of the current global change debate is leaning on such metabolic effects on ecosystem function. It seems much more likely that biodiversity changes will become major drivers of any alterations of ecosystem function. In other words, if climatic changes will leave fingerprints, they will in large materialise in the form of organismic presence and behaviour, and much less so in the form of a priori altered fluxes and pools of carbon.

CONCLUSIONS

Evidence from both plants and animals indicate that the period of thirty years of warmer temperatures at the end of the 20th century has affected the phenology and physiology of organisms, as well as the range and distribution of species. It is shown, that sensitive organisms and biota strongly respond to minute climatic changes.

- In the Antarctic, rapid increases in the numbers or extent of populations of existing species have been documented across a range of taxa. In the longer term, colonisation by new species will become increasingly important, resulting in increased competition and trophic complexity.
- Populations of pinnipeds and other marine mammals have responded measurably to changes in the physical environment in the Arctic.
- There is now evidence that poleward and altitudinal shifts of butterfly ranges in North America and Europe mimic patterns of regional warming over similar time frames, with little or no lag-times for most species.
- Dragonfly species show clear changes in the distribution patterns with the northward expansion of "southern" Odonata species, whereas negative effects are expected for "northern" and mooreland species, as they do not have an alternative area or biotope.
- Analyses of phenological observations clearly reveal a strong seasonal variation of changes in onset dates, with noteable advances of spring events and a lengthening of the growing season.
- Global warming has led to altitudinal and latitudinal shifts of native plant species during the last 30 years. The naturalisation of several exotic thermophilous species has been favoured by the rise in temperatures.
- Whereas many ecosystems will be facing gradual shifts in distribution and range, some others (e.g. coral reefs) may be facing abrupt changes triggered by physiological thresholds, that once reached, lead to a widespread collapse of ecosystem function.

The existence of consistent patterns for a broad range of taxonomic groups with diverse geographical distributions strongly suggest that biological systems are affected by climate change, in spite of the possibly confounding and conflicting influences of other anthropogenic forces (e.g. land use change or pollution). The contributions to this book provide collective evidence that organisms are responding to the recent warming trend of the past three decades and thus represent biological "fingerprints" of climatic change. These fingerprints of life may convince those, who find the 0.6 K mean global warming itself negligible.

REFERENCES

IPCC, 2001a, *Climate Change 2001: The scientific basis*. A report of the Working Group I of the Intergovernmental Panel on Climate Change. Cambridge University Press, Cambridge.

IPCC, 2001b, *Climate Change 2001: Impacts, adaptation and vulnerability*. A report of the Working Group II of the Intergovernmental Panel on Climate Change. Cambridge University Press, Cambridge.

Klötzli F., Walther G.-R., Carraro G. & Grundmann A., 1996, Anlaufender Biomwandel in Insubrien. *Verh. Ges. Ökol.* 26: 537-550.

Klötzli F. & Walther G.-R. (eds.), 1999, *Conference on recent shifts in vegetation boundaries of deciduous forests, especially due to general global warming*. Proceedings of the Centro Stefano Franscini, Monte Verità, Ascona. Birkhäuser, Basel, 342pp.

Körner Ch. & Stöcklin J., 1999, *Lorbeer, Bambus, Palmen in der Schweiz?* Collection of student reports of a field course in Locarno of the Institute of Botany, University of Basel, Switzerland.

Mai D.H., 1995, *Tertiäre Vegetationsgeschichte Europas*. Fischer, Jena, 691pp.

Möller I., Wüthrich C. & Thannheiser D., 2001, Changes of plant community patterns, phytomass and carbon balance in a high arctic tundra ecosystem under a climate of increasing cloudiness. In: C.A. Burga & A. Kratochwil (eds.) *Biomonitoring: General and applied aspects on regional and global scales*. Tasks for vegetation science 35, Kluwer Academic, Dordrecht, pp. 225-242.

Paulsen J., Weber U.M. & Körner Ch., 2000, Tree growth near treeline: Abrubt or gradual reduction with altitude? *Arct. Antarct. Alp. Res.* 32: 14-20.

Raich J.W. & Nadelhoffer K.J., 1989, Belowground Carbon Allocation in Forest Ecosystems: Global Trends. *Ecology* 70(5): 1346-1354.

Stocker O., 1935, Assimilation und Atmung westjavanischer Tropenbäume. *Planta* 24: 402-445.

WSL & BUWAL, 2001, *Lothar. Der Orkan 1999*. Ereignisanalyse. Eidg. Forschungsanstalt WSL, Birmensdorf; Bundesamt für Umwelt, Wald und Landschaft BUWAL, Bern. 365pp.

Index

abiotic 20, 51, 77, 101, 123, 140, 185, 186
absence 7, 19, 29, 60, 69, 81, 84, 145, 147, 157, 213, 284
absorption 31, 36, 229, 233, 242, 243, 245, 246
Acari 19, 40
acceptor 230
acclimation 204, 310, 313
Acer pseudoplatanus 154
acidification 102, 104, 106, 243
acidity 111, 209
Acropora 220, 221
Acyrthosiphon svalbardicum 30, 41
adaptation 9, 11, 20, 30, 40, 55, 89, 101, 161, 229-231, 241, 243, 246, 247, 310, 316
adventitious 286
aerosols 85, 88, 211
Aesculus hippocastanum 116, 127, 130
Aeshna affinis 94, 95, 96, 108
Aeshna caerulea 100, 105
Aeshna cyanea 102, 109
Aeshna subarctica 100, 103
Africa 10, 65, 67, 98, 111, 163, 164, 171, 180, 183, 210, 223, 227, 228
Afrotropical 98
Ageratina adenophora 163, 165, 166, 176, 177, 184
agriculture, agricultural 67, 90, 182, 186, 250, 251
Agrostis stolonifera 27
Alaska 43, 44, 46, 48, 51-54, 276, 314

Alaskozetes antarcticus 29, 37
albedo 44, 45, 54, 297
algae, algal 4, 19, 33, 42, 203, 205, 206, 220, 222, 227, 247
Algeria 98
alien 27, 28, 36, 39, 163, 165-167, 176, 180-182, 276
alkalinity 203, 208, 209
Alnus 118
alpine 13, 38-40, 106, 107, 111, 139, 140, 146-149, 197, 265, 266, 273, 274, 276-278, 281, 285, 292, 300, 313
Alps 10, 11, 14, 83, 120, 139, 140, 141, 144, 146-148, 152, 249-251, 258-260, 264, 273-277, 284, 300, 307, 308
Ambrosia 118
American 8, 53, 98, 179, 225
Ammophila arenaria 186
Amphibia 5, 106
Amphisorus 235, 236, 237, 238, 239, 241, 242
Amphisorus heimprichii 235, 237, 239, 242
Amphistegina 235, 241
Amphistegina lobifera 235
amplitude 124, 189, 236, 298, 306
Anax imperator 94, 96, 103, 108, 110
Anax parthenope 95, 96, 102, 108
Androsace alpina 143, 147
Anemone nemorosa 116, 117
animals 3, 4, 6, 10, 12, 13, 20, 29, 30, 36, 37, 45, 50, 53, 55, 60, 99, 123, 124,

139, 164, 165, 198, 204, 206, 222, 224, 227, 306, 307, 315
Anisoptera 97, 98, 107-109, 111
anomaly, -ies 2, 9, 11, 87, 169, 170, 213, 214, 216, 217, 221, 306
antagonistic 234, 247
Antarctica, Antarctic Peninsula 9-12, 14, 17-42, 45, 55, 73, 225, 315
Anthoxanthum odoratum 251
anthropogenic 24, 44, 59, 62, 166, 177, 208, 211, 222, 315
aphid 30
Apiaceae 195
appeared 61, 73, 93, 141, 238, 245, 284, 286-291
aragonite saturation 209, 220
Araschnia levana 70, 75, 76
Araujia sericifera 163, 165, 166, 177-179, 183
Archaeophytes 154
Arctic, arctic 11, 14, 20, 21, 24, 30, 31, 34, 37, 40-43, 45-48, 50, 52-55, 73, 148, 197, 264, 273, 274, 276-279, 315, 316
Arctocephalus gazella 28, 39
Arctotheca calendula 163, 165, 166, 180
Argynnis paphia 66
arrival dates 12, 61, 123
Artemisia 118
Artemisia genipi 265
arthropod 19, 30, 32, 36, 37
ascidians 234
Asclepideaceae 177
Asia 13, 82, 228
Australasia 210
Australia 18, 171, 203, 210, 212, 216, 217, 220, 224-226, 228
Austria, -n 10, 67, 88, 92, 93, 139, 148, 149, 263, 265
autecology 156
autochthonous, -y 91, 95, 96, 101
autotroph, -ic 22, 32, 34
autumn phases 121, 124, 126, 127
bacteria 19, 206
bear 43, 46, 48-51, 54, 55
beetle 28, 30, 36, 38, 103
behaviour 1, 3-5, 7, 10-12, 15, 47, 48, 52, 57, 59, 98, 214, 224, 229, 230-232, 234, 235, 238, 239, 241, 243, 245, 299, 305, 314

Belgium, -an 93, 96, 100, 194
Betula 118, 119
Betula pendula 125-127, 130
biochemistry, -al 17, 32, 33
biodiversity 11, 13, 19, 27, 74, 139, 147, 163, 182, 185-187, 194, 200, 204, 223, 224, 276, 305, 309, 314
bioindication, -r 74, 75, 104, 136, 160, 229, 231
biology, -cal 1-3, 5, 10, 12, 17, 18, 20-23, 25, 26, 28, 29, 34, 35, 37, 39, 41, 42, 50, 52, 54, 57, 68, 72, 89, 98, 102, 106, 107, 161, 186, 204, 209, 238, 246, 250, 272, 277, 282, 305, 307, 308, 315
biological consequences 10, 18, 22, 23, 34
biological responses 3, 5, 17, 34
biomass 20, 25, 42, 101, 234, 275, 277, 278
biomonitoring 8, 104, 123, 135, 148, 161, 261, 316
biophysical 229, 231, 232, 233, 238
biota 13, 17-19, 21, 24, 25, 35, 37, 54, 62, 75, 140, 307, 309, 315
biotic 20, 51, 58, 62, 67, 68, 77
biotope 89, 100, 102, 104, 105, 106, 186, 199, 245, 315
bird 5, 8, 14, 15, 19, 28, 52, 55, 60, 72, 74, 123, 146, 148, 164, 186, 199, 208, 307, 311
Bistorta vivipara 274, 278
bleaching 4, 9, 10, 12, 14, 203, 205, 213, 214, 216, 217, 220-229, 234, 235, 243, 245-247
bloom 69, 113, 116, 117, 125
Blysmus rufus 188-193
bog 101-104, 107
boreal 14, 75, 124, 194
Brachypodium pinnatum 251
Britain, British 9, 14, 17, 35, 53, 61, 62, 66, 71-73, 75, 77-79, 81, 84-88, 92-94, 96, 103, 109, 110, 193, 199, 201
Bromus erectus 251
bryophyte 19, 20, 22, 26, 32, 33, 38-41, 198, 292
buds 269
butterfly, -ies 9, 10, 13, 14, 57, 59-63, 66, 67, 69, 70, 72-88, 123, 201, 227, 309, 315
C_3 / C_4 164
calcification 203, 209, 225, 227

Index

calcium carbonate (CaCO$_3$) 207-209, 225, 227, 235
California 8-10, 13, 14, 54, 64, 69, 74, 171, 277
Canada 8, 9, 46, 53, 54, 63, 64, 73, 125, 136, 182
capacity 87, 160, 204, 229, 236, 243, 245, 247
Capra ibex 146
Capros aper 165
carbon (C) 74, 106, 164, 182, 203, 208-210, 225, 227, 266, 267, 272, 275, 276, 279, 310, 313, 314, 316
carbon dioxide (CO$_2$) 21, 40, 74, 106, 124, 125, 136, 164, 182, 203, 208-210, 213, 227, 229, 230, 235, 243, 246, 273-275, 277, 278, 306, 312-314
Cardamine resedifolia 144
Carex caryophyllea 251
Carex curvula 263, 265-267, 269, 274, 276
carnivores 44, 45
Carpinus 118
Carterocephalous palaemon 72
case study 12, 41, 229, 231, 310, 311, 313
Castanea 118, 152, 154
cats 27
cattle 27, 187
cell 17, 32, 52, 81-85, 206, 213, 225, 236, 238, 241, 243
Cerastium uniflorum 143
Cercion lindenii 94-96
Ceriagrion tenellum 102
chamaephyte 170
chamois 265
chemistry, -cal 40, 106, 225, 230, 233, 282-284, 305, 306
Chlorophyll (Chl) 229-232, 235, 243, 246, 247, 277
Ciconia ciconia 164
Cinnamomum glanduliferum 153, 157
clams 206
climate/-ic change 1-15, 17, 18, 21-24, 28-30, 34-40, 43, 45-47, 50-55, 57-59, 62, 63, 67, 68, 73, 74, 76-79, 85, 86, 88, 89, 97, 100, 102-106, 114, 121, 123, 124, 135, 136, 139, 140, 144, 146, 149, 151, 153, 154, 156, 160, 161, 163-165, 182, 185, 187, 189, 190, 191, 194, 195, 197-201, 203-205, 208, 211, 213, 216, 220, 221, 223, 226, 228, 229, 231, 250, 259, 261, 264, 276-278, 281, 285, 298, 300, 305-311, 314, 315
climate/-ic warming 1, 7-11, 21, 41, 58, 59, 68, 78, 85, 86, 139, 140, 144, 147, 160, 250, 305, 307, 313
climatic conditions 61, 91, 97, 152, 156, 157, 160, 189, 191, 193, 194, 197, 264, 291
climatic envelope 58, 185, 189, 190-196
climatology 200, 211
climax 15, 178
cloud cover 23, 216, 306
Cnidaria 206, 225, 234
coast, -al 12, 19, 23, 24, 41, 50, 54, 74, 96, 163, 165-167, 170, 172, 176, 178, 180-182, 185-191, 193-195, 197-200, 203, 209, 214, 223, 227, 286
Coenagrion armatum 103, 108
Coenagrion hastulatum 98
Coenonympha tullia 86
coenosis, -es 100-103, 107, 164
Colchicum autumnale 116
Coleoptera 9, 19, 36, 40, 73
Collembola 19, 31, 37, 42
Colobanthus quitensis 10, 25, 33, 38, 39, 41
colonisation 17, 20, 21, 25-29, 32, 35, 36, 57, 60, 68, 69, 71, 85, 89, 95, 104, 157, 159, 166, 176, 264, 272-274, 311, 315
colonise, -r 26, 35, 45, 66, 84, 86, 94, 95, 147, 151, 153, 164, 166, 179, 264, 273, 275
community structure 98, 250, 259
competition 2, 10, 25, 29, 98, 102, 156, 199, 264, 282, 308, 315
Compositae 170, 175, 176, 180, 182, 184
conservancy 100, 101, 107
conservation 41, 59, 75, 106, 186, 187, 227, 251
continentality 181, 282, 285
control plot 252, 263-272, 274, 275
cooling 1, 35, 210, 211, 213, 267
corals, coral reef 4, 10, 12, 13, 164, 203-209, 212-214, 216, 217, 220-229, 231, 234, 235, 241, 246, 247, 309, 315
Corylus 116, 118
Corylus avellana 116, 154
cover values 154, 284
cows 76, 297

Crithmum maritimum 195-197, 200
Crocothemis erythraea 89-94, 96, 100, 101, 104, 107-111
crop 38, 171, 249-255
cryological 264
Cryptopygus antarcticus 30
cultivation, cultures 26, 153, 234, 236, 237, 239, 241, 242, 245
Cupressaceae 118
cyanobacteria 19, 33, 40-42, 234, 247
Cyperaceae 189
Cytisus scoparius 154
Czech Republic 93
Danaus plexippus 60
dark phase 236
deciduous 8, 11, 84, 126, 130, 135, 152-154, 156, 157, 160, 161, 250, 311, 313, 316
decomposition 33, 40
deglaciation 264, 276
degradation 66, 208, 226, 264
Denmark 67, 74, 94, 96, 197, 277
Deschampsia antarctica 25, 26, 33, 38, 39, 41
desert 18, 38, 279
desiccation 20, 29, 31, 39, 42, 61
detritivore 33, 34
diaspores 146, 179, 199
die-back 31, 39
dinoflagellates 206, 207, 214, 224, 226, 227, 234, 246
Diptera 19, 36, 37
disappeared, -ance 78, 193, 209, 284, 286-291, 307, 309
disease 9, 10, 12, 59, 90, 165, 224
dislocation 185, 191, 195, 197
disperse, -al 15, 21, 60-62, 74, 75, 79, 96, 99, 104, 106, 114, 139, 146, 151, 160, 161, 171, 172, 176, 179, 194, 198, 199, 274, 311
distribution 1, 3, 6, 7, 9, 12, 13, 19, 25, 27, 29, 33, 35-38, 43, 46, 47, 49, 55, 57-63, 66, 67, 69, 72-74, 77-79, 81-87, 91-93, 96, 98, 103-105, 123, 126, 130, 135, 140, 142-146, 148, 149, 160, 161, 164-166, 168, 170, 176, 180, 181, 185-189, 191, 193-195, 197-201, 204, 205, 208, 209, 221, 224, 282-284, 287, 289, 293-295, 298, 313, 315

distribution limits 160, 188, 191, 193, 194, 197, 198
disturbance 2, 19, 38, 40, 53, 225, 226, 250, 259, 278, 286, 297
Dittrichia viscosa 163, 165, 166, 175, 176, 181
diversity 17, 20, 21, 25, 38, 42-45, 57, 67, 88, 100, 106, 107, 139, 140, 148, 186, 201, 224, 276, 292
dormancy 58, 313
Draba fladnizensis 143
dragonfly, -ies 89, 97, 107, 109-111, 307, 309, 315
driving force 147, 233, 234, 239, 241-243, 250
drought 8, 36, 69, 74, 152, 210, 230, 247, 250, 258, 259
dry meadows 66
Dryas octopetala 274, 275, 278, 279
dryness 197
duck 19
dune 185, 186, 188, 195, 197, 199, 200
Dutch 93, 96, 197
ecological consequences 2, 5, 15, 89, 226
ecological studies 45, 251, 311
ecology 36-38, 43, 45, 47, 52, 53, 55, 59, 79, 88, 90, 98, 140, 198, 207, 277
ecophysiological 20, 27, 156, 283, 311
ecosystem function 160, 250, 314, 315
ecosystem processes 42, 109, 313, 314
ecotone 7, 149, 307, 308
ecotypes 310
El Niño Southern Oscillation (ENSO) 23, 35, 45, 210, 212, 214, 225, 228, 306
Elaeagnus pungens 154
Elanus caeruleus 164
electron transport 229, 233, 236, 247
Ellenberg's index 281-285, 287-291, 293-295, 297, 298
Elymus farctus 186
emergence 69, 91, 95, 96, 98, 102
emission 106, 211, 213, 228
Empetraceae 197
Empetrum nigrum 188, 197, 199
Enallagma cyathigerum 102
endemic 38, 177, 182
England 78, 85, 87, 88, 92, 118, 197
Enhydra lutris 44

Index

environmental change 10, 11, 13, 20, 21, 28, 36, 39, 41, 42, 44, 45, 47, 51, 57, 106, 164, 204, 225, 275, 278, 279, 310
environmental conditions 1, 20, 26, 27, 37, 159, 160, 185, 204, 230, 290
environmental envelope 204
environmental factor 30, 123, 152, 160, 164, 282, 288, 295, 298
environmental variables 31, 32, 282, 284, 298
epiphyte 172
equator, -ial 164, 208, 210
Erebia aethiops 86
Erebia epiphron 86
Erignathus barbatus 46, 52
erosion 106, 142, 186, 187, 209, 286, 306, 309
Erythromma viridulum 94, 95, 102
Estonia 7, 94, 96, 125, 136
Eumetopias jubatus 44, 54
Euphydryas editha 63-70, 74, 75
Eurasiatic 286, 290
Europe, -an 8, 12, 18, 27, 38, 57, 60, 65-67, 70-72, 74, 77, 79, 81, 83-85, 87-91, 93, 94, 96, 98, 99, 105, 107, 110, 115, 116, 121, 123-125, 135-137, 139, 149, 152, 155, 160, 161, 164, 165, 177, 179, 188, 191, 193, 194, 196, 200, 201, 223, 260, 282, 298, 306, 307, 315
Eurosiberian 97, 104
eutrophication 101-103, 106, 208
evergreen 151-154, 156, 157, 159-161, 179, 250, 309, 311, 313
evergreen broad-leaved 151-153, 156, 157, 159-161, 309, 311, 313
evolution, -ary 50-53, 55, 58, 59, 86, 88, 101, 211, 217, 227, 246, 276, 290, 307, 310
Evonymus europaea 154
exotic 15, 17, 25, 28, 35, 102, 151-153, 155-161, 250, 286, 308-311, 313, 315
expansion 25-27, 29, 32, 35, 63, 71, 74, 79, 80, 85-90, 92-96, 98, 102, 104, 105, 110, 156, 158, 159, 163, 165-167, 171, 172, 176, 177, 181, 195, 197-199, 208, 220, 315
experiment, -al 2, 14, 17, 18, 21, 27-29, 31-35, 38, 39, 42, 59, 62, 91, 222, 227, 233, 235-238, 241, 243, 246, 249-253, 260, 263-266, 273-278, 282, 310, 313

exploitation 28, 52, 208
extinct, -ion 25, 57, 60, 64-70, 72-75, 146, 147, 195, 204, 222, 286, 307, 312
extreme 9, 13, 18-22, 28, 31, 32, 37, 47, 50, 57, 61, 67-70, 74, 75, 98, 161, 204, 259, 306, 309
Fabaceae 195
Fagus 116-118, 246
Fagus sylvatica 116, 117, 125, 127, 130, 246
farming 286, 292
fauna 8, 9, 19, 24, 27, 32, 39, 42, 55, 74, 87, 89, 97, 100-102, 105, 106, 205, 206
Fenno-scandia 71
ferns 19
fertilise, -ation 139, 146, 147, 220
fertility 102
Festuca tenuifolia 251
field 29, 31, 32, 34, 38-41, 51, 57, 59, 63, 68, 166, 220, 234, 245, 249-259, 263, 264, 272, 276, 277, 300, 311, 316
field studies 30, 34, 51, 59, 220
fingerprint 1, 5-7, 10-12, 15, 18, 50, 51, 123, 140, 195, 198, 199, 205, 229, 231, 250, 282, 305, 307, 308, 314, 315
Finland 62, 66, 71, 72, 96
fire 286, 291
fish 5, 9, 51, 52, 54, 55, 60, 101, 104, 123, 164, 206, 208, 223
fisheries 203, 223
fitness 15, 102, 123
flexibility 17, 21, 29, 35, 37
flies 28
flight period, flying season 98, 102
floods 90, 306
flora, floristic 11, 19, 22, 26, 27, 37, 100, 106, 153, 166, 177, 182-186, 191, 193, 195, 197, 199-201, 278, 281-292, 294, 297-300
flower 7, 19, 25, 31, 37, 69, 99, 113, 114, 123, 125-127, 130-132, 165, 170, 172, 175-180, 266, 267, 269-271, 274, 278, 313
fluctuation 36, 38, 55, 60-62, 88, 101, 108, 198, 199, 236, 306, 307
fluorescence 229-232, 234-239, 241, 242, 245-247, 310
flux 225-227, 229, 233, 241, 242, 245, 313, 314
foliation 113, 116

food chain, - web 12, 25, 33, 102, 104, 206, 208
foraminifer 229, 231, 234, 235, 237, 241, 243, 245-247
forb 249, 259, 260, 275
forest 7, 8, 10, 11, 13-15, 61, 73-75, 97, 106, 151-154, 159-161, 177, 183, 210, 250, 299, 309, 311-314, 316
forestry 59, 90
Forsythia suspensa 126, 127, 130
foxes 48
Frangula alnus 154
Fragilaria 235
fragmented, -ation 57, 59, 62, 75, 77, 78, 84, 85, 199, 306
France 67, 71, 91-93, 96, 108, 136, 194, 197
Fraxinus 118, 152
freezing 61, 157, 166, 167, 171, 175, 177-180
French 36, 91, 96, 212, 214, 225, 226
frequency 24, 26, 31, 57, 64, 130, 135, 151, 153, 154, 159, 160, 205, 214, 216, 224, 228, 250, 258, 284, 286, 291, 292, 294, 306
freshwater 51
frost 7, 156, 160, 179
fruit ripening 123, 127, 130-132, 135
fungi 19, 247
fungicides 252
Galanthus nivalis 127, 130
gas exchange 30
gene, -tic 199, 204, 217, 220, 222, 224, 249
generalists 102
generative 264, 265, 274
genotypes 217, 222, 310
Gentiana bavarica 143
geological 307
Georgia 18, 27, 36-39, 42, 160
German, -y 1, 8, 12, 67, 89-92, 94, 95, 97, 99, 100, 102, 106, 123, 125-127, 130, 135, 136, 151, 185-191, 193-195, 197-200, 210, 211, 217, 305
GIS 159, 189, 191
glacial 22, 23, 36, 107, 108
glacier 30, 36, 144, 263-266, 272-275, 277, 285, 297
Glaucium flavum 188, 197, 199

global change 3, 15, 53, 57, 58, 73, 114, 160, 161, 186, 200, 246, 261, 305, 306, 314
global warming 7-11, 13-15, 34, 36, 39, 52, 53, 57, 60, 71, 74, 89, 114, 119, 161, 164, 165, 187, 226, 281, 286, 292, 297, 305, 313, 315, 316
goats 146
gradient 21, 29, 98, 111, 140, 159, 191, 198, 295
graminoid 259, 273, 275
grass 20, 27, 28, 36, 40, 249, 250, 258, 274, 275, 286, 291
grassland 10, 66, 74, 76, 139, 140, 147, 249, 252, 258, 259, 261, 265, 266, 277, 278, 291
graze, -r 19, 59, 139, 146, 187, 252, 286, 291, 292, 297
greenhouse 31, 32, 39, 53, 74, 85, 88, 164, 185, 205, 208, 209, 212, 213, 228, 234, 272
Greenland 20, 55, 278, 313
growing season 5, 99, 125-127, 130-132, 135, 152, 155, 156, 258, 259, 267, 268, 272, 273, 278, 315
growth form 32
growth period 249, 258
growth rate 4, 59, 207
gulls 47
habitat conditions 101, 147, 250
habitat destruction 64, 67, 199
habitat requirements 57, 59, 62, 85, 86
harvest 53, 249, 252, 253, 255-258, 260, 276
health 12, 13, 165, 203, 208, 213, 223, 227
heath 38, 86, 197, 278
Hedera helix 152
Hemianax ephippiger 94-96, 111
herb 153, 158, 252, 274, 278
herbivores 33, 75, 265
Hesperidae 78-80
Holocene 36, 40, 53, 58, 73, 74
horses 265
host 62, 63, 66, 67, 69, 82, 206, 214, 222, 225, 229, 234, 245
human 2, 13, 18, 27, 28, 47, 66, 73, 75, 100, 164, 165, 179, 205, 208, 223, 224, 249, 251, 258, 281, 286, 308

Index 323

humidity 30, 31, 197, 249, 251, 254-260, 306
Hungary 15, 96, 125, 137, 282
ice 10-12, 18, 19, 22-24, 30, 31, 37, 40, 41, 43-55, 58, 74, 91, 106, 123, 136, 208, 211, 212, 264, 265, 273
Ice Age 23
Ilex aquifolium 152, 154, 165, 188
immigrant 41, 61, 94, 100, 147
immigration 96, 101, 139, 194, 195
impacts 1-3, 5, 7, 9, 10, 12, 13, 19, 21, 22, 24, 25, 27-33, 35, 41, 45-47, 49, 54, 57, 58, 66, 73, 74, 97, 100, 104, 106, 123, 146, 156, 185, 198, 200, 203-205, 208, 209, 213, 223, 225, 226, 228, 277, 251, 300, 305, 308, 313
in vivo 229, 231, 234, 236, 245, 247
Indian 216, 217, 221, 223, 225, 228
indicator, -ion 12, 27, 29, 35, 38, 41, 42, 44, 46, 59, 73, 74, 79, 98, 100, 104, 115, 116, 123, 124, 126, 135, 144, 151, 157, 161, 165, 191, 198, 199, 229, 231, 243, 281-284, 288, 290-292, 298-300, 310
indicator value 281, 283, 284, 288, 292, 295, 297, 298, 310
indigenous 40, 45, 100, 152-154, 159-161, 213, 308, 313
insect 4, 19, 31, 36, 37, 39, 59, 60, 69, 70, 73, 78, 87, 100, 108, 109
insecticides 252
interaction 10, 20, 22, 30, 31, 37, 53, 74, 78, 86, 164, 182, 204, 234, 250, 270, 282, 298, 306, 311
inter-annual 49, 258, 259
interglacial 108, 210
introduced, -tions 27, 28, 36-39, 53, 87, 113, 153-155, 160, 163, 177, 179, 181, 182, 231, 233, 286
inundation 186, 187
invade, -r 139, 146, 147, 159, 161, 182, 309, 311
invasibility 182
invasion, -ve 38, 62, 75, 94-96, 101, 102, 153-155, 159, 160, 166, 176, 180, 182, 230, 234, 245, 273, 306, 308, 311
invertebrate 4, 19, 27-29, 31, 32, 37, 39, 42, 49, 75, 90, 203, 207
IPCC 1, 2, 3, 5, 11, 73, 74, 88, 147, 148, 152, 154, 161, 168, 183, 189, 190, 200, 204, 205, 208, 210-213, 226, 250, 263, 264, 277, 306, 307, 309, 316
Ischnura elegans 102, 103
Ischnura pumilio 95, 102
isolation 17, 21, 22, 27, 35, 37
Italy, -ian 12, 266, 283, 284, 298
IUCN 93, 100
Jamaica 74, 211, 225, 226
Japan, -ese 98, 110, 160, 274, 278
JIP-test 229, 231-236, 238, 241, 245-247
Juncaceae 191
Juncus maritimus 188, 191, 192, 197, 198, 201
key groups 197, 213, 216, 220, 309, 314
Kobresia myosuroides 265
K-strategists 103
La Niña 1
laboratory 32, 59, 243, 245
lag time 59, 61, 204, 315
lag-effects 258
Lagopus mutus 146
LAI 311, 312
land management 13, 59
land use 62, 66, 250, 261, 292, 298, 305, 306, 315
landscape 2, 7, 57, 75, 87, 107, 187, 199, 201
Langustis angusticollis 30
large-scale 17, 44, 58, 68, 148, 205, 221, 226, 250
Larix decidua 116
Latvija 96
laurophyllisation 8, 157, 159, 310, 311
laurophyllous 151-153, 155-161, 306, 308, 309, 311, 312
Laurus nobilis 154
leaf, leaves 9, 40, 41, 113, 115-117, 123, 125-127, 130-132, 135, 246, 263, 266, 267, 269-272, 275, 278, 311, 313
leaf area 266, 267, 270-272, 311
leaf colouring 113, 116, 117, 123, 126, 127, 130-132, 135
leaf fall 113, 116, 123
leaf unfolding 126, 127, 130-132, 135
lengthening 30, 126, 130, 315
Lepidoptera 9, 27, 36, 59-61, 73, 75, 76
Leptostilos crumeniferus 165
Lestes barbarus 96
Lestes dryas 103, 110
lethal limits 35, 204

Leucanthemopsis alpina 144
Leucanthemum vulgare 116
Leucorrhinia albifrons 103
liana 177
Libellula depressa 94, 96
lichen 19, 22, 26, 33, 40, 292
life cycle 25, 29, 38, 58
life history 21, 29, 35-37, 52, 199
light phase 236, 238, 241, 242
Ligustrum vulgare 154
limiting factor 39, 120, 165, 199, 259
Linaria alpina 265
liverwort 33
long-term 1, 7, 8, 17, 30, 34, 36, 49, 58, 59, 61, 62, 68, 75, 79, 99, 101, 104, 106, 148, 155, 156, 241, 249, 250, 251
loss 10, 24, 25, 30, 38, 40, 62, 63, 66, 67, 72, 73, 77, 78, 84, 104, 107, 139, 148, 191, 194, 206, 209, 210, 216, 223, 224, 229, 234, 243, 306, 309, 313, 314
Luxembourg 93, 96, 100
Luzula spicata 144
macroclimatic 22
Maculinea 67
magnitude 23, 24, 32, 57, 60, 64, 69, 79, 85, 214, 305, 308
maize 246, 249, 251-258, 260
Maldives 9, 221, 223, 225
Malus domestica 127, 130
mammals 4, 28, 36, 43-47, 52, 53, 55, 60, 73, 74, 139, 208, 286, 315
man-induced, -made 106, 114
map 15, 82, 85, 88, 105, 126, 130, 135, 142-145, 171, 175, 177-180, 189, 191, 199, 231, 292
marine ecosystem 55, 203-205, 208
marsh 66, 185-188, 193, 198, 200
Mauritius 235
maxima, -um 24, 41, 44, 130, 135, 152, 167-169, 190, 203, 214, 230-233, 236, 259, 268, 288, 298, 306
meadow 67, 249-252, 255-257, 259, 260
mean temperature 1, 7, 82, 97, 99, 103, 154, 155, 163, 167-170, 181, 189, 191, 194, 195, 205, 210, 214, 250
mechanistic links 57, 68
Mediterranean 8, 81, 89, 90, 94-97, 100, 101, 104, 152, 163, 165, 166, 170, 172, 176, 181, 182, 197, 235, 241, 281, 284-286, 290, 310

Melanargia galathea 82, 83
melt 22, 23, 26, 31, 33, 35, 43, 44, 48, 49, 64, 208, 264, 285, 297, 313
metabolism 106, 164, 222, 226, 227, 246
metapopulation 70, 75
meteorological 15, 22, 26, 74, 75, 87, 114, 151, 155, 163, 165, 167, 168, 179, 251, 261, 312
methane (CH_4) 164, 213
Mexico 33, 61, 64, 69, 171, 176, 182
mice 27, 36
micro habitats 139, 147
microbe, -ial 19, 22, 32, 41, 42, 306
microclimate, -ic 22, 23, 34, 41, 42, 61, 66, 74, 99, 121, 259, 263, 264, 266, 267, 272, 275
microhabitat 30, 31, 35, 37
migrant, migrate, -ory 8, 27, 58, 60-62, 73, 90, 94, 123, 139, 144, 164, 198, 274, 310
migration 2, 10, 14, 39, 60, 61, 86, 108, 140, 142, 159, 198, 200, 250, 310
mineralisation 101, 226
minima, -um 2, 21, 50, 114, 142, 152, 156, 157, 160, 163, 167-169, 171-176, 178-182, 194, 208, 252, 259, 264, 268, 306
mite 20, 28, 29, 37, 42
mobility 60, 86
model 1-3, 23, 26, 29, 31, 34, 35, 39, 43, 44, 57-59, 74, 77, 78, 81, 82, 84, 85, 87, 88, 115, 168, 189, 190, 191, 193-196, 199, 200, 203, 204, 210-213, 217, 220, 224, 225, 228, 242-244, 247, 253, 255-257, 282, 284, 295, 298, 310
moisture 82, 266-268, 272, 273, 282, 285, 287, 289, 294, 297, 300
molluscs 234
monitor 105, 199
monitoring 14, 42, 46, 47, 48, 52, 59, 60, 62, 66, 75, 88, 98, 105, 113-115, 121, 123, 136, 142, 148, 149, 165, 199, 201, 221, 228, 263, 266
mooreland 100, 101, 315
Morocco 82
morphological 34, 52
mortality 9, 98, 203, 204, 216, 217, 220-227, 234, 247
moss 29, 32, 33, 37, 38, 40
moth 15, 27

Index

mountain 10, 13, 64, 69, 86, 93, 95, 97, 103, 127, 139, 140, 144, 146-149, 172, 198, 200, 201, 258, 264, 274, 275, 277, 278, 300, 309
mufflon 286, 291
Mycteria ibis 165
national park 64, 73, 186, 281, 284, 286, 292
native 9, 15, 27, 28, 38, 152, 163, 165, 166, 170, 176, 181, 200, 276, 315
natural habitat 77, 78, 176, 241, 243
naturalise, -ation 28, 153, 165, 177, 179-181, 315
negative trends 115, 116, 118, 119, 120, 126, 127
Nematoda 19, 32
neophytes, -ic 154, 199, 311
neotropic 163, 165, 177, 178
neozoon 100
New Zealand 15, 171
nitrogen (N) 59, 106, 109, 170, 199, 200, 213, 225, 226, 247, 252, 266, 267, 272, 273, 275-278, 300, 310
nival 139-141, 147, 149
non-invasive 230, 234, 245
North America 9, 13, 14, 51, 52, 57, 63, 65, 70, 72, 74, 123, 124, 135, 137, 183, 315
North Atlantic Oscillation (NAO) 259, 261
Northern Hemisphere 23, 47, 51, 70, 124, 136, 205
northward 14, 61, 64, 67, 68, 70-73, 79, 84, 85, 89, 90, 94, 103, 104, 193, 198, 249, 315
Norway 96, 193, 197, 274
Nucifraga caryocatactes 146
nutrient 19, 31, 76, 147, 164, 206, 207, 227, 277, 282, 311, 313
Nymphalidae 75, 78
observation 1, 25, 27, 30, 39, 41, 45-47, 54, 55, 59, 60, 68, 70, 74, 91-93, 96, 98, 102, 113, 114, 118, 121, 126, 130, 135, 139, 140, 142, 148, 166, 170, 180, 213, 214, 221, 222, 251, 259, 275, 292, 297, 298, 315
Odobenus rosmarus 45, 52
Odonata 89-91, 94-99, 101, 102-104, 106-111, 315
Odontites litoralis 188, 193, 194

O-J-I-P 229-232, 234, 236, 237, 247
Oligocene 45, 51
open-top chamber (OTC) 263-275, 277
ornamental 153, 155, 177, 179
Orthetrum brunneum 93-96
Otarioidea 44
oxygen 98, 236, 243, 246
ozone 22, 24, 32, 33, 35, 38, 40, 42, 73, 229, 230, 246
ozone depletion 24, 32, 40, 42
ozone hole 22, 24, 33, 35, 40
Palaearctic 82
Palau 214, 216, 217, 221, 224
palm 153, 160, 172, 311, 312
Panama 222
panmixis 217, 220
Papaveraceae 197
PAR 22, 24
Pararge aegeria 78, 84, 85, 86
pastures 187, 251, 297
performance index 229, 233, 238, 241, 246
permafrost 264, 277
Peru 33, 179
phanerogam 32, 33, 165
phenology, -al 3, 4, 7-9, 12-15, 49, 53, 66, 88, 98, 99, 102, 110, 113-116, 118, 120, 121, 123-126, 130, 135-137, 155, 166, 269, 273, 278, 279, 310, 315
phenophase 113-116, 118, 120, 121, 123, 130, 135, 137
Phoca hispida 45, 52-55
Phoca vitulina richardsi 44, 54
Phocoidea 44
Phoenix canariensis 172
photochemical 230, 236, 242
photoinhibition 214
photon 245
photosynthesis, -tic 22, 33, 36, 38, 40, 42, 157, 164, 182, 203, 206, 207, 224, 229-231, 234, 235, 238, 241-243, 245-247, 279, 311, 312
photosystem (PS) II 229-232, 234-236, 238, 239, 241, 243, 245, 247
physical 9, 18, 20, 22, 26, 30, 44, 63, 165, 186, 204, 209, 282, 283, 305, 308, 315
physiology, -cal 3, 17, 20, 21, 29, 30, 35, 37, 52, 55, 59, 197, 204, 207, 220, 221, 234, 241, 242, 244, 278, 315

phytomass 252, 263, 264, 270-272, 278, 316
phytoplankton 4, 12, 106
Picea 74, 116, 118
Picea abies 74, 116, 127
pigment 33, 213, 229, 234, 243
pinniped 11, 43-47, 49-52, 309, 315
Pinus 12, 118
pioneer 42, 140, 175, 264, 265, 273, 274, 297
plant 3, 4, 6, 7, 9, 10, 12-14, 19, 20, 25, 26, 28, 31, 32, 37-42, 60, 62, 63, 66, 67, 69, 73, 76, 82, 87, 99, 106, 113, 114, 120, 123-126, 130, 135, 136, 139, 140, 142, 144, 146-149, 154, 155, 160, 161, 163-166, 170, 175, 177, 179, 181-183, 185-189, 197, 198, 200, 201, 204, 206, 214, 222, 227, 228, 231, 232, 234, 235, 238, 243, 246, 247, 249, 250, 252, 253, 258, 259, 261, 263-267, 269-279, 281-284, 292, 297-300, 306, 307, 311-313, 315, 316
Plantago 118
Platanus 118
Pleistocene 22, 45, 58, 73, 74
Plutella xylostella 27, 36
Poa 27, 28, 38, 143, 147, 263, 265-267, 270-275
Poa alpina 263, 265-267, 270-275
Poa annua 27
Poa laxa 143, 147
Poa pratensis 28
Poaceae 118
Pocillopora 207, 220, 221, 228
poikilothermic 78, 98
Poland 93, 96, 103, 107-109, 282, 300
polar 22, 24, 43, 44
pole 164
poleward 5, 57, 58, 72, 73, 78, 205, 208, 315
pollen 113-115, 118, 119, 121, 276
pollution 24, 59, 62, 139, 147, 246, 298, 315
Polygonia c-album 78, 88
Polygonum bistorta 273, 278
population density 12, 32, 101, 226, 278
population dynamics 45, 57, 62, 68, 69, 74, 75, 276
population size 69, 89, 139, 143, 199, 225
Populus 118

Populus tremuloides 125
Porites 220, 221
Portugal 181, 183
positive trend 115, 116, 118, 120, 126, 127
precipitation 17, 18, 22, 23, 31, 34, 35, 37, 41, 59, 69, 73, 101, 187, 189, 190, 199, 249-251, 254-259, 281, 285, 306
predation, -or 15, 30, 38, 43, 45, 46, 49, 50, 52, 54
presence 9, 18, 21, 22, 27, 35, 47, 60, 63, 81, 84, 92, 143, 145, 165, 166, 170, 171, 177, 185-187, 193, 236, 284, 291, 310, 314
prey 46, 48, 51
primary production 22, 208
probability 5, 82, 84, 233, 250, 253, 293, 295, 296, 306
producer 206, 208
productivity 20, 38, 49, 52, 164, 186, 187, 206, 210, 229, 234, 245, 249-251, 258, 259
prolongation 115, 121, 156, 273
propagule 26, 32, 40, 41
protection 33, 49, 107, 203, 209, 223
protists 204
Prunus avium 127
Prunus laurocerasus 154
Prunus serotina 154
Pueraria hirsuta 154
Pyronia tithonus 78
Pyrrophyta 206
Quaternary 73, 74, 86, 87, 228
Quercus 40, 118, 152, 154, 183, 184
Quercus ilex 178
Quercus robur 127, 130
rabbits 27
radiation 21, 22, 24, 25, 31-33, 35-42, 164, 225
rainfall 152, 164, 189, 210, 285
rainforests 204
range 1, 3-5, 7, 9-15, 19, 21-29, 31-33, 51, 52, 57, 58, 60-75, 77-82, 84-88, 91, 98, 99, 105, 114, 124, 127, 140, 146, 152, 156, 158, 160, 163, 165, 171, 176, 181, 182, 185, 187-189, 193-196, 198, 199, 201, 204, 208-210, 213, 217, 220, 227, 241, 243, 250, 253, 282, 283, 292, 293, 299, 305, 307, 308, 310, 314, 315

range boundary/limit/margin 62-64, 66-68, 70, 79, 82, 84, 86, 159
range shift 52, 57, 58, 67, 68, 70, 72, 79, 159, 189, 194, 198
Ranunculus glacialis 143, 147
rats 27
ravens 47
reaction centres 229, 230, 233, 241
recruitment 10, 39, 52, 157, 220
Red List 93, 100
reduction 13, 33, 46, 50, 54, 74, 106, 165, 210, 229, 230, 233, 236, 243, 245, 292, 306, 316
regeneration 47, 220, 227
reindeer 27, 42
relevé 153, 157, 159, 283, 293, 295-297
relic 51, 107
reproduction, -ive 15, 21, 29, 32, 37, 38, 50, 52, 53, 57, 58, 61, 78, 92, 94-96, 220, 228, 264, 274, 276, 278, 279, 311
reptile 5
reserve 61, 106, 186, 187
respiration 164, 314
restoration 57, 59
reversibility, -ble 229, 238, 245
rhizosphere 234
root 266, 267, 270-272
Rotifera 19
r-strategists 103
Rubus 154
ruderal 171, 179, 184
Russia 15, 193
salinity, -ne 186, 193, 197, 226, 243, 246
Salix 118, 188, 277
salt 185, 186, 188, 193, 198, 200, 282
Sambucus nigra 116, 127, 130, 154
saprotrophic 32
Satyrinae 78, 79
Saxifraga aizoides 265
Saxifraga bryoides 143, 147
Saxifraga exarata 144
Saxifraga oppositifolia 143, 147, 265, 278
Saxifraga stellaris 274, 278
Scandinavia 74, 84, 103
Scotland 12, 78, 85, 88, 193
Scrophulariaceae 193, 199
sea level 36, 64, 94, 95, 120, 166, 179, 186, 208
sea lions 44, 54
sea otters 44, 52

sea temperature 2, 9, 203, 205, 208-214, 216, 217, 220, 224
seal 19, 28, 39, 41, 43-49, 51-55
seawater 48, 203, 208, 209, 225, 235, 243
sedentary species 60, 61, 85
sedimentation 186, 187, 208
seed 26, 40, 139, 146, 177, 179, 198, 199, 274, 278, 291, 311
semi-natural 78, 249, 251, 252, 258, 259
Seychelles 203, 216, 221, 224, 228
sheep 27, 146, 187, 265, 297
shoot 127, 263, 266, 267, 269-273, 277, 278
short-term 22, 36, 68, 70, 94, 217
shrub 124, 152-154, 158, 200, 250, 273-275, 277, 278
Sieversia pentapetala 274, 278
snow 18, 22, 23, 26, 30, 33, 43, 47-50, 52, 53, 69, 106, 140, 264, 267, 313
snow cover 24, 48, 49
snowmelt 49, 54, 64, 263, 272, 273, 313
socio-economic 223, 225, 228
soil 14, 19, 24, 26, 30-32, 37, 38, 40-42, 63, 152, 170, 176, 186, 252, 258, 265-267, 272-274, 276, 278, 282, 285, 286, 297, 300, 306, 309, 313, 314
Somatochlora alpestris 103, 105, 111
Somatochlora arctica 100, 103
Somatochlora sahlbergi 105
Sonchus tenerrimus 163, 165, 166, 170-172, 181
Sorites 235, 241
Sorites variabilis 235
South Africa 98, 163, 171, 180, 228
South America 26, 27, 36, 177
southward 61, 67, 68, 71, 72
Spain 71, 72, 92, 96, 163, 164, 170, 181
species abundance 142, 143, 153, 159, 195
species composition 10, 15, 100, 146, 160, 182, 216, 220, 221, 250, 259, 278, 282
species pool 139, 147, 194
species richness 20, 139-144, 146, 147, 286
sponges 206, 234
spores 26, 198
spread 65, 90-96, 113, 152-155, 157, 170, 178, 180, 181, 194, 221, 286, 292, 309, 311

spring phases 121, 124, 126, 130
springtail 20, 28, 30, 36
stability 21, 62, 74, 75, 185, 241
steady state 230
storm 69, 186, 259, 306
stress 8, 12, 20, 26, 27, 29-31, 33, 35, 59, 151, 186, 203, 208, 214, 217, 221, 222, 224-226, 229, 230, 231, 234, 235, 241-243, 245-247, 258, 259
subalpine 13, 292, 297
subnival 139-141, 146, 147
subtropics, -al 98, 153, 161, 165, 166, 177, 307
succession, -al 20, 41, 200, 252, 263-265, 273-276
sulfur 211, 228
survival 37, 38, 42, 47, 48, 70, 78, 82, 95, 156, 157, 161, 178, 180, 227, 241, 267, 311
Sweden 11, 71, 96, 98
Swiss 11, 15, 51, 113, 120, 151, 153, 156, 246, 251, 253, 258, 260, 273, 275, 276, 298
Switzerland 11, 15, 92, 93, 96, 98, 109, 113-116, 118, 120, 125, 136, 139, 140, 144, 146, 149, 151, 152, 155, 160, 161, 229, 249-251, 259, 261, 305, 309, 316
Symbiodinium 226, 235, 246
symbiont 203, 205, 207, 217, 221, 222, 228, 229, 231, 234, 235, 241-243, 245, 247
symbioses, -tic 203, 206, 207, 209, 214, 220, 224-229, 234, 245, 246, 273
Sympetrum flaveolum 96
Sympetrum fonscolombii 94-96, 102, 109
Sympetrum meridionale 95
Sympetrum sanguineum 101
Sympetrum strioliatum 101, 104
Sympetrum vulgatum 101, 104
synanthropic 183, 286
synecology, -cal 101, 282, 283
synergism, -tic 106, 234, 243
Taraxacum 116
Tardigrada 19
Taxus 118, 154
temperate 10, 14, 20, 34, 124, 152, 153, 155, 160, 161, 205, 210, 229, 234, 235, 241, 247, 250, 278, 307
terrestrial ecosystems 14, 17, 18, 28, 30, 34, 41, 277

Thailand 212
The Netherlands 9, 67, 71, 76, 88, 93, 96, 195, 197, 212, 246, 247
thermolability 235
thermophilic, -ous 12, 91, 98, 100, 102, 163, 165, 166, 170, 181, 182, 290, 291, 310, 315
thermoprotection 229, 236, 241, 242, 245
therophyte, -ic 170, 180, 286, 291
threshold 2, 5, 7, 21, 58, 82, 84, 182, 212, 224, 243, 259, 281, 282, 298, 308, 315
Thymelicus sylvestris 78-81
Tilia 116, 127, 152
Tilia cordata 116
Tilia platyphyllos 116, 127
time scale 14, 50, 51, 217, 230-232, 237, 239, 307
time series 43, 50, 59, 61, 62, 85, 113, 114, 115, 118, 121, 124, 154, 249, 251, 253, 258
tolerance 20, 33, 35, 36, 38, 40, 42, 101, 203, 214, 216, 220, 221
tourism 146, 187, 203, 223, 225, 228, 292
Trachycarpus fortunei 153, 154, 311, 312
transect 60, 166, 167, 172, 293, 307, 308
transformation 242, 260
transition 235, 238, 241-243
transplantation 27, 28, 222, 263, 274
tree 9, 11, 13, 15, 59, 73, 74, 124-126, 130, 135, 137, 152, 153, 157, 158, 178, 200, 250, 307, 308, 313
treeline 13, 307, 308, 316
treering 59, 74, 157, 307
trend analysis 114, 115, 121, 123, 126, 130, 135
Trifolium 188, 263, 265-267, 270-275, 278
Trifolium alpinum 273
Trifolium pallescens 263, 265-267, 270-273, 275
Trithemis annulata 96, 105
Tropaeolaceae 179
Tropaeolum majus 163, 165, 166, 179, 180
trophic 7, 10, 20, 25, 28, 30, 34, 315
tropic, -al 1, 13, 20, 98, 165, 166, 176, 177, 182, 203, 205, 206, 208-211, 213, 214, 216, 223-226, 228, 245, 246, 313
tundra 14, 273, 276-278, 316
Turdus torquatus 146

Index 329

turn-over 13, 111, 220, 236, 238, 251, 286, 291, 313
turtles 208
UK 8, 61, 77-79, 84, 85, 87, 96, 121, 247, 277
Ukraine 94, 96
Ulex europaeus 188, 195, 196
upward 1, 13, 30, 57, 58, 64, 70, 139, 140, 144, 146, 147, 149, 210, 213, 221, 222, 250, 292, 297
urban 59, 108, 124, 125, 137, 178, 285
Ursus maritimus 46
Urtica 118
USA 7, 8, 10, 11, 57, 73, 74, 125, 136, 210, 224-226, 228, 246, 277
UV 22, 24, 25, 31-33, 35, 36, 38-40
variety 2, 51, 60, 147, 153, 217, 221, 222, 229, 231, 249-254
vascular 9, 27, 38, 39, 42, 139, 140, 182, 183, 185, 188, 198, 282, 287, 289, 300
vegetation 4, 7, 8, 11, 15, 19, 28, 32, 37, 39, 42, 63, 88, 98, 104, 105, 113-115, 121, 136, 139, 149, 151-154, 156, 157, 160, 161, 164, 165, 178, 185, 186, 249, 251, 258, 261, 265, 277, 278, 281, 282, 284, 291-293, 297, 298, 300, 306, 309, 316
vegetative 26, 264, 265, 270, 271, 273, 274, 279, 285, 297
vertebrates 19, 27, 73
vine 97, 99, 125
vintage 99, 115, 116
Viscum album 152
Vitis vinifera 116
volcanic 210, 211
walrus 43-47, 49-52, 54, 307
water availability 22, 30, 258, 259, 285
wavelength 24, 31, 32
weed 27, 75, 171, 178
wetlands 108, 165
whales 10, 28
wildlife 51, 55, 57, 58, 68
wind 18, 23, 30, 31, 47, 171, 179, 186, 198, 210, 249, 258, 259, 260, 272
wood 7, 84, 85, 116, 177, 286
worm 28, 32, 206
Zea mays 251
zooplankton 4
zooxanthellae 206, 207, 213, 216, 247
Zygoptera 94, 98